LONDON MATHEMATICAL SOCIETY LECTURE NOTE SERIES

Managing Editor: Professor J.W.S. Cassels, Department of Pure Mathematics and Mathematical Statistics, University of Cambridge, 16 Mill Lane, Cambridge CB2 1SB, England

The titles below are available from booksellers, or, in case of difficulty, from Cambridge University Press.

34 Representation theory of Lie groups, M.F. ATIYAH *et al*
36 Homological group theory, C.T.C. WALL (ed)
39 Affine sets and affine groups, D.G. NORTHCOTT
46 p-adic analysis: a short course on recent work, N. KOBLITZ
50 Commutator calculus and groups of homotopy classes, H.J. BAUES
59 Applicable differential geometry, M. CRAMPIN & F.A.E. PIRANI
66 Several complex variables and complex manifolds II, M.J. FIELD
69 Representation theory, I.M. GELFAND *et al*
76 Spectral theory of linear differential operators and comparison algebras, H.O. CORDES
77 Isolated singular points on complete intersections, E.J.N. LOOIJENGA
83 Homogeneous structures on Riemannian manifolds, F. TRICERRI & L. VANHECKE
86 Topological topics, I.M. JAMES (ed)
87 Surveys in set theory, A.R.D. MATHIAS (ed)
88 FPF ring theory, C. FAITH & S. PAGE
89 An F-space sampler, N.J. KALTON, N.T. PECK & J.W. ROBERTS
90 Polytopes and symmetry, S.A. ROBERTSON
92 Representation of rings over skew fields, A.H. SCHOFIELD
93 Aspects of topology, I.M. JAMES & E.H. KRONHEIMER (eds)
94 Representations of general linear groups, G.D. JAMES
95 Low-dimensional topology 1982, R.A. FENN (ed)
96 Diophantine equations over function fields, R.C. MASON
97 Varieties of constructive mathematics, D.S. BRIDGES & F. RICHMAN
98 Localization in Noetherian rings, A.V. JATEGAONKAR
99 Methods of differential geometry in algebraic topology, M. KAROUBI & C. LERUSTE
100 Stopping time techniques for analysts and probabilists, L. EGGHE
101 Groups and geometry, ROGER C. LYNDON
103 Surveys in combinatorics 1985, I. ANDERSON (ed)
104 Elliptic structures on 3-manifolds, C.B. THOMAS
105 A local spectral theory for closed operators, I. ERDELYI & WANG SHENGWANG
106 Syzygies, E.G. EVANS & P. GRIFFITH
107 Compactification of Siegel moduli schemes, C-L. CHAI
108 Some topics in graph theory, H.P. YAP
109 Diophantine analysis, J. LOXTON & A. VAN DER POORTEN (eds)
110 An introduction to surreal numbers, H. GONSHOR
113 Lectures on the asymptotic theory of ideals, D. REES
114 Lectures on Bochner-Riesz means, K.M. DAVIS & Y-C. CHANG
115 An introduction to independence for analysts, H.G. DALES & W.H. WOODIN
116 Representations of algebras, P.J. WEBB (ed)
117 Homotopy theory, E. REES & J.D.S. JONES (eds)
118 Skew linear groups, M. SHIRVANI & B. WEHRFRITZ
119 Triangulated categories in the representation theory of finite-dimensional algebras, D. HAPPEL
121 Proceedings of *Groups - St Andrews 1985*, E. ROBERTSON & C. CAMPBELL (eds)
122 Non-classical continuum mechanics, R.J. KNOPS & A.A. LACEY (eds)
125 Commutator theory for congruence modular varieties, R. FREESE & R. MCKENZIE
126 Van der Corput's method of exponential sums, S.W. GRAHAM & G. KOLESNIK
127 New directions in dynamical systems, T.J. BEDFORD & J.W. SWIFT (eds)
128 Descriptive set theory and the structure of sets of uniqueness, A.S. KECHRIS & A. LOUVEAU
129 The subgroup structure of the finite classical groups, P.B. KLEIDMAN & M.W.LIEBECK
130 Model theory and modules, M. PREST
131 Algebraic, extremal & metric combinatorics, M-M. DEZA, P. FRANKL & I.G. ROSENBERG (eds)
132 Whitehead groups of finite groups, ROBERT OLIVER
133 Linear algebraic monoids, MOHAN S. PUTCHA
134 Number theory and dynamical systems, M. DODSON & J. VICKERS (eds)
135 Operator algebras and applications, 1, D. EVANS & M. TAKESAKI (eds)
136 Operator algebras and applications, 2, D. EVANS & M. TAKESAKI (eds)
137 Analysis at Urbana, I, E. BERKSON, T. PECK, & J. UHL (eds)
138 Analysis at Urbana, II, E. BERKSON, T. PECK, & J. UHL (eds)
139 Advances in homotopy theory, S. SALAMON, B. STEER & W. SUTHERLAND (eds)
140 Geometric aspects of Banach spaces, E.M. PEINADOR & A. RODES (eds)
141 Surveys in combinatorics 1989, J. SIEMONS (ed)

142 The geometry of jet bundles, D.J. SAUNDERS
143 The ergodic theory of discrete groups, PETER J. NICHOLLS
144 Introduction to uniform spaces, I.M. JAMES
145 Homological questions in local algebra, JAN R. STROOKER
146 Cohen-Macaulay modules over Cohen-Macaulay rings, Y. YOSHINO
147 Continuous and discrete modules, S.H. MOHAMED & B.J. MÜLLER
148 Helices and vector bundles, A.N. RUDAKOV *et al*
149 Solitons, nonlinear evolution equations and inverse scattering, M. ABLOWITZ & P. CLARKSON
150 Geometry of low-dimensional manifolds 1, S. DONALDSON & C.B. THOMAS (eds)
151 Geometry of low-dimensional manifolds 2, S. DONALDSON & C.B. THOMAS (eds)
152 Oligomorphic permutation groups, P. CAMERON
153 L-functions and arithmetic, J. COATES & M.J. TAYLOR (eds)
154 Number theory and cryptography, J. LOXTON (ed)
155 Classification theories of polarized varieties, TAKAO FUJITA
156 Twistors in mathematics and physics, T.N. BAILEY & R.J. BASTON (eds)
157 Analytic pro-*p* groups, J.D. DIXON, M.P.F. DU SAUTOY, A. MANN & D. SEGAL
158 Geometry of Banach spaces, P.F.X. MÜLLER & W. SCHACHERMAYER (eds)
159 Groups St Andrews 1989 volume 1, C.M. CAMPBELL & E.F. ROBERTSON (eds)
160 Groups St Andrews 1989 volume 2, C.M. CAMPBELL & E.F. ROBERTSON (eds)
161 Lectures on block theory, BURKHARD KÜLSHAMMER
162 Harmonic analysis and representation theory, A. FIGA-TALAMANCA & C. NEBBIA
163 Topics in varieties of group representations, S.M. VOVSI
164 Quasi-symmetric designs, M.S. SHRIKANDE & S.S. SANE
165 Groups, combinatorics & geometry, M.W. LIEBECK & J. SAXL (eds)
166 Surveys in combinatorics, 1991, A.D. KEEDWELL (ed)
167 Stochastic analysis, M.T. BARLOW & N.H. BINGHAM (eds)
168 Representations of algebras, H. TACHIKAWA & S. BRENNER (eds)
169 Boolean function complexity, M.S. PATERSON (ed)
170 Manifolds with singularities and the Adams-Novikov spectral sequence, B. BOTVINNIK
171 Squares, A.R. RAJWADE
172 Algebraic varieties, GEORGE R. KEMPF
173 Discrete groups and geometry, W.J. HARVEY & C. MACLACHLAN (eds)
174 Lectures on mechanics, J.E. MARSDEN
175 Adams memorial symposium on algebraic topology 1, N. RAY & G. WALKER (eds)
176 Adams memorial symposium on algebraic topology 2, N. RAY & G. WALKER (eds)
177 Applications of categories in computer science, M. FOURMAN, P. JOHNSTONE, & A. PITTS (eds)
178 Lower K- and L-theory, A. RANICKI
179 Complex projective geometry, G. ELLINGSRUD *et al*
180 Lectures on ergodic theory and Pesin theory on compact manifolds, M. POLLICOTT
181 Geometric group theory I, G.A. NIBLO & M.A. ROLLER (eds)
182 Geometric group theory II, G.A. NIBLO & M.A. ROLLER (eds)
183 Shintani zeta functions, A. YUKIE
184 Arithmetical functions, W. SCHWARZ & J. SPILKER
185 Representations of solvable groups, O. MANZ & T.R. WOLF
186 Complexity: knots, colourings and counting, D.J.A. WELSH
187 Surveys in combinatorics, 1993, K. WALKER (ed)
188 Local analysis for the odd order theorem, H. BENDER & G. GLAUBERMAN
189 Locally presentable and accessible categories, J. ADAMEK & J. ROSICKY
190 Polynomial invariants of finite groups, D.J. BENSON
191 Finite geometry and combinatorics, F. DE CLERCK *et al*
192 Symplectic geometry, D. SALAMON (ed)
193 Computer algebra and differential equations, E. TOURNIER (ed)
194 Independent random variables and rearrangement invariant spaces, M. BRAVERMAN
195 Arithmetic of blowup algebras, WOLMER VASCONCELOS
196 Microlocal analysis for differential operators, A. GRIGIS & J. SJÖSTRAND
197 Two-dimensional homotopy and combinatorial group theory, C. HOG-ANGELONI, W. METZLER & A.J. SIERADSKI (eds)
198 The algebraic characterization of geometric 4-manifolds, J.A. HILLMAN
199 Invariant potential theory in the unit ball of C^n, MANFRED STOLL
200 The Grothendieck theory of dessins d'enfant, L. SCHNEPS (ed)
201 Singularities, JEAN-PAUL BRASSELET (ed)
202 The technique of pseudodifferential operators, H.O. CORDES
203 Hochschild cohomology of von Neumann algebras, A. SINCLAIR & R. SMITH
204 Combinatorial and geometric group theory, A.J. DUNCAN, N.D. GILBERT & J. HOWIE (eds)
207 Groups of Lie type and their geometries, W.M. KANTOR & L. DI MARTINO (eds)
208 Vector bundles in algebraic geometry, N.J. HITCHIN, P. NEWSTEAD & W.M. OXBURY (eds)

London Mathematical Society Lecture Note Series. 212

Groups '93 Galway / St Andrews

Galway 1993

Volume 2

Edited by

C.M. Campbell
University of St Andrews

T.C. Hurley
University College, Galway

E.F. Robertson
University of St Andrews

S.J. Tobin
University College, Galway

J.J. Ward
University College, Galway

Published by the Press Syndicate of the University of Cambridge
The Pitt Building, Trumpington Street, Cambridge CB2 1RP
40 West 20th Street, New York, NY 10011-4211, USA
10 Stamford Road, Oakleigh, Melbourne 3166, Australia

© Cambridge University Press 1995

First published 1995

Library of Congress cataloguing in publication data available

British Library cataloguing in publication data available

ISBN 0 521 47750 6 paperback

Transferred to digital printing 2004

CONTENTS

Volume II

Contents of Volume I

PREFACE

This book is the second of two volumes of the Proceedings of Groups 1993 Galway / St Andrews. There is a full contents of both volumes and papers written by authors with the name of the first author lying in the range H-Z. Contained in this part are the papers of the main speakers A Lubotzky, J Neubüser and E I Zel'manov, and we would like especially to thank them for their contributions both to the conference and to this volume. There is also a collection of problems contributed by conference participants included at the end of the volume. There is some perturbation in strict alphabetical order at the join between the volumes and the first volume contains papers written by authors with the name of the first author lying in the range A-K.

INTRODUCTION

An international conference 'Groups 1993 Galway / St Andrews' was held at University College, Galway, Ireland during the period 1 to 14 August 1993. This followed in the main the successful format developed in 1981, 1985 and 1989 by C M Campbell and E F Robertson. They invited T Hurley, S Tobin and J Ward to join them and continue the series in 1993 in Galway. Serious planning got under way when the five organisers met at a Warwick conference in March 1991, and decided to invite as principal speakers J L Alperin (Chicago), M Broué (Paris), P H Kropholler (London), A Lubotzky (Jerusalem) and E I Zel'manov (Madison). All of these agreed to give courses of about five lectures each, and articles based on these courses form a valuable part of these Proceedings - particularly so as in some cases they are strongly complementary in subject matter. Also, as it transpired, one speaker was awarded a Fields Medal exactly one year later at the 1994 ICM in Zurich; the organisers have great pleasure in congratulating Professor Zel'manov most heartily - and hope perhaps that this may augur well for future speakers in the series!

An invitation to J Neubüser (Aachen) to arrange a workshop on Computational Group Theory and the use of GAP was taken up so enthusiastically by him that the workshop became effectively a fully-fledged parallel meeting throughout the second week, with over thirty hours of lectures by experts and with practical sessions organised by M Schönert (Aachen). These Proceedings contain an article by Professor Neubüser based on a lecture he gave in the first week of the conference.

In addition there were sixteen invited one-hour lectures, and a very great number of research talks, which reflected the fact that the meeting was attended by 285 mathematicians (who were accompanied by about 150 family members) from 35 countries around the world. The articles in these Proceedings, other than those of the main speakers, are based on one-hour invited lectures and other research talks.

As in 1989 there are two volumes of the Proceedings, with the contributions arranged according to author's names in alphabetical order from A in Volume One to Z in Volume Two, except for a minor perturbation in alphabetical order at the join between the two volumes. They will form a stimulating record of recent work as well as being a valuable reference source, due to the wealth of material covered in the main courses. We thank all our contributors, and regret that we could not - because of space restrictions - accept several other worthy papers.

An unusual feature of this conference was the setting aside of one day for a special programme of lectures to honour the 65th birthday of Professor K W Gruenberg, in recognition of his many contributions to group theory. A second feature was the publication by the Galway Mathematics Department of a splendid memoir by Professor Lubotzky on Subgroup Growth, prepared by him as a background for his course of lectures. Another feature was the videoing of all the talks of the main speakers which form a lovely record of the events. These videos may be obtained from the Galway Mathematics Department.

There are many who helped to make the conference a memorable occasion; in particular we thank the staff of the Computing Services in University College Galway, our colleagues and students, members of the Secretariat in both Galway and St Andrews, and members of our own families who cheerfully helped out in various ways. We extend our thanks to the Administration of University College Galway for the Reception in the Aula Maxima; and we are grateful to the Mayor of Galway who gave a special reception in the Corporation's Council Chamber.

We gratefully acknowledge financial support for the academic program from the London Mathematical Society, the Royal Irish Academy, the Irish Mathematical Society, An Bord Gais, Irish Shell, Allied Irish Banks, Bank of Ireland, Bord Failte, the International Science Foundation, University College Galway, the University of St Andrews and the Deutsche Forschungsgemeinschaft.

As our final word, an focal scuir, we wish to thank Cambridge University Press for help with these Proceedings and Nik Ruškuc for undertaking so willingly the enormous task of reformatting files in many flavours of TEX. As in previous volumes, the present editors have endeavoured to produce a measure of uniformity - hopefully without distorting individual styles. For any inconsistencies in, or errors introduced by, our editing we of course accept responsibility.

Colin M Campbell
Thaddeus C Hurley
Edmund F Robertson
Sean J Tobin
James J Ward

AN ARMY OF COHOMOLOGY AGAINST RESIDUAL FINITENESS

P.R. HEWITT

University of Toledo, Toledo OH 43606, U.S.A.

1. Phylogeny

In [10] we demonstrate the existence of a group of shape $\mathbb{Z}^3.\mathcal{SL}_3(\mathbb{Z})$ that fails to be residually finite. This is a locally graded group in which the normal closure of every finite set is finitely generated. H. Smith had shown that the finite residual of a group with these properties is abelian; and that if such a group is locally solvable, then its finite residual is central. He sought an instance where the finite residual is noncentral. Groups of the above shape– which are perforce perfect– induce nearly the full automorphism group on the finite residual.

Unfortunately, [10] does not provide the desired example explicitly. Rather, it offers cohomological reconnaissance of residual finiteness for arbitrary extensions, and then rolls out a potent cohomological arsenal in order to attack groups of shape $\mathbb{Z}^d.\mathcal{SL}_d(\mathbb{Z})$. It is disappointing not to have secured a concrete example after such a high-powered assault, but the battle lines are drawn rather neatly:

$$\text{residual finiteness} \iff \text{linearity} \iff \text{cohomological torsion}$$

The proof of this triple equivalence calls up Lubotzky's Criterion (see [13]) and succeeds for arithmetic subgroups of essentially all Chevalley groups (see [11]).

Cohomology enters the fray since split extensions of residually finite groups are residually finite; as are virtually split extensions; as are residually virtually split extensions. In fact, the converse is true: residually finite extensions of residually finite groups must be residually virtually split. As was known to Schur, the second cohomology group controls splitting.

Not available to Schur were the generic cohomology of algebraic groups (see [8, 9, 12]), which enables application of Lubotzky's Criterion, by first constraining the residual splitting of a 2-class to finitely many primes; nor Soulé's explicit reduction theory for $\mathcal{SL}_3(\mathbb{Z})$ (see [21]), which displays nontorsion 2-cohomology for $\mathcal{SL}_3(\mathbb{Z})$ on $\mathbb{Z}^3$; nor Borel's stable real cohomology (see [3]), which shows that $\mathcal{SL}_3(\mathbb{Z})$ is the exception, not the rule: $\mathrm{H}^2(\mathcal{SL}_d(\mathbb{Z}), \mathbb{Z}^d)$ is finite when $d \geq 7$.

We exhibit an explicit presentation for an anomalous extension $\mathbb{Z}^3.\mathcal{SL}_3(\mathbb{Z})$, and then explore a geometric representation– one we hope will prove to be flexible enough to reflect back on some of the cohomology. Ultimately we hope to provide group-theoretic rationale for Borel's weight-theoretic hypotheses.

2. Ontogeny

The proper context in which to study extensions of a given shape $M.\Gamma$ is that of an outer action, $\omega: \Gamma \to Out(M)$. To simplify matters a bit, we suppose that all groups under discussion are finitely generated. Suppose that Γ and M are residually finite. If there is a split extension in this context– *i.e.*, if Γ acts on M– then it is residually finite. In fact, split extensions are the basic building block out of which all residually finite extensions are constructed. This is because, on the one hand, virtually residually finite groups are residually finite, as are residually residually finite groups. Hence, residually virtually split extensions are residually finite. The converse is just as straightforward.

Incidentally, this theorem is well-known in the world of Galois cohomology, at least when M is finite and abelian (see [17, p. 100]). However, the example in [7] shows that the general case cannot be deduced from this special one, at least not merely by taking limits.

To see that the introduction of cohomology is not entirely frivolous, note a useful corollary of the above characterization. Suppose that M is free abelian, and that there are (up to equivalence) only finitely many extensions of shape $M.\Gamma$. If E is any one of these extensions, then its cohomology class has finite order, whence E is residually finite (see [10]). So, if we set $M_{\mathbb{Q}} = M \otimes \mathbb{Q}$, then we find the non-residually-finite extensions of shape $M.\Gamma$ among the nonzero elements of $\mathrm{H}^2(\Gamma, M_{\mathbb{Q}})$.

Now let $\Gamma = \mathcal{SL}_d(\mathbb{Z})$, and $M = \mathbb{Z}^d$. The ideas above show that there is something distinctive about the case when $d = 3$. To begin with, if $d = 2$ then Γ is virtually free, and so every extension $M.\Gamma$ is residually finite. When d is even, $-I \in \Gamma$. Since $-I$ acts freely on M, a Frattini argument shows that $\mathrm{H}^2(\Gamma, M_{\mathbb{Q}}) = 0$ (see [10]). Finally, Borel's results (see [3]) can be applied to show that $\mathrm{H}^2(\Gamma, M_{\mathbb{Q}}) = 0$ when $d \geq 7$. (I am at a loss when $d = 5$.)

For groups and modules of the sort we are considering– arithmetic subgroups of algebraic groups and most of their simple rational modules– one can show that every nontorsion element of $\mathrm{H}^2(\Gamma, M)$ determines a non-residually-finite extension. Owing to the above characterization, this is a consequence of the following: for each prime p and each finite-index subgroup Δ, the restriction map $\mathrm{H}^2(\Gamma, M/p) \to \mathrm{H}^2(\Delta, M/p)$ is faithful. As we may assume that $\Delta = \Gamma(n)$, for some n (see [1, 14][1]), the spectral sequence of the inclusion

[1]Matsumoto's result is applied incorrectly in [11], in the case of the ring of integers in a totally imaginary number field K. In this case the congruence kernel is isomorphic with the group μ_K of roots of unity. The hole is patched simply by considering only those

$\Delta < \Gamma$ reduces this latter statement to the computation of $\mathrm{H}^k(\Gamma/p, X)$, where $\Gamma/p = \mathcal{SL}_3(\mathbb{F}_p)$; $k \leq 2$; and X is a 'smallish' module. To convey the idea behind 'small', we remark that the largest such X we need to consider is $\mathrm{Hom}(\otimes^2\mathfrak{L}, M/p)$, where $\mathfrak{L}$ is the p-modular Lie algebra.

The Classification era taught us that sporadic cohomology is most often linked to sporadic simple groups. For example, $\mathrm{Hom}_{\Gamma/p}(\otimes^2\mathfrak{L}, M/p) = 0$ except when $p = 2$ and $d = 3$. The exception gives rise to a curious 2-local parabolic in Rudvalis' simple group. We don't need the Classification, except philosophically, but it does hint strongly that the kernel of the restriction map is 0, usually.

This we can verify, using the 'generic cohomology' of finite algebraic groups (see [8, 9, 12, 22]; or, for direct computation in the case of $\mathcal{SL}_d(\mathbb{Z})$, see [2]). We find that restriction is faithful whenever $p > 5$. Because of the Pigeon Hole Principle, we deduce as a corollary that a residually finite extension E of shape $M.\Gamma$ and of infinite cohomological order, must contain a normal series $E_1 > E_2 > E_3 > \cdots$ whose sections E_k/E_{k+1} are boundedly finite p-groups, where $p = 2$, 3, or 5.

Now Lubotzky's Criterion comes to the rescue, and delivers a contradiction. Indeed, an extension containing such a series would be linear, and hence contain such a series for almost every prime. This is in conflict with the cohomological calculations, and so E fails to be residually finite.

Now we are left with the task of finding nontorsion 2-cohomology. It turns out that there is a 'reduction theory', due to Soulé, which gives the cohomology of $\mathcal{SL}_3(\mathbb{Z})$ about as explicitly as one is likely to see (see [21]). It works best when 6 is invertible on the coefficient module, but we can arrange this in our case, by extending scalars to $\mathbb{Q}$. One computes that $\mathrm{H}^k(\Gamma, M_{\mathbb{Q}}) = 0$ whenever $k \neq 2$, deducing that the Euler characteristic of $M_{\mathbb{Q}}$ is equal to the dimension of $\mathrm{H}^2(\Gamma, M_{\mathbb{Q}})$. On the other hand, the Euler characteristic can be computed at the chain level, using Soulé's reduction theory: The answer is 1.

3. Presentation

Armed with the answer, let us try to hunt down these elusive extensions exactly. As before, set $\Gamma = \mathcal{SL}_3(\mathbb{Z})$ and $M = \mathbb{Z}^3$. We will show that the following presentation— based on the Steinberg relations (see [22]) and the action of Γ on M— defines an extension that fails to be residually finite.

$$E = \Big\langle x_{ij}, e_k \ (1 \leq i,j,k \leq 3;\ i \neq j) \Big| \quad [e_k, x_{ij}] = \begin{cases} e_j, & \text{if } k = i; \\ 1, & \text{if } k \neq i; \end{cases}$$
$$[e_k, e_{k'}] = 1; \quad [x_{ij}, x_{kj}] = 1; \quad [x_{ij}, x_{jk}] = e_j^2 e_k, \text{ if } k \neq i. \Big\rangle$$

primes that do not divide $|\mu_K|$, and then applying the Transfer one more time.

We do this by finding an explicit 2-cocycle for the extension. I should note, however, that I do not yet know how to prove that this extension fails to be residually finite, except to invoke the results from [10].

Since $\mathcal{GL}_2(\mathbb{Z})$ is virtually free, some Baer power of E is generated by copies of the root subgroups Λ_1 and Λ_2, which intersect in the diagonal subgroup Δ:

$$\Pi_1 = \begin{bmatrix} * & * & 0 \\ * & * & 0 \\ * & * & * \end{bmatrix} \qquad \Delta = \begin{bmatrix} * & 0 & 0 \\ 0 & * & 0 \\ 0 & 0 & * \end{bmatrix} \qquad \Pi_2 = \begin{bmatrix} * & 0 & 0 \\ * & * & * \\ * & * & * \end{bmatrix}$$

$$\Lambda_1 = \begin{bmatrix} * & * & 0 \\ * & * & 0 \\ 0 & 0 & * \end{bmatrix} \qquad \Lambda_2 = \begin{bmatrix} * & 0 & 0 \\ 0 & * & * \\ 0 & * & * \end{bmatrix}$$

Now any Baer power of a non-residually-finite extension of this shape must itself fail to be residually finite, since any multiple of a cocycle of infinite order must itself have infinite order. So, without loss of generality, our extension is generated by the Λ_i.

However, if E fails to be residually finite, then it cannot be generated by copies of the parabolic overgroups Π_i (see [15, 16]– more on this anon). This fact allows us to find the cocycle explicitly, by restricting to the Π_i. In fact, since every pair of root elements x_{ij} lie together in a conjugate of Π_2, it suffices to consider the cocycle on this subgroup.

Since Π_2 is simply a split extension $U_2.\Lambda_2$, where $U_2 \cong \mathbb{Z}^2$, the neatest description of the cohomological character of the parabolics in such extensions would be given by the characteristic classes of Charlap and Vasquez (see [5, 6]), which were explored by Sah (see [18]). However, the cohomology is easy to compute directly.

To do so, we apply the spectral sequence. So, first we compute $\mathrm{H}^*(U_2, M_\mathbb{Q})$, by means of the cohomology sequence associated to the decomposition

$$0 \to [M_\mathbb{Q}, U_2] \to M_\mathbb{Q} \to M_\mathbb{Q}/[M_\mathbb{Q}, U_2] \to 0.$$

The induced long exact sequence

$$\begin{aligned} 0 \longrightarrow \mathrm{H}^0(U_2, [M_\mathbb{Q}, U_2]) &\xrightarrow{\cong} \mathrm{H}^0(U_2, M_\mathbb{Q}) \xrightarrow{0} \mathrm{H}^0(U_2, M_\mathbb{Q}/[M_\mathbb{Q}, U_2]) \xrightarrow{\cong} \\ \mathrm{H}^1(U_2, [M_\mathbb{Q}, U_2]) &\xrightarrow{0} \mathrm{H}^1(U_2, M_\mathbb{Q}) \longrightarrow \mathrm{H}^1(U_2, M_\mathbb{Q}/[M_\mathbb{Q}, U_2]) \longrightarrow \\ \mathrm{H}^2(U_2, [M_\mathbb{Q}, U_2]) &\xrightarrow{0} \mathrm{H}^2(U_2, M_\mathbb{Q}) \xrightarrow{\cong} \mathrm{H}^2(U_2, M_\mathbb{Q}/[M_\mathbb{Q}, U_2]) \longrightarrow 0. \end{aligned}$$

is Λ_2-equivariant, and we find that

$$\mathrm{H}^1(U_2, M_\mathbb{Q}) \cong S_2(\mathbb{Q}) = \{2 \times 2 \text{ symmetric matrices}\}; \quad \mathrm{H}^2(U_2, M_\mathbb{Q}) \cong \mathbb{Q}^2.$$

Indeed, a matrix $\delta\colon \mathbb{Z}^2 \to \mathbb{Q}^2$ lifts to a 1-cocycle $\tilde{\delta}\colon \mathbb{Z}^2 \to M_\mathbb{Q}$ precisely when there is a function $q\colon \mathbb{Z}^2 \to \mathbb{Q}$ satisfying the following identity:

$$q(v + w) = q(v) + q(w) + \delta(v) \cdot w,$$

where $\cdot$ denotes inner product. This means that δ is a symmetric matrix, and q is its associated quadratic form: $q(v) = \frac{1}{2}\delta(v) \cdot v$. In particular, the cocycle restricts to 0 on U_2.

Moreover, Λ_2 acts through similarity on $\mathrm{H}^1(U_2, M_{\mathbb{Q}})$; and through multiplication by the determinant on $\mathrm{H}^2(U_2, M_{\mathbb{Q}})$. Again, Λ_2 is virtually free, and so the spectral sequence tells us that

$$\mathrm{H}^2(\Pi_2, M_{\mathbb{Q}}) = \mathrm{H}^1(\mathcal{GL}_2(\mathbb{Z}), S_2(\mathbb{Q})).$$

It remains, then, to identify $\mathrm{H}^1(\mathcal{GL}_2(\mathbb{Z}), S_2(\mathbb{Q}))$.

We apply the Mayer-Vietoris sequence ([4, p. 37]; but compare with [23]), since $\mathcal{SL}_2(\mathbb{Z})$ is simply a freely amalgamated product: $\mathcal{SL}_2(\mathbb{Z}) = \langle s \rangle *_{\pm I} \langle u \rangle$, where $s = \begin{bmatrix} 0 & 1 \\ -1 & 0 \end{bmatrix}$ and $u = \begin{bmatrix} 1 & 1 \\ -1 & 0 \end{bmatrix}$. Thus,

$$\begin{aligned} 0 \longrightarrow\ & S_2(\mathbb{Q})^s \oplus S_2(\mathbb{Q})^u \longrightarrow S_2(\mathbb{Q}) \xrightarrow{p} \mathrm{H}^1(\mathcal{SL}_2(\mathbb{Z}), S_2(\mathbb{Q})) \\ \xrightarrow{q}\ & \mathrm{H}^1(\langle s \rangle, S_2(\mathbb{Q})) \oplus \mathrm{H}^1(\langle u \rangle, S_2(\mathbb{Q})) \longrightarrow 0. \end{aligned}$$

The sum of the fixed-point subspaces– which are indicated by the superscripts– is 2-dimensional. Hence the image of p is 1-dimensional, and $\mathrm{H}^1(\mathcal{SL}_2(\mathbb{Z}), S_2(\mathbb{Q}))$ is 5-dimensional.

If $H = \begin{bmatrix} 0 & 1 \\ 1 & 0 \end{bmatrix}$, then $\mathrm{H}^1(\mathcal{GL}_2(\mathbb{Z}), S_2(\mathbb{Q})) \cong \mathrm{H}^1(\mathcal{SL}_2(\mathbb{Z}), S_2(\mathbb{Q}))^H$. Since the Mayer-Vietoris sequence is equivariant for H, we find that

$$\mathrm{H}^1(\mathcal{SL}_2(\mathbb{Z}), S_2(\mathbb{Q}))^H = \left\{ s \mapsto \begin{bmatrix} 0 & a \\ a & 0 \end{bmatrix}; \ u \mapsto \begin{bmatrix} b & b \\ b & 0 \end{bmatrix} \right\}$$

Indeed, H is nontrivial on $S_2(\mathbb{Q})/(S_2(\mathbb{Q})^s + S_2(\mathbb{Q})^u)$; and the map q is given by composing differentials $\pm s \mapsto \begin{bmatrix} c & a \\ a & -c \end{bmatrix}$ and $\pm u \mapsto \begin{bmatrix} b & b+d \\ b+d & d \end{bmatrix}$, with surjections $\mathcal{SL}_2(\mathbb{Z}) \to \langle s \rangle / \pm I$ and $\mathcal{SL}_2(\mathbb{Z}) \to \langle u \rangle / \pm I$, respectively.

Let us translate these 1-cocycles for Λ_2 into 2-cocycles for Π_2. Write the elements of $\mathrm{H}^1(\Lambda_2, S_2(\mathbb{Q}))$ as $x \mapsto \delta_x$, where $\delta_x \in S_2(\mathbb{Q})$. So, we can regard each δ_x as a homomorphism from U_2 to $\mathbb{Q}^2$. For each x in Λ_2, let $q_x \colon U_2 \to \mathbb{Q}$ be the quadratic form associated to δ_x; this lifts δ_x to a differential $\tilde{\delta}_x \colon U_2 \to M_{\mathbb{Q}}$:

$$\tilde{\delta}_x(u) = (q_x(u), \delta_x(u)).$$

The formula for the associated 2-cocycle φ is now easy to write down:

$$\varphi(xv, yw) = \tilde{\delta}_y(v^y) \cdot w.$$

To finish, we need to determine which of these 2-cocycles are Γ-invariant. This is quite straightforward. Set $t^+ = \begin{bmatrix} 1 & 0 \\ 1 & 1 \end{bmatrix}$ and $t^- = \begin{bmatrix} 1 & 1 \\ 0 & 1 \end{bmatrix}$. We have that $t^+ = s^{-1}u$ and $t^- = su^{-1}$, whence

$$\delta_{t^+} = \begin{bmatrix} b-2a & b-a \\ b-a & 0 \end{bmatrix}; \qquad \delta_{t^-} = \begin{bmatrix} 0 & b-a \\ b-a & b-2a \end{bmatrix}.$$

The above formula tells us that $\varphi(x_{21}, x_{23}) = 0$. But the pair x_{21}, x_{23} is conjugate in Γ to the pair $x_{23}, x_{21}{}^{-1}$, which tells us that $a = b$ if φ is to be invariant. If we choose $a = -2$, then $\varphi(x_{31}, x_{23}) = (1, 0, 2)$. From this value we can obtain the above presentation, by means of conjugation in Γ.

I have done the above calculations in absolutely excruciating detail in order to remove any further need for cohomology in determining E 'explicitly'. However, having E in hand brings me no closer to establishing its residual finiteness, unless I revert to cohomology. In the next section, I outline a geometric representation of E that I hope will shed new light.

4. Representation

Let $\mathfrak{K}$ be a 'nice' topological space admitting the group Γ. If π' is a Γ-invariant normal subgroup of $\pi_1(\mathfrak{K})$, then the elements of Γ lift to automorphisms of the cover $\mathfrak{G}$ corresponding to π', by the Homotopy Lifting Property. The ambiguity of each lift is the group of covering transformations, $\pi_1(\mathfrak{K})/\pi'$. In particular, for each Γ-module M we have a map $\mathrm{Hom}_\Gamma(\mathrm{H}_1(\mathfrak{K}), M) \to \mathrm{H}^2(\Gamma, M)$.

Let us specialize to the situation at hand: $\Gamma = \mathcal{SL}_3(\mathbb{Z})$, $M = \mathbb{Z}^3$, and E is an extension that is split over the root subgroups Λ_i. The free product can be viewed as automorphisms of a tree $\mathfrak{B}$, whose vertices are the Γ-conjugates of the Λ_i (see [19]). Thus, any completion of this amalgam embeds as automorphisms of the corresponding quotient of $\mathfrak{B}$:

$$\begin{array}{ccccc} \Lambda_1 *_\Delta \Lambda_2 = F & \to & E & \to & \Gamma \\ \cup & & \cup & & \\ \pi_1(\mathfrak{K}) = \pi & \to & M & & \\ \cup & & & & \\ \pi' & & & & \end{array} \qquad \begin{array}{ccccc} \mathfrak{B} & \to & \mathfrak{G} & \to & \mathfrak{K} \\ & & \| & & \| \\ & & \mathfrak{B}/\pi' & & \mathfrak{B}/\pi \\ & & \underbrace{\phantom{\mathfrak{B}/\pi' \quad \mathfrak{B}/\pi}}_{M} & & \end{array}$$

Note that $\mathfrak{K}$ embeds in the graph $\mathfrak{K}_\mathbb{Q}$ on point-line antiflags in $\mathbb{P}(M_\mathbb{Q})$, where a pair of antiflags determine an edge when their geodesic closure in the Tits building is an apartment. Alternatively, the graph $\mathfrak{K}_\mathbb{Q}$ is simply the commuting graph on reflections in $\mathcal{GL}_3(\mathbb{Q})$, and was studied by A. Wagner (see [24]).

Since $\mathrm{H}_1(\mathfrak{K})$ is the abelianization of the covering transformations π, we wish to find an explicit homomorphism $\mathrm{H}_1(\mathfrak{K}_\mathbb{Q}) \to M$ that is nontrivial on $\mathrm{H}_1(\mathfrak{K})$. Each such homomorphism defines an extension E of the right shape.

Note that if E were generated by the Π_i, then we obtain an analogous graph, which is simply be the Tits building of $\mathbb{P}(\mathbb{Q}^3)$. (Since $\mathbb{Z}$ is a 'Bezout ring', Γ is flag-transitive on the building, and so it does not matter whether we work over $\mathbb{Z}$ or $\mathbb{Q}$.) Its homology is the Steinberg module, and there are no nonzero homomorphisms from the Steinberg module to M (see [15, 16]), whence E would be split. This justifies our restriction φ to the Π_i, in the previous section.

Let us return to $\mathfrak{K}_{\mathbb{Q}}$. Now $\Gamma_{\mathbb{Q}}$ has rank 8 on antiflags: a basepoint $*$; a single suborbit at distance 1; 5 suborbits at distance 2, of which 3 contain points joined uniquely to $*$ by a geodesic; and a single suborbit at distance 3. If $* = p, L$, and p', L' is a second antiflag, then we label p', L' according to the following scheme:

$$
\begin{array}{llll}
*: & p' = p;\ L' = L & \delta: & p' \notin L;\ p \notin L';\ l \cap L' \notin p + p' \\
\alpha: & p' \in L - \{p\};\ p \in L' - \{p'\} & \epsilon: & p' = p;\ L' \neq L \\
\beta: & p' \notin L;\ p \in L' - \{p'\} & \zeta: & p' \neq p;\ L' = L \\
\gamma: & p' \in L - \{p\};\ p \notin L' & \eta: & p' \notin L;\ p \notin L';\ L \cap L' \in p + p'
\end{array}
$$

So, there are two types of quads: $* \to \alpha \to \epsilon \to \alpha \to *$ and $* \to \alpha \to \zeta \to \alpha \to *$.

Following a suggestion of D. Evans, we view the oriented edges in $\mathfrak{K}_{\mathbb{Q}}$ as corresponding naturally to ordered frames. Thus, $\mathrm{H}_1(\mathfrak{K}_{\mathbb{Q}})$ is a homomorphic image of the permutation module Φ on frames $p.q.r$. The boundaries $\mathrm{B}_1(\mathfrak{K}_{\mathbb{Q}})$ are sums of terms of the form $p.q.r + r.q.p$, and so we may replace Φ by the module induced from the sign character of $\mathfrak{S}_3$.

An oriented triangle is represented in Φ as $p.q.r + q.r.p + r.p.q$, which is left invariant by a subgroup isomorphic to $\mathfrak{A}_4$. Since this subgroup leaves invariant no nonzero vector in M, we deduce that the portion of the homology generated by the triangles maps to 0 under the homomorphism we seek.

An oriented quad

is represented in Φ as $p_1.x.p_2 + p_2.x.p_3 + p_3.x.p_4 + p_4.x.p_1$, where $x = L_1 \cap L_2$. Let this quad be denoted $\phi(x; p_1, p_2, p_3, p_4)$. On the other hand, an oriented quad of dual type is easier to describe through a dual description of the group of chains. So, an ordered triple of noncollinear points $p.p'.p''$ corresponds to an ordered triple of noncoïncident lines $L.L'.L''$, where $p = L' \cap L''$, $p' = L'' \cap L$, and $p'' = L \cap L'$. Thus, the oriented quad of dual type correspond to sums of the form $L_1.X.L_2 + L_2.X.L_3 + L_3.X.L_4 + L_4.X.L_1$, where $X = p_1 + p_2$.

The portion of Φ generated by the first type of quad is induced from the span of $\{\phi(x; p_1, p_2, p_3, p_4) | p_i \neq x\}$, considered as a module for the stabilizer of a point Γ_x. Let the span of the quads and triangles be denoted Θ; and let the span of the quads based at x be denoted Φ_x. Choose a generator v for the projective point x. Take a basis of Φ_x consisting of quads, and map any part of this basis to v, the rest to 0. This gives a Γ_x-equivariant map $\Phi_x \to M$, from which we can induce to a Γ-equivariant map $\Theta \to M$. (Recall that the triangles map to 0.) These maps span all equivariant maps $\Theta \to M$, so if all yield residually finite extensions, then the explicit homomorphism we seek is defined modulo the triangles and quads.

This idea runs out of gas at the pentagons, since these have trivial centralizers. One needs to address the pentagons, since one can show that the triangles and quads generate a proper submodule. On the other hand, the pentagons and triangles *do* generate all of the homology. Moreover, there does not appear to be any invariant complement to Θ in $\mathrm{H}_1(\mathfrak{K}_{\mathbb{Q}})$.

I expect that the above homomorphisms extend to all of the homology, but even so I have as yet no means of deciding whether they result in residually finite extensions, other than to verify that the 2-cocycles they yield have infinite order. Nevertheless, this approach seems promising, and appears to be tailor-made for the low-rank groups that Borel's theory does not cover.

References

[1] H. Bass, J. Milnor, and J.P. Serre, Solution of the congruence subgroup problem for SL_n $(n \geq 3)$ and Sp_{2n} $(n \geq 2)$, *Publ. Math. Inst. Hautes Études Sci.* **33**(1967), 59–137.

[2] G.W. Bell, On the cohomology of the finite special linear groups I–II, *J. Algebra* **54**(1978), 216–238, 239–259.

[3] A. Borel, Stable and L^2-cohomology of arithmetic groups, *Bull. Amer. Math. Soc.* (N.S.) **3**(1980), 1025–1027. [Œuvres/Collected Papers, v. III. New York: Springer-Verlag, 1983; pp. 614–616.]

[4] K.S. Brown, *Cohomology of Groups* (Springer-Verlag, New York, 1982).

[5] L.S. Charlap and A.T. Vasquez, The cohomology of group extensions, *Trans. Amer. Math. Soc.* **124**(1966), 24–40.

[6] L.S. Charlap and A.T. Vasquez, Characteristic classes for modules over groups, *Trans. Amer. Math. Soc.* **137** (1969), 533–549.

[7] D.M. Evans and P.R. Hewitt, Counterexamples to a conjecture on relative categoricity, *Ann. Pure Appl. Logic* **46** (1990), 201–209.

[8] E.M. Friedlander and B.J. Parshall, On the cohomology of algebraic and related finite groups, *Invent. Math.* **74** (1983), 85–117.

[9] E.M. Friedlander and B.J. Parshall, Cohomology of infinitesimal and discrete groups, *Math. Ann.* **273** (1986), 353–374.

[10] P.R. Hewitt, Extensions of residually finite groups, *J. Algebra*, to appear.

[11] P.R. Hewitt, An application of Lubotzky's criterion to extensions of arithmetic groups over their rational modules, in *Group Theory: Proceedings of the Bi-*

ennial Ohio State-Denison Conference (S. Sehgal and R. Solomon (eds.), NJ: World Scientific, River Edge, 1993).

[12] J.C. Jantzen, *Representations of Algebraic Groups* (New York: Academic Press, 1987).

[13] A. Lubotzky, A group theoretic characterization of linear groups, *J. Algebra* **113**(1988), 207–214.

[14] H. Matsumoto, Sur les sous-groupes arithmétiques des groupes semi-simples déployés, *Ann. Sci. École Nor. Sup.* **2**(1969), 1–62.

[15] M. Reeder, The Steinberg module and the cohomology of arithmetic groups, *J. Algebra* **141**(1991), 287–315.

[16] M. Reeder, Modular symbols and the Steinberg representation (Springer L.N.M. **1447** (1992)), 287–302.

[17] L. Ribes, *Introduction to Profinite Groups and Galois Cohomology* (Queen's Papers in Pure and Applied Mathematics **24**, Kingston, Ont.: Queen's University, 1970).

[18] Ch.H. Sah, Cohomology of split group extensions, I–II, *J. Algebra* **29**(1974), 255–302; and **45**(1977), 17–68.

[19] J.P. Serre, *Trees* (Springer-Verlag, New York, 1980).

[20] J.P. Serre, Cohomologie des groupes discrets, *Ann. of Math. Stud.* **70**(1971), 77–169.

[21] Ch. Soulé, The cohomology of $SL_3(Z)$, *Topology* **17**(1978), 1–22.

[22] R. Steinberg, Générateurs, relations et revêtements de groupes algébriques, in *Colloque sur la Théorie des Groupes Algébriques* (Bruxelles, 1962), 113–127; and Generators, relations and coverings of algebraic groups, II, *J. Algebra* **71**(1981), 527–543.

[23] O. Taussky and H. Zassenhaus, On the 1-cohomology of the general and special linear groups, *Aequationes Math.* **5**(1970), 129–201.

[24] A. Wagner, Determination of the finite primitive reflection groups over an arbitrary field of characteristic not two, I, II, III, *Geom. Dedicata* **9**(1980), 239–253; **10**(1981), 191–203, 475–523.

ON SOME QUESTIONS CONCERNING SUBNORMALLY MONOMIAL GROUPS

E. HORVÁTH

Department of Mathematics, Faculty of Mechanical Engineering, Budapest University of Technology, 1521 Budapest, Hungary

In his papers [4], [5] and [6] Guan-Aun How described some properties of SM (subnormally monomial) groups. He proved that the class of SM-groups is the intersection of the class of CSF (chiefly sub-Frobenius) and the class X of those solvable groups G, for which for all primes p and for all subgroups A, $O^p(A)$ has no central p-factor. Among other things, he proved that the class of SM-groups is closed under taking direct products, factor groups and subgroups. First we consider the relation between SM-groups, subgroup-closed M-groups and supersolvable groups, showing that these classes are all distinct. For the class of Frobenius groups the first two classes coincide, and they properly contain the class of supersolvable Frobenius groups. For Frobenius complements the three classes are equal. The class SM is not closed under extensions. We show that even the split extension of an abelian group with an SM-group can be non-SM. On the basis of the notion of relative M-groups, we introduce the notion of relative SM-groups. We investigate whether some results which are known to be true for relative M-groups, remain true for relative SM-groups or not. Some of them remain true: we show that every SM-group is relative SM with respect to every abelian normal subgroup. According to [7], if G/N is supersolvable, then G is relative M-group with respect to N. The analogous statement is not true for SM: it may happen that G/N is supersolvable, but G is not relative SM with respect to N. However, if G/N is nilpotent, then G is relative SM with respect to N. Every Frobenius SM-group is relative SM with respect to its kernel, moreover, this property characterizes Frobenius SM-groups. We also show that a Frobenius group is SM if and only if its complement is SM. So fixed point free extension with an SM-group is always SM. Finally we show an algorithm which tests whether a group is SM or not. This algorithm can be implemented in GAP, it has been tested on some groups already (e.g. on Examples 1.5 and 1.6). It is of polynomial complexity as a function of the order of the group, provided that the multiplication table of the group is known. Throughout this paper all groups are finite and all characters are complex-valued.

1.

Definition 1.1. An irreducible character of a group G is called *monomial*

if it can be induced from a linear character of a subgroup of G. A group G is called a *monomial group (M-group)* if each of its irreducible characters is monomial.

Definition 1.2. An irreducible character of a group G is called *subnormally monomial* character *(SM-character)* if it can be induced from a linear character of a subnormal subgroup of G. A group G is called *subnormally monomial* group *(SM-group)* if each of its irreducible characters is an SM-character.

Definition 1.3. A group is called a *subgroup-closed M-group (MU-group)* if all of its subgroups are M-groups.

Remark 1.4. It is well-known that the class of M-groups is not closed under taking subgroups. The class of MU-groups was characterized by Price [8]. In his papers Guan-Aun How shows that SM-groups are MU-groups. However not every MU-group is SM, as we see from the following example.

Example 1.5. In their paper [1] Brewster and Yeh described a non-M-group T which is the subdirect product of two MU-groups T/T_1 and T/D where T_1 and D are minimal normal subgroups of T. Actually $T/T_1 \cong T/D$ and this group cannot be SM as then T would be SM, too. Let us denote T/T_1 by H. Its structure is the following: $H \cong P split\langle\alpha\rangle$, where P is a 2-group of order 2^9 with automorphism α of order 5 such that:

(i) P' is elementary abelian of order 2^5

(ii) $Z(P)$ has order 2

(iii) α acts irreducibly on P/P' and on $P'/Z(P)$

Its irreducible characters of degree 16 cannot be induced from any subnormal subgroup. According to [2], a similar construction is possible for odd order groups, namely instead of P one can take a group of order 3^9 with an automorphism α of order 5.

Supersolvable groups are MU-groups, but as the above H shows, not all MU-groups are supersolvable. The class of SM-groups is also different from the class of supersolvable groups. A_4 gives a simple example of a nonsupersolvable group which is subnormally (even normally) monomial. The next example shows that supersolvability does not imply SM.

Example 1.6. Let

$$G = \langle x_1, x_2, z, t | x_1^3 = x_2^3 = z^3 = t^2 = 1,$$
$$[x_1, x_2] = z, [x_1, t] = x_1, [x_2, t] = x_2, [z, t] = 1\rangle$$

be an extraspecial group P of order 3^3 of exponent 3, extended by an automorphism of order 2, acting reducibly on $P/Z(P)$. This group is supersolvable but not SM, as its degree 3 characters cannot be induced from any subnormal subgroup. Supersolvable groups are chiefly sub-Frobenius (that is for each chief-factor K/L and each $k \in K$, $C_G(kL)$ is subnormal (even normal) in G), but they need not belong to class X, mentioned in the introduction, as the above example shows.

2.

It is easy to see that the split extension of SM-groups by SM-groups need not be SM-groups, even not M-groups, e.g. Q_8 *split* $Z_3 \cong SL(2,3)$. In [3] we proved that the split extension of an abelian group by an MU-group is always an M-group. The analogous statement for SM-groups is not true, as the following example shows:

Example 2.1. Let

$$G = (Z_2 \wr Z_2^2) \wr Z_3 \cong Z_2^{12} \textit{ split } (Z_2^6 \textit{ split } Z_3).$$

Here Z_2^{12} is abelian, Z_2^6 *split* Z_3 is abelian by nilpotent, so by [5] it is SM, but as $Q_8 \wr Z_3 \leq (Z_2 \wr Z_2^2) \wr Z_3$, so G cannot be SM, in fact not even MU.

The necessary and sufficient condition for the split extension of an abelian group by an SM-group to be SM, is the following

Proposition 2.2. *If $G = A$ split B with $A \lhd G$ abelian and B an SM-group, then G is SM, if and only if, for every $\tau \in Irr(A)$, $I_G(\tau)$ is subnormal in G.*

PROOF. If G is SM, then by Theorem 3.4 in [5], for every $\tau \in Irr(A)$, $I_G(\tau)$ is subnormal in G. Let us suppose now that for every $\tau \in Irr(A)$, $I_G(\tau)$ is subnormal in G. We prove that G is SM. Let G be a counterexample of minimal order with $\chi \in Irr(G)$, a non-SM character. Then $Ker(\chi) \not\geq A$. Let $\tau \in Irr(A)$ be such that $(\chi_A, \tau) \neq 0$. Then τ is linear. If $I_G(\tau) = G$ then there exists a $\tilde{\tau} \in Irr(G)$ such that $\tilde{\tau}_A = \tau$. Then we have $\tau^G = \sum_{\omega \in Irr(G/A) = Irr(B)} \omega(1)(\omega\tilde{\tau})$. So there exists an $\omega \in Irr(G/A)$ such that $\chi = \omega\tilde{\tau}$. As G/A is SM, there exists a subnormal subgroup H containing A and a linear character λ of H such that $\omega = \lambda^G$. Then $\omega\tilde{\tau} = (\lambda\tilde{\tau}_H)^G$, which is a contradiction. So we may assume that $I_G(\tau) < G$. By induction $I_G(\tau) = A$ *split* $(I_G(\tau) \cap B)$ is SM. But $\chi = \vartheta^G$ for some $\vartheta \in Irr(I_G(\tau))$ and by assumption $I_G(\tau)$ is subnormal in G. By induction $\vartheta = \lambda^{I_G(\tau)}$ for some linear character λ of a subnormal subgroup of $I_G(\tau)$. By transitivity of induction and that of subnormality, χ is an SM-character. This contradicts to the assumption. □

The notion of relative M-characters is well-known, see e.g. [7]. We introduce analogously the notion of relative SM-character

Definition 2.3. A character $\chi \in Irr(G)$ is said to be a *relative SM-character* with respect to a normal subgroup N if it can be induced from a character $\xi \in Irr(H)$ for a subnormal subgroup H of G containing N, so that $\xi_N \in Irr(N)$. A group G is said to be *relative SM-group* with respect to normal subgroup N if every irreducible character of G is relative SM-character with respect to N.

It is well-known that every irreducible character of an M-group is relatively monomial with respect to any abelian normal subgroup. We prove analogous statement for SM-groups:

Proposition 2.4. *Let G be SM and $A \lhd G$ abelian. Then G is relative SM with respect to the normal subgroup A.*

PROOF. Let $\chi \in Irr(G)$. Then there exists a subnormal subgroup H in G and a linear character $\lambda \in Irr(H)$ such that $\lambda^G = \chi$. So $\lambda^{AH} \in Irr(AH)$. Then $\lambda_{A\cap H}$ extends to a $\tilde{\lambda} \in Irr(A)$, and so

$$(\lambda^{AH})_A = (\lambda_{A\cap H})^A = \sum_{\beta \in Irr(A/A\cap H)} (\tilde{\lambda}\beta).$$

Let $\alpha = \tilde{\lambda}\beta$ for some $\beta \in Irr(A/A \cap H)$. Then there exists a unique $\vartheta \in Irr(I_{AH}(\alpha))$ such that $\vartheta^{AH} = \lambda^{AH}$ and $(\vartheta_A, \alpha) = ((\lambda^{AH})_A, \alpha) = 1$. As α is $I_{AH}(\alpha)$-invariant, $\vartheta_A = \alpha$, and so ϑ is linear. According to Th. 3.4. in [5], as AH is SM and $A \lhd AH$, and for $\alpha \in Irr(A)$, $A/Ker(\alpha)$ is nilpotent, we have that $I_{AH}(\alpha)$ is subnormal in AH, which is subnormal in G. As $\chi = \vartheta^G$ for a linear character $\vartheta \in Irr(I_{AH}(\alpha))$ and $A \leq I_{AH}(\alpha) \lhd \lhd G$, χ is a relative SM-character with respect to A, and we are done.

Corollary 2.5. *Let G be a group, $A \lhd G$ abelian, then G is SM if and only if it is relative SM with respect to A.*

PROOF. One direction follows from Proposition 2.4, the other direction is trivial. □

Remark 2.6. The if part of Corollary 2.5 is true also locally, namely if χ is relative SM with respect to the abelian normal subgroup A, then χ is SM. We could not prove the local statement for the other direction. However it is easy to see that $\chi \in Irr(G)$ is SM if and only if it is relative SM with respect to 1. If G is a relative SM-group with respect to a normal subgroup N then G/N is obviously an SM-group. On the other hand, it may happen that G/N is an SM-group, but G is not relative SM with respect to N. E.g. if $G = SL(2,3)$, $N = Z(G)$, then $G/N \cong A_4$, which is SM, but G cannot be relative SM with respect to N, as then G would be SM, too, but G is not even an M-group. Each non-SM M-group having a nontrivial abelian normal subgroup N, is a relative M-group but not a relative SM-group with respect to N (See Example 1.6 or Example 2.1). A relative SM-group can be non-SM, as each solvable group is relative SM with respect to every maximal normal subgroup of it. In [7] it is proved that if $N \triangleleft G$, and G/N is supersolvable, then G is a relative M-group with respect to N. The analogous statement is not true for SM, as the group G of Example 1.6 with normal subgroup $N = \{1\}$ shows. However if G/N is nilpotent, then the analogous statement for SM is true:

Proposition 2.7. *Let G be a group with normal subgroup N such that G/N is nilpotent. The G is relative SM with respect to N.*

PROOF. The proof is similar to that of Theorem 6.22 in [7]. □

It is well-known, see e.g. [9], that a Frobenius group is an M-group if and only if it is already an MU-group. We consider when a Frobenius group is SM.

Proposition 2.8. *Let G be a Frobenius group with kernel N and complement H. Then the following statements are equivalent:*

(i) *G is SM*

(ii) *G is relative SM with respect to N*

(iii) *H is SM*

(iv) *H is an M-group*

(v) *H' is cyclic*

(vi) *H is supersolvable*

(vii) *G is MU*

(viii) *H is MU*

(ix) *G is an M-group*

PROOF. *(i)* → *(ii)* If $\chi \in Irr(G)$ then either $\chi = \xi^G$ for a $\xi \in Irr(N)$, and then χ is relative SM with respect to N, or $Ker(\chi) \geq N$. In the second case, as $\chi = \lambda^G$ for some linear character λ of a subnormal subgroup H of G, we have that $N \leq Ker(\chi) = Ker(\lambda^G) \leq Ker(\lambda)$, and $\lambda_N = 1_N$, so χ is also relative SM with respect to N.

(ii) → *(iii)* As $H \cong G/N$, the statement is trivial from Remark 2.6.

(iii) → *(i)* Let $\chi \in Irr(G)$. If $Ker(\chi) \geq N$ then $\chi \in Irr(G/N) = Irr(H)$, so it is SM. If $\chi = \eta^G$ for $\eta \in Irr(N)$, then as N is nilpotent, η is SM, and by transitivity of subnormality and induction, χ is SM.

(iii) → *(iv)* This is trivial.

(iv) → *(v)* This follows from Lemma 2.10 in [6].

$(v) \to (vi)$ This is trivial.

$(vi) \to (iii)$ If H is supersolvable then H is an M-group, so, as we have seen, it implies that H' is cyclic. Therefore H is metabelian, which implies that H is SM.

$(i) \to (vii)$ This follows from the results of Guan-Aun How, mentioned in the introduction.

$(vii) \to (viii)$ This is trivial, as MU is inherited by factors.

$(viii) \to (iv)$ This is trivial.

$(ix) \to (iv)$ This is trivial, as M is inherited by factors.

$(iv) \to (ix)$ This is easy to prove, see e.g. [3].

Thus we have shown that all the statements $(i) - (ix)$ are equivalent. □

Corollary 2.9. *The classes SM, MU and that of supersolvable groups coincide for Frobenius complements. For Frobenius groups SM and MU coincide, and they properly contain the class of supersolvable Frobenius groups.*

PROOF. The first two statements are obvious. If G is a supersolvable Frobenius group, then its complement H is also supersolvable, so according to the above proposition, G is SM and MU. Howeve A_4 is a Frobenius group, which is SM and MU, but not supersolvable, so the containment of this class is proper. □

Corollary 2.10. *A fixed point free extension of a group by an SM-group is again SM.*

PROOF. A fixed point free extension by an SM-group means taking a Frobenius group with an SM complement. Then by Proposition 2.8, G is SM, too. □

3.

Finally we give an algorithm which determines whether a group G is SM or not. The algorithm is the following:

For $\chi \in Irr(G)$ find a maximal normal subgroup N of G, such that $\chi_N \notin Irr(N)$. Then take a $\vartheta \in Irr(N)$ such that $(\chi_N, \vartheta) \neq 0$. If $\vartheta(1) > 1$, then find a maximal normal subgroup N_1 in N such that $\vartheta_{N_1} \notin Irr(N_1)$ etc. We state that G is SM, if and only if this algorithm stops for each irreducible character χ of G, only at a linear character of a subnormal subgroup of G. Then χ can be induced from this character.

PROOF. If G is SM, then for $\chi \in Irr(G)$ with $\chi(1) > 1$, there exists a subnormal subgroup L of G and a linear character $\lambda \in Irr(L)$, such that $\lambda^G = \chi$. But then there exists a maximal normal subgroup N in G which contains L, so λ^N is an irreducible constituent of χ_N, and χ is induced from it, so χ_N cannot be irreducible. Let us denote an irreducible constituent of χ_N by ϑ. If G is SM, then as SM is inherited by subgroups, N is SM, too. So we see similarly as above, that if $\vartheta(1) > 1$ then there exists a maximal normal subgroup in N, such that the restriction of ϑ to it is not irreducible, and it can be induced from the irreducible components. So if G is SM then for each irreducible character this algorithms stops only when it reaches a linear character of a subnormal subgroup. On the other hand, if for each irreducible character of a group the algorithms reaches a linear character of a subnormal subgroup, then the group is SM, as we found the linear characters of subnormal subgroups, from where the irreducible characters of G are induced. □

We need not use this algorithm, if we already know that the group is nilpotent, as then we get automatically that G is SM. Otherwise we have to do computations of the following complexity:

If the order n of the group G is given together with the multiplication table of the group, then in polynomial time in n we get the character table of G, with the Dixon-Schneider algorithm. The number of maximal normal subgroups at each step is bounded by the number of irreducible characters of the given subgroup, which is at most n. The classes belonging to a maximal normal subgroup are determined by the character table. From their representatives one gets in polynomial time in n the maximal normal subgroups. With each irreducible character we may take at most $\log_2(n)$ steps down in a subnormal series, and at each step we have to consider at most n^2 characters of at most n maximal normal subgroups of the actual subnormal subgroup, whether our character, got from the previous step, is induced from them, or not. At each level, we calculate the character table of the subnormal subgroup, which is done in polynomial time in the order of the subgroup, that is again at most n. Summing up all these, we still get polynomial time, in the order of the group.

Acknowledgements. This paper was written during my 3 months stay in Aachen at Lehrstuhl D für Mathematik with a DAAD fellowship. I would like to express my gratitude to Professor Neubüser and his colleagues, for the possibility to work with them and to learn the GAP system. My participation in the conference was supported by the SOROS Foundation, the Technical University of Budapest and the Hungarian National Science Foundation Grant No. T 7441. Thanks for their support. I am also grateful to Professor L. G. Kovács for sending me Guan Aun How's papers, and for

Gábor Ivanyos for consulting on some questions. Finally, I am grateful for the referee's comments, which were very useful for me.

References

[1] B. Brewster, G.Yeh, Closed Subclasses of M-groups, *J. Algebra* **146**(1992), 18–29.

[2] T. Breuer, Construction of certain p-groups, manuscript.

[3] E. Horváth, On a problem of finite solvable groups (in Hungarian), *Matematikai Lapok*, **34** 1-3 (1987), 99–108.

[4] G.A. How, Special classes of monomial groups I, *Chinese J. Math.* **12** 2 (1984).

[5] G.A. How, Special classes of monomial groups II, *Chinese J. Math.*, **12** 2 (1984).

[6] G.A. How, Special classes of monomial groups III, *Chinese J. Math.* **12** 3 (1984).

[7] I.M. Isaacs, *Character Theory of Finite Groups* (Acad. Press, New York, 1976).

[8] D.T. Price, Character ramification and M-groups, *Math. Z.* **130**(1973), 326–337.

[9] G.M. Seitz, M-groups and the supersolvable residual, *Math. Z.* **110**(1969), 101-122.

A CONJECTURE CONCERNING THE EVALUATION OF PRODUCTS OF CLASS-SUMS OF THE SYMMETRIC GROUP

JACOB KATRIEL

Department of Chemistry, Technion - Israel Institute of Technology, Haifa 32000, Israel
E-mail: chr09kt@technion

Abstract

A conjecture is proposed concerning the explicit form of the product of a class-sum consisting of a single active cycle and an appropriate number of fixed points, $[(p)]_n \equiv [(1)^{n-p}(p)]_n$, and an arbitrary class-sum $[*]_n \equiv [(1)^{\ell_1}(2)^{\ell_2}\ldots(n)^{\ell_n}]_n$ in the symmetric group algebra. This conjecture specifies the set of class-sums C_i present in the expansion of $[(p)]_n \cdot [*]_n$ as well as the value of the coefficient of each class-sum present, denoted by $[(p)]_n \cdot [*]_n\Big|_{C_i}$. The operation of $[(p)]_n$ on an arbitrary class-sum is represented by means of a certain linear combination of operators whose forms are specified. The coefficients of these operators, which do not depend on n, are referred to as the reduced class-coefficients (RCCs). Explicit expressions for a certain subset of RCCs and recurrence relations for RCCs involving embedded cycles are presented. It is pointed out that, for bridging cycles, one can obtain elimination rules that involve symmetrization over sets of RCCs with common cycle-structures but inequivalent index distributions. Consequently, some of these RCCs remain individually inaccessible.

1. Introduction

The class-algebra of the symmetric group has attracted considerable attention because of its utility in constructing the representations of the group. It has also found applications in graph-theory [8], the quantum-mechanical many-body problem, and other areas. The product of a pair of class-sums C_1 and C_2 is a linear combination with non-negative integral coefficients $C_1 \cdot C_2\Big|_{C_i}$ of class-sums C_i, which we write in the form

$$C_1 \cdot C_2 = \sum_{C_i \in CS_n} C_1 \cdot C_2\Big|_{C_i} C_i \qquad (1)$$

These coefficients are sometimes referred to as the structure constants of the class-algebra. The complete set of coefficients is equivalent to the set of characters of the corresponding symmetric group. Most of the results obtained on this topic ([1], [6], [7], [8], [9], [17]) use the representation theoretic relation

$$C_1 \cdot C_2\Big|_{C_i} = \frac{|C_1|\,|C_2|}{n!} \sum_{\Gamma} \frac{\chi_1^\Gamma \chi_2^\Gamma \chi_i^\Gamma}{|\Gamma|}\,, \tag{2}$$

where Γ ranges over the irreducible representations of S_n (*i.e.*, over the partitions of n), χ_i^Γ is the character corresponding to the class C_i in the irreducible representation Γ, and $|C|$, $|\Gamma|$ stand for the order of the class C and the dimension of the irreducible representation Γ, respectively. Several attempts were reported to derive these coefficients using their combinatorial significance [2], [3], [4], [5], [16].

A combinatorial study of the product of the class of transpositions with an arbitrary class of the symmetric group [15], followed by a similar study of the class of three-cycles [10], motivated the formulation of a conjecture concerning the form of the product of a single-cycle class-sum of arbitrary length with an arbitrary class-sum in the symmetric group algebra [10], [11], [12], [13]. This conjecture is reviewed in the present article.

2. Representation of $[(p)]_n \cdot [*]_n$ in terms of reduced class-sums

The product of a class-sum with cycle-structure $[(p)]_n \equiv [(p)(1)^{n-p}]_n$ and an arbitrary class-sum $[*]_n \equiv [(1)^{\ell_1}(2)^{\ell_2}\ldots(n)^{\ell_n}]_n$ is given by an expression of the form:

$$\begin{gathered}
[(p)]_n \cdot [*]_n = \\
\sum_{\substack{r_1,\ldots,r_p \\ (r_1\cdots+r_p\leq n)}} \ \sum_{\substack{k,\ell \\ (k+\ell+p=\text{odd})}} \ \sum_{(p_1,\ldots,p_k)\vdash p} \ \sum_{(p'_1,\ldots,p'_\ell)\vdash p} \ \sum_{\substack{\text{distinct connected} \\ \text{distributions}}} \\
C < (\overbrace{r_1+\ldots+r_{p_1}}^{p_1})(\overbrace{r_{p_1+1}+\ldots}^{p_2})\ldots(\overbrace{\ldots+r_p}^{p_k});(\overbrace{\cdots}^{p'_1})(\overbrace{\cdots}^{p'_2})\ldots(\overbrace{\cdots}^{p'_\ell}) > \\
\cdot < (\overbrace{r_1+\ldots+r_{p_1}}^{p_1})(\overbrace{r_{p_1+1}+\ldots}^{p_2})\ldots(\overbrace{\ldots+r_p}^{p_k});(\overbrace{\cdots}^{p'_1})(\overbrace{\cdots}^{p'_2})\ldots(\overbrace{\cdots}^{p'_\ell})* >
\end{gathered} \tag{3}$$

To interprete this expression we make the following remarks:

1. The symbol $(p_1,\ldots,p_k) \vdash p$ stands for a partition of p into k non-vanishing parts $p_1 \geq p_2 \geq \ldots \geq p_k > 0$ $(p_1+p_2+\ldots+p_k = p)$.

2. The expression for $[(p)]_n \cdot [*]_n$ involves a sum over p positive integral indices, $r_1, r_2, \ldots, r_p$ which satisfy $r_1 + r_2 + \ldots + r_p \leq n$. The summand is a linear combination of terms, each of which is a product of an operator, to which we refer as the reduced class-sum, and a corresponding rational and non-negative numerical coefficient. The numerical coefficient will be referred to as the *reduced class-coefficient* (RCC).

3. The reduced class-sum is denoted by a symbol of the form

$$< (\overbrace{r_1 + \ldots + r_{p_1}}^{p_1})(\overbrace{r_{p_1+1} + \ldots}^{p_2})\ldots(\overbrace{\ldots + r_p}^{p_k}); (\overbrace{\ldots}^{p'_1})(\overbrace{\ldots}^{p'_2})\ldots(\overbrace{\ldots}^{p'_\ell})* > \qquad (4)$$

where the asterisk indicates that the reduced class-sum operates on the class-sum $*$. The RCC corresponding to it is denoted by prefixing the letter C to the symbol denoting the reduced class-sum, and suppressing the asterisk, on which the RCC does not depend. As the symbol suggests, the reduced class-sum and its RCC are specified by distinct distributions of the indices $r_1, r_2, \ldots, r_p$ within two sets of cycles, which are separated by a semicolon in the symbol. The indices $r_1, r_2, \ldots, r_p$ are distributed in each of the two sets of cycles in a non-repetitive way. The length of each cycle is equal to the sum of a certain subset of indices. The number of indices whose sum specifies the length of a given cycle will be referred to as this cycle's *index–length.* Thus a cycle of index-length k, say $(r_1 + r_2 + \ldots + r_k)$, has an actual length that is at least k (when $r_1 = r_2 = \ldots = r_k = 1$) and at most $n - (p - k)$ (which is possible only if $r_{k+1} = r_{k+2} = \ldots = r_p = 1$). A cycle of index-length k will be denoted by $((k))$, when convenient.

4. For a particular set of values of the indices $r_1, r_2, \ldots, r_p$, the reduced class-sum, (4), generates a linear combination of two class-sums from the original class-sum $*$. In one of them, a set of cycles of the lengths specified by one of the two sets of index-sums is eliminated and a set specified by the second is inserted. The other class is generated by interchanging the roles of the two sets of cycles. If either set contains a cycle of a length that does not appear in the original class $*$, the term involving the elimination of that set vanishes.

5. When the two sets of indexed cycles are related to one another by means of a permutation of indices that occupy equivalent positions, the term is referred to as self-associated and is defined to operate only once on the original class $*$.

6. The reduced class-sum is defined so as to imply that each class generated in the procedure described above should be multiplied by

$$\prod_{i=1}^{n} \left[i^{\lambda_i} \cdot \lambda_i! \binom{\ell_i + \lambda_i - \mu_i}{\lambda_i} \right]$$

where λ_i is the number of cycles of actual length i inserted, μ_i the number of cycles of that length eliminated and ℓ_i the number of cycles of the same length that were present in the original class $*$.

7. In order to specify the set of reduced class-sums which appear in the product $[(p)]_n \cdot [*]_n$ we note:

a) A cycle of length λ_i has parity $\lambda_i + 1$. The elimination of one set of cycles and the insertion of another conserves parity if the difference between the number of cycles eliminated and that of the cycles inserted is even, and reverses it if this difference is odd. Thus, the above difference should be even if p is odd and odd if p is even. This is the origin of the condition $k+\ell+p=$ odd specified in eq. (3).

b) The distributions of $r_1, r_2, \ldots, r_p$ within the two sets of cycles comprising any reduced class-sum should satisfy the following connectedness property: For any partitioning of these indices into two subsets there is at least one pair of indices - one from each subset - that appears in a common cycle.

We note in passing that the set of reduced class-sums specified by these two conditions is in general overcomplete (*cf.* [11]).

8. Making use of appropriately chosen (low n) special cases, the RCCs can be determined by equating the products obtained using the expressions in terms of reduced class-sums with the products obtained using the representation-theoretic relation, eqs. (1, 2). This is the procedure that was originally used to determine the RCCs corresponding to single-cycle class-sums with $p \leq 8$ ([10], [11], [12]). An incomplete representation-free algorithm is presented below.

For RCCs in which (at least) one of the two sets consists of a single cycle of index-length p, a closed form expression was derived in [11] using a result due to Boccara [4]. In this case the distribution of indices is unique and the reduced class-sums can be fully specified by the index-lengths of the cycles involved. The explicit expression for RCCs containing a single cycle on at least one side of the semicolon is

$$C < ((p)); ((p_1))((p_2))\ldots((p_k)) > \\ = \frac{(p-1)!}{(p+1)\left(\prod_{i=1}^{k} p_i\right)\left(\prod_{j=1}^{p} \ell_j!\right)} \sum_{m=0}^{k}(-1)^m \sum_{I_m \subseteq K} g(I_m) \tag{5}$$

where ℓ_j is the number of cycles of index-length j on the right-hand-side of the semicolon, I_m is an m-element subset of $K=\{1, 2, \ldots, k\}$, $g(I_m)= (-1)^{S(I_m)}\binom{p}{S(I_m)}^{-1}$ and $S(I_m)=\sum_{i\in I_m} p_i$.

3. Elimination of fully embedded cycles of a given length

Recurrence relations have been formulated for terms containing cycles consisting of indices all of which are contained in the same cycle of the complementary set ([11], [12], [13]). From now on r_i will be abbreviated to i within the cycles specifying an RCC and Greek letters $\alpha, \beta, \ldots$ will be used to highlight particular sets of indices. To formulate the recurrence relations we define two cycles of the same index-length which appear on the same side of the semicolon to be equivalent if their indices are distributed in an equivalent way within the cycles on the other side of the semicolon.

The simplest recurrence relation, involving the elimination of a cycle of unit index-length, is

$$C\langle(\alpha+1+2+\ldots+k)A;(\alpha)B\rangle$$
$$= \frac{k}{\gamma_1(B)+1}\cdot\frac{\gamma_k(A)+1}{\gamma_{k+1}(A)+1}C\langle(1+2+\ldots+k)A;B\rangle \tag{6}$$

where $\gamma_1(B)$ is the number of cycles of unit index-length in B whose indices appear in $(\alpha+1+2+\ldots+k)$; $\gamma_{k+1}(A)$ is the number of cycles of index-length $k+1$ in A that are equivalent to $(\alpha+1+2+\ldots+k)$; $\gamma_k(A)$ is the number of cycles of index-length k that are equivalent to $(1+2+\ldots+k)$.

As was explained in [11], the use of the closed form expression for the single cycle RCCs and of the recurrence relation for RCCs with a cycle of unit index-length enables the determination of all the RCCs needed for the evaluation of the product of $[(p)(1)^{n-p}]_n$ with any class-sum of the form $[(q)(1)^{n-q}]_n$. If $q<p$ it may be more efficient to evaluate the same product using the q-index expression for $[(q)]_n\cdot[(p)]_n$.

The conjecture described in the previous section, along with the recurrence relation just specified, was used recently [14] to evaluate the following set of coefficients:

$$[(2)]_n^k\Big|_{[(2)^{\ell_2}(3)^{\ell_3}\ldots(n)^{\ell_n}]_n}(\textstyle\sum_{i=2}^n(i-1)\ell_i=k),\quad [(p)]_n\cdot[(2)^k]_n\Big|_{[(2)^q]_n},$$
$$[(p)]_n\cdot[(q)]_n\Big|_{[(2)^k]_n}\ \text{and}\ [(p)]_n\cdot[(q)]_n\Big|_{[(r)]_n}$$

The recurrence relation

$$C\langle(\alpha+\beta+1+2+\ldots+k)A;(\alpha+\beta)B\rangle=$$
$$=\sum_{\substack{k_1\le k_2\\(k_1+k_2=k)}}\frac{k_1k_2}{\gamma_2(B)+1}\sum_{\tau_{k_1,k_2}(B)}\frac{\gamma_{\tau_{k_1,k_2}}(A)+1}{\gamma_{k+2}(A)+1}C\langle\tau_{k_1,k_2}(B)A;B\rangle$$

involves the elimination of an embedded cycle of index-length two. The symbol τ_{k_1,k_2} stands for a pair of cycles, of index-lengths k_1 and k_2 respectively, consisting of a particular distribution of the indices 1, 2, ..., k. The sum runs over all index distributions which yield distinct RCCs. Note that the range of this sum depends on the manner in which the k_1+k_2 indices are distributed in B. $\gamma_{\tau_{k_1,k_2}}(A)$ is the number of pairs of cycles on the side of the semicolon containing A which are equivalent to the pair specified by τ_{k_1,k_2}. Note that the same cycle can appear in more than one pair. The other factors are defined in analogy with the previous recurrence relation. One immediate consequence of this recurrence relation is that RCCs in which a cycle of index-length two is embedded in a cycle of index-length three - vanish.

For an embedded cycle of index-length three

$$\begin{aligned}
&C\langle(\alpha+\beta+\gamma+1+2+\ldots+k)A;(\alpha+\beta+\gamma)B\rangle \\
&=\binom{k+2}{3}\cdot\frac{1}{\gamma_3(B)+1}\cdot\frac{\gamma_k(A)+1}{\gamma_{k+3}(A)+1}\cdot C\langle(1+2+\ldots+k)A;B\rangle \\
&+2\cdot\sum_{\substack{k_1\le k_2\le k_3\\(k_1+k_2+k_3=k)}}\sum_{\tau_{k_1k_2k_3}}\frac{k_1k_2k_3}{\gamma_3(B)+1}\cdot\frac{\gamma_{\tau_{k_1k_2k_3}}(A)+1}{\gamma_{k+3}(A)+1}\cdot C\langle\tau_{k_1k_2k_3}A;B\rangle
\end{aligned}$$

and for an embedded cycle of index-length four

$$\begin{aligned}
&C\langle(\alpha+\beta+\gamma+\delta+1+2+\ldots+k)A;(\alpha+\beta+\gamma+\delta)B\rangle \\
&=2\sum_{\substack{k_1\le k_2\\(k_1+k_2=k)}}\sum_{\tau_{k_1k_2}(B)}\frac{k_1(k_1+1)k_2(k_2+1)}{\gamma_4(B)+1}\cdot\frac{\gamma_{\tau_{k_1k_2}}(A)+1}{\gamma_{k+4}(A)+1}\cdot C\langle\tau_{k_1k_2}A;B\rangle \\
&+12\sum_{\substack{k_1\le k_2\le k_3\le k_4\\(k_1+k_2+k_3+k_4=k)}}\sum_{\tau_{k_1k_2k_3k_4}(B)}\frac{k_1k_2k_3k_4}{\gamma_4(B)+1}\cdot\frac{\gamma_{\tau_{k_1k_2k_3k_4}}(A)+1}{\gamma_{k+4}(A)+1}\cdot C\langle\tau_{k_1k_2k_3k_4}A;B\rangle
\end{aligned}$$

One immediate consequence is that RCCs in which a cycle of index-length four is embedded in a cycle of index length five vanish even when they satisfy the parity and connectedness criteria. Thus, while $C\langle(1+2+3+4+5);(1+2+3+4)(5)\rangle$ vanishes by the parity condition and the vanishing of $C\langle(1+2+3+4+5)(6+7);(1+2+3+4)(5+6+7)\rangle$ is accounted for by the fact that a cycle of index-length two is embedded in a cycle of index-length three, the RCC $C\langle(1+2+3+4+5)(6+7+8+9);(1+2+3+4)(5+6+7+8+9)\rangle$ should vanish by the elimination rule for an embedded cycle of index length four.

While we are not yet able to write down the elimination rule for an embedded cycle of arbitrary length in an embedding cycle of arbitrary length, the leading term can be written as follows

$$\begin{aligned}
&C\langle(\alpha_1+\alpha_2+\cdots+\alpha_\ell+1+2+\cdots+k)A;(\alpha_1+\alpha_2+\cdots+\alpha_\ell)B\rangle \\
&=\begin{cases}
\dfrac{2\cdot(\ell-1)!}{\ell+1}\dbinom{k-1+\ell}{\ell}\cdot\dfrac{1}{\gamma_\ell(B)+1}\cdot\dfrac{\gamma_k(A)+1}{\gamma_{k+\ell}(A)+1} \\
\qquad\cdot C\langle(1+2+\cdots+k)A;B\rangle+\cdots & \ell\text{ odd}\\
\dfrac{2\cdot\ell!}{\ell+2}\displaystyle\sum_{\substack{k_1\le k_2\\(k_1+k_2=k)}}\dbinom{k_1-1+\frac{\ell}{2}}{\frac{\ell}{2}}\dbinom{k_2-1+\frac{\ell}{2}}{\frac{\ell}{2}}\cdot\dfrac{1}{\gamma_\ell(B)+1} \\
\qquad\cdot\dfrac{\gamma_{\tau_{k_1k_2}}(A)+1}{\gamma_{k+\ell}(A)+1}\cdot C\langle\tau_{k_1k_2}A;B\rangle+\cdots & \ell\text{ even}
\end{cases}
\end{aligned}$$

In certain special cases, some of which are non-trivial, this leading term is sufficient.

4. Fully embedded cycles: elimination rules of the second kind

The recurrence relations presented in the previous section involve the elimination of a cycle of a given index-length, which is embedded in another cycle of arbitrary index-length. In the present section we still consider the elimination of a fully embedded cycle, but specify an alternative definition of the relation between the cycle to be eliminated and the cycle in which it is embedded.

Instead of formulating rules for the elimination of a cycle of a given index-length, we attempt to formulate rules for the elimination of a cycle whose index-length differs from that of the embedding cycle by some specific integer. It is obvious that a complete set of elimination rules of this type overlaps with a complete set of elimination rules of the first kind. However, given a finite subset of elimination rules of the first kind, they are usefully complemented by a finite subset of elimination rules of the second kind.

The simplest elimination rule of the second kind is

$$C\langle(1+2+\ldots+k+\alpha)A;(1+2+\ldots+k)B\rangle$$
$$=\begin{cases} C\langle(\alpha)A;B\rangle\dfrac{2\cdot(k-1)!}{(k+1)(1+\delta_{k,1}\gamma_1(B))}\cdot\dfrac{\gamma_1(A)+1}{\gamma_{k+1}(A)+1} & k \text{ odd}\\ 0 & k \text{ even}\end{cases}$$

The next elimination rule of the second kind involves the elimination of a cycle of index-length k, embedded in a cycle of index-length $k+2$:

$$C\langle 1+2+\ldots+k+\alpha+\beta)A;(1+2+\ldots+k)B\rangle$$
$$=\begin{cases} C\langle(\alpha)(\beta)A;B\rangle\cdot\dfrac{2\cdot k!}{(k+2)(1+\delta_{k,2}\cdot\gamma_2(B))}\cdot\dfrac{\gamma_{1,1}(A)+1}{\gamma_{k+2}(A)+1} & k \text{ even}\\ C\langle(\alpha+\beta)A;B\rangle\cdot\dfrac{2\cdot(k-1)!}{1+\delta_{k,1}\cdot\gamma_1(B)}\cdot\dfrac{\gamma_2(A)+1}{\gamma_{k+2}(A)+1} & k \text{ odd}\end{cases}$$

The last elimination rule of the second kind to be considered explicitly is

$$C\langle(1+2+\ldots+k+\alpha+\beta+\gamma)A;(1+2+\ldots+k)B\rangle$$
$$=\begin{cases} C\langle(\alpha+\beta+\gamma)A;B\rangle\cdot\dfrac{(k-1)!(k+2)}{\gamma_k(B)+1}\cdot\dfrac{\gamma_3(A)+1}{\gamma_{k+3}(A)+1} & \\ +C\langle(\alpha)(\beta)(\gamma)A;B\rangle\cdot\dfrac{k!(k-1)}{k+3}\cdot\dfrac{1}{\gamma_k(B)+1}\cdot\dfrac{\gamma_{1,1,1}(A)+1}{\gamma_{k+3}(A)+1} & k \text{ odd}\\ k!\cdot\sum_{\tau_{2,1}}C\langle\tau_{2,1}(\alpha,\beta,\gamma)A;B\rangle\cdot\dfrac{1}{\gamma_k(B)+1}\cdot\dfrac{\gamma_{\tau_{2,1}}(A)+1}{\gamma_{k+3}(A)+1} & k \text{ even}\end{cases}$$

where $\tau_{2,1}$ ranges over the distinct distributions of the indices α,β,γ in a cycle of index-length 2 and a cycle of unit index-length.

5. Elimination rules for fully bridging cycles

A cycle will be referred to as fully bridging if each one of the indices specifying its length belongs to a different cycle in the complementary set. The simplest type of term involving bridging cycles is of the form

$$C\langle(\alpha+1+2+\ldots+k)(\beta+1'+2'+\ldots+k')A;(\alpha+\beta)B\rangle \tag{7}$$

By a straightforward parity argument we conclude that elimination of the indices α and β leads to the RCC $C\langle(1+2+\ldots+k+1'+2'+\ldots+k')A;\ B\rangle$. However, if the $k+k'$ indices comprising the amalgamated cycle can be distributed in the complementary set in ways that are not equivalent to the one in our original RCC, eq. (7), the following symmetrized relation

$$\sum_{\tau_{1,2,\ldots,k'}(B)} C\langle(\alpha+1+2+\ldots+k)(\beta+1'+2'+\ldots+k')A;(\alpha+\beta)B\rangle$$
$$\cdot(\gamma_2(B)+1)\cdot\frac{\gamma_{k+1,\,k'+1}(A)+1}{\gamma_{k+k'}(A)+1}$$
$$=\frac{k+k'}{1+\delta_{k,\,k'}}C\langle(1+2+\ldots+k+1'+2'+\ldots+k')A;B\rangle$$

where the sum runs over the distinct distributions of $1,\ 2,\ \ldots\ k,\ 1',\ 2',\ \ldots,\ k'$ in B, is found to hold.

The available data suggests that the general expression for the elimination of a bridging cycle of length ℓ is

$$\sum C\langle\left(\prod_{j=1}^{\ell}(\alpha_j+1_j+2_j+\ldots+k_j)\right)A;(\alpha_1+\alpha_2+\ldots+\alpha_\ell)B\rangle$$
$$\cdot(\gamma_\ell(B)+1)\cdot\frac{\gamma_{k_1+1,\,k_2+1,\,\ldots,\,k_\ell+1}(A)+1}{\gamma_{k_1+k_2+\ldots+k_\ell}(A)+1}$$
$$=(\ell-1)!\frac{k_1+k_2+\ldots+k_\ell}{[\#(k_1,\,k_2,\,\ldots,\,k_\ell)]!}\cdot C\langle\left(\sum_{j=1}^{\ell}(1_j+2_j+\ldots+k_j)\right)A;B\rangle$$

The sum on the left hand side ranges over non-equivalent distributions of the indices $1_1,\ 2_1,\ \ldots,\ k_\ell$.

6. Simple illustrations

The simplest class-sum is the sum of transpositions, $[(2)]_n$. The product of this class-sum with an arbitrary class-sum of S_n is given in terms of a single reduced class-sum specified by the pair of partitions $((2))$ and $((1))^2$. This pair satisfies the parity condition $k + \ell + p =$ odd. The unique distribution $(1+2);(1)(2)$ satisfies the connectedness condition as well. Thus,

$$[(2)]_n \cdot [*]_n = \frac{1}{2} \sum_{\substack{r_1, r_2 \\ (r_1+r_2 \leq n)}} \langle (1+2);(1)(2)* \rangle \tag{8}$$

where the RCC $C < (1+2);(1)(2) >= \frac{1}{2}$ can be determined either by use of the general expression for RCCs containing a cycle of index-length p (in this case, $p = 2$), or by comparison of eq. (8) with the trivial relation $[(2)]_2 \cdot [(2)]_2 = 1$. A combinatorial proof of eq. (8) was given in [15].

For $p = 3$ we have to consider the following partitions: $((3))$, $((2))((1))$ and $((1))((1))((1))$. Pairs of partitions which satisfy the parity condition are $((3));((3))$, $((3));((1))((1))((1))$ and $((2))((1));((2))((1))$. The first two pairs give rise to the reduced class-sums $< (1+2+3);(1+2+3)* >$ and $< (1+2+3);(1)(2)(3)* >$, both of which trivially satisfy the connectedness criterion. The only reduced class-sum satisfying the connectedness criterion which arises from the third pair of partitions is $< (1+2)(3);(1)(2+3)* >$. Using the explicit expression for RCCs with a single cycle on at least one side of the semicolon, eq. (5), we evaluate the first and second RCCs. The recurrence relation involving the elimination of a cycle of unit index-length (eq. 6) can be used to evaluate the third (as well as the second) RCC. The final result is

$$[(3)]_n \cdot [*]_n = \sum_{\substack{r_1, r_2, r_3 \\ (r_1+r_2+r_3 \leq n)}} \left\{ \frac{1}{3} < (1+2+3);(1+2+3)* > + \right.$$

$$\left. + \frac{1}{3} < (1+2+3);(1)(2)(3)* > + < (1+2)(3);(1)(2+3)* > \right\}$$

A combinatorial proof of this result was given in [10].

7. Conclusions

An expression has been proposed for the operation of the single-cycle class-sum $[(p)]_n$ on an arbitrary class-sum, in terms of a linear combination of reduced class-sums. Each reduced class-sum is specified by a unique distribution of two sets of p indices in a pair of partitions of p. Legitimate reduced class-sums satisfy both a parity and a connectedness condition. Coefficients of reduced class-sums involving a cycle of index-length p can be evaluated explicitly, and additional coefficients can be obtained using a set of recurrence relations and, more generally, elimination rules.

The set of elimination rules presently available enables the evaluation of a very large set of RCCs, forming what will hopefully prove to be a significant step towards the complete representation-free treatment of the class-algebra of the symmetric group.

Three remaining major problems are:

a. The formulation of elimination rules for both embedded and bridging cycles of higher index-length. In principle, this seems to be a more-or-less straightforward problem.
b. The formulation of additional rules in order to evaluate RCCs which the presently available rules generate only within linear combinations with other RCCs. From a practical point of view this is the most significant open problem. Its resolution will actually provide the complete representation-free procedure for the evaluation of products of class-sums within the symmetric group algebra.
c. From a rigorous mathematical point of view it should be stressed again that almost everything that was stated in the present article is actually a conjecture. While it appears to be supported by a significant body of data, a satisfactory proof of the various components of this conjecture would certainly be desirable.

Acknowledgements. Helpful discussions with Professors David Chillag, Gordon James, Arie Juhasz and Ian Macdonald are gratefully acknowledged.

References

[1] F. Bédard and A. Goupil, The poset of conjugacy classes and decomposition of products in the symmetric group, *Canad. Math. Bull.* **35**(1992), 152–160.

[2] E. A. Bertram and V. K. Wei, Decomposing a permutation into two large cycles: an enumeration, *SIAM J. Alg. Disc. Meth.* **1**(1980), 450–461.

[3] G. Boccara, Decompositions d'une permutation d'un ensemble fini en produit de deux cycles, *Discrete Math.* **23**(1978), 189–205.

[4] G. Boccara, Nombre de representations d'une permutation comme produit de deux cycles de longueurs donnees, *Discrete Math.* **29**(1980), 105–134.

[5] G. Boccara, Cycles comme produit de deux permutations de classes donnees, *Discrete Math.* **38**(1982), 129–142.

[6] A. Goupil, On products of conjugacy classes of the symmetric group, *Discrete Math.* **79**(1989/90), 49–57.

[7] A. Goupil, Decomposition of certain products of conjugacy classes of S_n, *J. Combin. Theory A*, to appear.

[8] D. M. Jackson, Counting semiregular permutations which are products of a full cycle and an involution, *Trans. Amer. Math. Soc.* **305**(1988), 317–331.

[9] G. James and A. Kerber, *The Representation Theory of the Symmetric Group* (Addison-Wesley, Reading, MA., 1981).

[10] J. Katriel, Products of class operators of the symmetric group, *Int. J. Quantum Chem.* **35**(1989), 461–470.

[11] J. Katriel, A partial recurrence relation for reduced class coefficients of the symmetric group, *Int. J. Quantum Chem.* **39**(1991), 593–604.

[12] J. Katriel, Products of class-sums of the symmetric group: elimination of two-index cycles, *Israel J. Chem.* **31**(1991), 287–295.

[13] J. Katriel, Products of class-sums of the symmetric group: Generalizing the recurrence relations, *Int. J. Quantum Chem.*, to appear.

[14] J. Katriel, Explicit expressions for some structure constants in the class-algebra of the symmetric group, to appear.

[15] J. Katriel and J. Paldus, Explicit expression for the product of the class of transpositions with an arbitrary class of the symmetric group, in *Group Theoretical Methods in Physics* (R. Gilmore (ed.), World Scientific, Singapore, 1987), 503–506.

[16] A. Machí, Sur les représentations des permutations impaires, *European J. Combin.* **13**(1992), 273–277.

[17] R. P. Stanley, Factorization of permutations into n-cycles, *Discrete Math.* **37**(1981), 255–262.

AUTOMORPHISMS OF BURNSIDE RINGS[1]

W. KIMMERLE and K.W. ROGGENKAMP

Mathematisches Institut B, University of Stuttgart, Pfaffenwaldring 57, D–70550 Stuttgart, Germany

The object of this article is the Burnside ring $\Omega(G)$ for a finite group G. If G is soluble we show how the group of ring automorphisms $\mathrm{Aut}(\Omega(G))$ may be calculated purely from the knowledge of the subgroup lattice of G. The results on $\mathrm{Aut}(\Omega(G))$ for an abelian group G by Krämer [6] and results of Lezaun [7] for dihedral and some special metacyclic groups follow as special cases. For general facts on Burnside rings we refer to [2, §80].

The paper is organized as follows: In Section 1 we recall some more or less well known facts about the table of marks, the ghost ring of $\Omega(G)$ and $\mathrm{Aut}(\Omega(G))$. We introduce the group of normalized automorphism $\mathrm{Aut}_n(\Omega(G))$ as the automorphisms which stabilize the regular $G - set \;\; G/1$. In (2.4) we reduce the study of $\mathrm{Aut}(\Omega(G))$ for soluble groups to the study of $\mathrm{Aut}_n(\Omega(G))$, extending a result of Krämer [6] from nilpotent groups to soluble groups. Any automorphism σ of $\Omega(G)$ induces in a natural way – via the ghost ring – a bijection σ_* on the set $V(G)$ of conjugacy classes of subgroups of G. In (3.1) it is shown that this is indeed an automorphism of $V(G)$ considered in a special way as a partially ordered set, provided σ is augmented and G is soluble. In (3.2) we characterize those $\sigma \in \mathrm{Aut}(\Omega(G))$, which send transitive G-sets to transitive G-sets, in terms of the map σ_*. In (3.6) we give an explicit constructive description of $\mathrm{Aut}_n(\Omega(G))$ in terms of the automorphisms of $V(G)$ as a partially ordered set; i.e. for a fixed group it allows one to compute explicitly $\mathrm{Aut}_n(\Omega(G))$ with the help of a computer algebra system. This result will be used in Section 4 to describe $\mathrm{Aut}(\Omega(G))$ for special classes of groups.

Before we state our results we have to introduce the basic notations and the fundamental results.

1. Preliminaries

The Burnside ring $\Omega(G)$ is the Grothendieck ring of the category of (left) G-sets, which is thus the free abelian group on the isomorphism classes of transitive G-sets G/U for subgroups U of G. The addition is given by the disjoint union and the multiplication by the cartesian product of G-sets; the identity element is the G-set G/G. For a subgroup U of G we denote by $[U]$ its G-conjugacy class; for the set of all conjugacy classes of subgroups of G we write $V(G)$. If M is a transitive (left) G-set, then M is isomorphic to

[1]This research was partially supported by the Deutsche Forschungsgemeinschaft.

the (left) cosets G/U for some subgroup U of G, and two transitive G-sets G/U and G/V are isomorphic if and only if U and V are conjugate in G. We denote by $G/[U]$ the isomorphism class of the transitive G-set G/U. A general element in $\Omega(G)$ will be written as

$$\omega = \sum_{[V]\in V(G)} z_{[V]} \cdot G/[V], z_{[V]} \in \mathbb{Z}.$$

The Burnside ring $\Omega(G)$ is not only a $\mathbb{Z}$-algebra; it is also an *augmented* $\mathbb{Z}$-algebra, where the augmentation $\eta_G : \Omega(G) \longrightarrow \mathbb{Z}$ is given by

$$\eta_G : \sum_{[V]\in V(G)} z_{[V]} \cdot G/[V] \longrightarrow \sum_{[V]\in V(G)} z_{[V]} \cdot |G|/|V|.$$

It is easily checked that this is a ring homomorphism with kernel freely generated over $\mathbb{Z}$ by

$$\{G/[V] - \frac{|G|}{|V|} \cdot G/[G]\}_{[V]\in V(G)\setminus\{[G]\}}.$$

1.1. The table of marks. We number the elements $[U_i]$ of $V(G)$ in increasing order — with respect to the order of the subgroups. A matrix $M(G) = (m_{ik})$ with $m_{ik} = (G/[U_i])^{[U_k]}$, the fixed points of U_k on the G-set G/U_i, is called a table of marks; $M(G)$ is clearly a lower triangular matrix. We note that m_{ik} coincides with the number of G-homomorphisms from G/U_k to G/U_i, and thus is independent of the chosen representatives.

The entries of the table of marks may be calculated as follows. For $[U]$, $[V] \in V(G)$ we denote by

1.2. $\#([V],[U])$ the number of different G-conjugates of V in U. Then

1.3. $(G/[U])^{[V]} = (|N_G(V)|/|U|) \cdot \#([V],[U])$.

1.4. Normalized automorphisms. A ring automorphism α of $\Omega(G)$ is called augmented or normalized, if it commutes with the augmentation η_G. This is equivalent to the condition $\alpha(G/[1]) = G/[1]$ (cf. 1.5). The group of normalized ring automorphisms will be denoted by $\mathrm{Aut}_n(\Omega(G))$.

Since the set $V(G)$ is partially ordered by inclusion, we consider $V(G)$ as a partially ordered set (poset) and by $\mathrm{Aut}(V(G))$ we denote the group of poset automorphisms of $V(G)$.

1.5. $\Omega(G)$ as order and the ghost ring. For each $U \in V(G)$ the map

$$\nu_{[U]} : \Omega(G) \longrightarrow \mathbb{Z}$$

defined by

$$\nu_{[U]} : \sum_{[V]\in V(G)} z_{[V]} \cdot G/[V] \longrightarrow \sum_{[V]\in V(G)} z_{[V]} \cdot (G/[V])^{[U]}$$

is a ring homomorphism, whose kernel is a minimal prime ideal of $\Omega(G)$.

Note that $\nu_{[1]}$ is exactly the augmentation.

The map

$$\nu := \prod_{[U]\in V(G)} \nu_{[U]} : \Omega(G) \longrightarrow gh(G) := \prod_{[U]\in \nu(G)} \mathbb{Z}_{[U]}$$

is an injective ring homomorphism, and hence $\Omega(G)$ is a $\mathbb{Z}$-order. Following [4] the maximal order containing $\nu(\Omega(G))$ is called the ghost ring $gh(G)$ of $\Omega(G)$. The elements of $gh(G)\backslash \mathrm{Im}(\nu)$ are called ghosts. We shall often identify $\Omega(G)$ with its image under ν. We point out that

$$\nu(G/[1]) = (0, ..., 0, |G|_{[1]}),$$

which proves the statement in (1.4).

A more conceptual way of describing the ghost ring $gh(G)$ is as follows: Let $\mathcal{G}$ be the set of subgroups of G, on which G acts via conjugation , and let $\mathbb{Z}$ be the trivial G - set. Then $gh(G)$ is isomorphic to $\mathrm{Hom}_G(\mathcal{G}, \mathbb{Z})$. The map ν is induced from $[G/[U]] \longrightarrow (\phi_U : V \to V^U)$.

1.6. The congruences of Dress [3], [4]. Let $x = (x_{[U]})$ be an element of the ghost ring $gh(G)$. Then $x \in \Omega(G)$ if and only if x satisfies for each class $[U]$ and for each rational prime p the following congruences:

$$\sum_V x_{[V]} \equiv 0 \mod(|N_G(U) : U|),$$

where the sum is taken over all subgroups V of $N_G(U)$ containing U and such that V/U is a cyclic subgroup of a Sylow p-subgroup of $N_G(U)/U$.

1.7. Idempotents and quasi-idempotents. We denote by $e_{[U]}$ the primitive idempotents of the ghost ring Γ corresponding to the component $\mathbf{Z}_{[U]}$. Then there exists a unique minimal natural number $\lambda_{[U]}$ such that $q_{[U]} := \lambda_{[U]} \cdot e_{[U]} \in \Omega(G)$, which we shall call the index with respect to U and $q_{[U]}$ is called the *quasi-idempotent* with respect to $[U]$.

For a finite H we denote by $\pi(H)$ the set of distinct rational prime divisors of $|H|$, and $\bar{H}$ stands for the commutator factor group. The congruences in 1.6 show that

$\lambda_{[U]} = |N_G(U)/U| \cdot \prod_{q\in\pi(\bar{U})} q,$ if U is not perfect, and
$\lambda_{[U]} = |N_G(U)/U|$ otherwise.

The table of marks 1.1 shows that

$$\nu(G/[U]) = \sum_{[V]\leq[U]} |N_G(V)|/|U| \cdot \#([V],[U]) \cdot e_{[V]}.$$

Its inversion formula using the Moebius function μ of the subgroup lattice is given as

$$e_{[U]} = |N_G(U)|^{-1} \cdot \sum_{V\leq U} \mu(V,U) \cdot |V| \cdot \nu(G/[V]),$$

cf. [5] or [10]. The above formula for the quasi-idempotents shows that for any commutative ring R, in which the prime divisors of $|G|$ are invertible,

1.8. $$R \otimes_{\mathbb{Z}} \nu(\Omega(G)) = R \otimes_{\mathbb{Z}} \Gamma.$$

At this stage we like to recall an interesting result of Dress, characterizing finite soluble groups [3]:

1.9. $R \otimes_{\mathbb{Z}} \nu(\Omega(G))$ contains no non-trivial idempotent if and only if G is soluble and no prime divisor of $|G|$ is a unit in R.

2. Reduction to normalized automorphisms

Proposition 2.1. *(i) Let $\sigma \in \mathrm{Aut}(\Omega(G))$. Then there exists a unique ring homomorphism $\sigma^* : gh(G) \to gh(G)$, making the following diagram commute.*

$$\begin{array}{ccc} \Omega & \xrightarrow{\nu} & gh(G) = \prod \mathbb{Z}_{[U]} \\ \sigma\downarrow & & \downarrow\sigma^* \\ \Omega & \xrightarrow{\nu} & gh(G) = \prod \mathbb{Z}_{[U]} \end{array}$$

σ^* induces a bijection σ_* of $V(G)$ as follows. Let $[U] \in V(G)$, then for each $y \in \Omega(G)$

$$y^{[U]} = \sigma(y)^{\sigma_*([U])}.$$

For brevity we also write $[U_*]$ for $\sigma_*[U]$– note though that U_* is not a well defined subgroup of G, only its G-conjugacy class is well defined.

(ii) The map

$$\kappa : \mathrm{Aut}(\Omega(G)) \longrightarrow \mathrm{Sym}(V(G)) \text{ given by } \sigma \longrightarrow \sigma_*$$

is an injective group anti-homomorphism (here $\mathrm{Sym}(V(G))$ is the group of the bijective maps on $V(G)$).

(iii) Let $U \in [U]$ and $U_* \in \sigma_*([U])$, then

$$|N_G(U)/U| \cdot \prod_{q\in\pi(\bar{U})} q = |N_G(U_*)/U_*| \cdot \prod_{q\in\pi(\bar{U}_*)} q,$$

where we use the convention that the product $\prod$ is 1, if the associated subgroup is perfect.

(iv) If $U = 1$, then U_* is an abelian normal subgroup of G of square free order.

PROOF. (i) Clearly σ induces an automorphism also denoted by σ of $\mathrm{Im}(\nu)$, which extends uniquely to an automorphism σ^* of the maximal order $gh(G)$, making the above diagram commute. Moreover, σ^* permutes the primitive idempotents $e_{[U]}$ of $gh(G)$, which are in 1 - 1 correspondence with the elements in $V(G)$.

Define $\sigma_* : V(G) \longrightarrow V(G)$ by $\sigma_*([U]) = [V]$, if $\sigma^*(e_{[U]}) = e_{[V]}$; i.e. $e_{[U]} = \sigma^{*^{-1}}(e_{[\sigma_*([U])]})$. Then the above commutative diagram shows that $\sigma(y)^{\sigma_*([U])} = y^{[U]}$ for $y \in \Omega(G)$.

(ii) Clearly the map $\kappa : \sigma \longrightarrow \sigma_*$ is an anti-homomorphism. Since the only ring automorphism of $\mathbb{Z}$ is the identity, κ is injective.

(iii) Since $\sigma^{*-1}(e_{[U]}) = e_{\sigma_*([U])}$, it follows that

$$\sigma(z \cdot e_{[U]}) = z \cdot e_{[\sigma_*^{-1}([U])]} \in \Omega(G) \quad \text{if and only if} \quad z \cdot e_{[U]} \in \Omega(G).$$

Hence $\lambda_{[U]} = \lambda_{[U_*]}$ and (iii) follows from 1.6.

(iv) If $U = 1$, then (iii) shows that $|G| = |N_G(U_*)/U_*| \cdot \prod_{q \in \pi(\bar{U}_*)} q$. The right hand side is less than or equal to $|N_G(U_*)|$ and coincides with $|N_G(U_*)|$ if and only if U_* is abelian of square free order. So (iv) follows immediately. □

Next we derive an important property of the above map σ_*.

Proposition 2.2. *Let $\sigma \in \mathrm{Aut}_n(\Omega(G))$. Then the corresponding subgroup classes $[U]$ and $\sigma_*([U])$ have representatives of the same order provided U is soluble. Moreover, in that case $\sigma_*([U])$ is a class of soluble groups and*

$$\sigma(G/[U]) = G/\sigma_*([U]) + \sum_{|U_\nu|<|U|} a_\nu \cdot G/[U_\nu].$$

PROOF. Since σ is normalized, the proposition holds for the trivial class [1]. We use induction on $|U|$. Assume that $\sigma(G/[U]) = \Sigma_\nu a_\nu \cdot G/[U_\nu]$.

Claim 1. *There is exactly one maximal $[U_\mu]$ with $a_\mu \neq 0$ and $|U_\mu| \geq |U|$; in fact $[U_\mu] = \sigma_*([U])$. Moreover*

$$\sigma(G/[U]) = G/\sigma_*([U]) + \sum_{|U_\nu|<|U|} a_\nu \cdot G/[U_\nu].$$

PROOF. For every such maximal $[U_\mu]$ with $a_\mu \neq 0$, let $[V_\mu]$ be its inverse image under σ_*. By Proposition 2.1 (i) we obtain that $G/[U]^{[V_\mu]} \neq 0$; thus $[V_\mu] \leq [U]$. Since $\sigma(G/[U])^{\sigma_*([U])} \neq 0$, the first part of the claim follows by induction.

Taking fixed points with respect to $[U]$ and $\sigma_*([U])$ we conclude

$$a_\mu \cdot |N_G(U_\mu)/U_\mu| = |N_G(U)/U|$$

Consider now the $\mathbb{Z}$-sublattice L of $gh(G)$ generated by $\{G/[V] \mid [V] \leq [U]\}$. Then L has index

$$\prod_{[V]\leq[U]} |N_G(V)/V)|$$

in the $\mathbb{Z}$-lattice M generated by $\{e_{[V]} | [V] \leq [U]\}$. The index of the image under σ of L in $\sigma(M)$ is therefore $\prod_{[V]\leq\sigma_*([U])} |N_G(V)/V)|$. By induction $|N_G(U_\mu)/U_\mu| = |N_G(U)/U|$, and so $a_\mu = 1$. This proves the second part of the claim. □

By Proposition 2.1 (iii) we see that U_μ can not be perfect. Consequently there exists a prime p such that U_μ has a normal subgroup of index p. Assume that $|U_\mu| > |U|$ and denote by $M(p)$ the set of all subgroups of U_μ of index p. According to the induction hypothesis, we know that for each $M \in M(p)$ we have $G/[U]^{\sigma_*^{-1}([M])} = 0$, otherwise $|U_\mu| = |U|$. Note also that $|M|$ cannot be a proper divisor of $|U|$ and so we conclude $\sigma(G/[U])^{[M]} = 0$. Hence

$$G/[U_\mu]^{[M]} + a_{[M]} \cdot G/[M]^{[M]} = 0.$$

By 1.4 we get

$$\#([M], [U_\mu]) \cdot (|N_G(M)|/|U_\mu|) + a_{[M]} \cdot |N_G(M)/M| = 0. \qquad (*)$$

Since $|U_\mu| = |M| \cdot p$ we deduce from (*)

$$\#([M], [U_\mu]) + a_{[M]} \cdot p = 0.$$

We shall show however that $\#([M], [U_\mu])$ cannot be congruent to zero mod(p) for each M. This follows from the following general fact.

Claim 2. *Let G be a finite group. Then the number of maximal subgroups of index p is congruent to zero* mod(p) *if and only if G has no normal subgroup of index p.*

PROOF. Let $S \in Syl_p(G)$. Then S acts by conjugation on the maximal subgroups of index p. The only orbits of length not divisible by p are the fixed points of this action, i.e. the normal subgroups of index p. Let N be the intersection of all these fixed points – i.e. the maximal normal subgroups of G of index p – then the quotient G/N is an elementary abelian p-group. The fixed points are in bijection to the maximal subgroups of G/N, the number of which is $(p^n - 1)/(p - 1)$, if $|G/N| = p^n$. □

Therefore the claim shows that $|M(p)|$ is not divisible by p. This contradiction forces $|U_\mu| = |U|$. We may apply our arguments also in this situation.

The arguments show that U_μ has in fact a normal subgroup V of index p such that $G/[U_\mu]^{[V]} \neq 0$. By induction $[V]$ corresponds via σ_* to a class of subgroups of U and is therefore soluble. Hence U_μ is soluble. This completes the proof of Proposition 2.2. □

2.3. Examples of non-normalized automorphisms. In that case of integral group rings, it is easy to modify a given automorphism to make it commute with the augmentation, the reason being that the group elements have augmentation one. Here the situation is much more involved, since $\eta_G(G/[V]) = |G|/|V|$ is not a unit in $\mathbb{Z}$, in general.

Next we shall construct non-normalized automorphisms of $\Omega(G)$, which essentially will allow us to describe $\mathrm{Aut}(\Omega(G))$ for a soluble group in terms of $\mathrm{Aut}_n(\Omega(G))$ and these special automorphisms.

(i) Let G be a finite group having a unique subgroup P of prime order p. Then Nicolson [8, Proof of Prop. 3.4] has shown that $\Omega(G)$ has an automorphism σ with the following properties on the quasi-idempotents:

$$\begin{aligned} q_{[U]} &\longrightarrow q_{[U]}, \text{ if } p^2 \,|\, |U|; \\ q_{[U]} &\longrightarrow q_{[U\cdot P]}, \text{ if } (p, |U|) = 1; \\ q_{[U]} &\longrightarrow q_{[W]}, \text{ if } U = P \cdot W \text{ and } (p, |W|) = 1. \end{aligned}$$

We note that these conditions describe σ uniquely. In particular

$$q_{[1]} \longrightarrow q_{[P]} \text{ and } q_{[P]} \longrightarrow q_{[1]},$$

and so this automorphism is not normalized. Clearly the automorphism σ has order 2.

(ii) Let G be a finite group such that G has a normal subgroup P of order 2 and every subgroup of G of order 4 contains P, then $\Omega(G)$ has an automorphism σ depending on P with the following properties for the quasi-idempotents:

$$\begin{aligned} q_{[U]} &\longrightarrow q_{[U\cdot P]}, && \text{if } |U| \text{ is odd;} \\ q_{[U]} &\longrightarrow q_{[W]}, && \text{if } U = W \cdot P \text{ and } |W| \text{ is odd;} \\ q_{[U]} &\longrightarrow q_{[U]} && \text{in all other cases.} \end{aligned}$$

Again σ has order 2 and $\sigma_*([1]) = [P]$, and so σ is not normalized.

Note that Klein's four group, $G = C_2 \times C_2$, has precisely three such normal subgroups P of order two. In contrast to (i) we do not assume that G has a unique normal subgroup of order 2. E.g. $G = C_{2^m} \times C_2$ has precisely one such subgroup P if $m > 1$.

(iii) Denote the proper subgroups of Klein's four group by 1, 2, 3 and 4, where 1 (as usual) denotes the trivial subgroup. The three non-normalized automorphisms of $\Omega(C_2 \times C_2)$ described in (ii) correspond then to the permutations (on the natural basis of $\Omega(C_2 \times C_2)$) (1,2), (1,3) and (1,4). These permutations generate the symmetric group S_4 and an easy calculation shows that S_4 is isomorphic to $\mathrm{Aut}(\Omega(C_2 \times C_2))$.

We next describe $\mathrm{Aut}(\Omega(G))$ for soluble groups in terms of $\mathrm{Aut}_n(\Omega(G))$.

Theorem 2.4. *Let G be a finite group and let $\mathrm{Aut}_n(\Omega(G))$ be the group of normalized automorphisms of $\Omega(G)$.*

a) If $G \cong C_2 \times C_2 \times \mathcal{O}$, where $\mathcal{O}$ is a group of odd order, then

$$\mathrm{Aut}(\Omega(G)) \cong S_4 \times \mathrm{Aut}(\Omega(\mathcal{O})).$$

b) Suppose that G is soluble. Assume that G is not isomorphic to $C_2 \times C_2 \times \mathcal{O}$ for some group $\mathcal{O}$ of odd order. Denote by m the number of rational primes p such that G has a unique subgroup of order p, p odd. Let $\epsilon = 1$, if G has a normal subgroup of order 2 contained in each subgroup of order 4. Otherwise put $\epsilon = 0$. Then

$$\mathrm{Aut}(\Omega(G)) \cong \mathrm{Aut}_n(\Omega(G)) \rtimes C_2^{m+\epsilon}.$$

The semidirect product[2] is direct, if $[U] \leq [V]$ implies $\sigma_*([U]) \leq \sigma_*([V])$ for each $\sigma \in \mathrm{Aut}_n(\Omega(G))$.

Remark. Note that by the odd order theorem of Feit and Thompson part a) of Theorem 2.4 is a statement for soluble groups.

First we establish some ingredients for the proof of Theorem 2.4. The following group theoretical fact explains the exceptional role of the Klein four group.

Lemma 2.5. *Let G be a finite group. Suppose that G has more than one normal subgroup of order two such that each subgroup of order 4 contains each of these normal subgroups. Then G is isomorphic to $C_2 \times C_2 \times \mathcal{O}$ for some group $\mathcal{O}$ of odd order.*

PROOF. Clearly 4 divides $|G|$. The classification of the groups of order 8 shows that G is not of order 8. Sylow's theorem yields that 8 does not divide $|G|$. Moreover G has a normal Sylow 2-subgroup V isomorphic to the Klein four group. At least two subgroups of order 2 are by assumption normal. Hence all subgroups of order 2 are normal and each element of odd order centralizes V. Since $\mathrm{H}^2(G/V, V) = 0$, the lemma follows. □

[2]In [7] it is shown that for dihedral groups the semidirect product is in general not direct.

The next result describes the (normalized) automorphisms of the Burnside ring of a product of soluble groups of coprime order in terms of the automorphisms of the Burnside rings of the factors. Let us first recall a fact for automorphisms of orders:

Let Λ and Γ be $\mathbb{Z}$-orders in $\mathbb{Q}\Lambda = \Pi_{i=1}^{s}\mathbb{Q}_i$ and $\mathbb{Q}\Gamma = \Pi_{i=1}^{t}\mathbb{Q}_i$ respectively with $\mathbb{Q}_i = \mathbb{Q}$. By $C(\Lambda) = \{x \in \mathbb{Z} | x \cdot (\Pi_{i=1}^{s}\mathbb{Z}_i) \subset \Lambda\}$ we denote the conductor of Λ.

Lemma 2.6. *Let Λ and Γ be as above, and assume that both are indecomposable, and that the conductors $C(\Lambda)$ and $C(\Gamma)$ are relatively prime. Then*

$$\mathrm{Aut}(\Lambda \otimes_{\mathbb{Z}} \Gamma) = \mathrm{Aut}(\Lambda) \times \mathrm{Aut}(\Gamma),$$

where $\mathrm{Aut}(\Lambda)$ *denotes the group of ring automorphisms of* Λ.

PROOF. By $\{\varepsilon_i\}$ and $\{\eta_j\}$ we denote the primitive idempotents in $\mathbb{Q}\Lambda$ and $\mathbb{Q}$ respectively. It should be noted that these are unique. Since the only ring automorphism of $\mathbb{Q}$ is the identity, every automorphism β of Λ is uniquely determined by a permutation $\sigma_\beta \in S_s$, the symmetric group on the s idempotents $\{\varepsilon_i\}$. Now let α be an automorphism of $\Lambda \otimes_{\mathbb{Z}} \Gamma$, which then corresponds to a permutation of the primitive idempotents $\{\varepsilon_i \otimes \eta_j\}_{1 \leq i \leq s, 1 \leq j \leq t}$ of $\mathbb{Q}(\Lambda \otimes_{\mathbb{Z}} \Gamma)$, and thus can be written as $\sigma_\alpha \times \tau_\alpha$, with $\sigma_\alpha \in \mathrm{Map}(\{1,...,s\} \times \{1,...,t\}, \{1,...,s\})$ and $\tau_\alpha \in \mathrm{Map}(\{1,...,s\} \times \{1,...,t\}, \{1,...,t\})$; i.e.,

$$\alpha(\varepsilon_i \otimes \eta_j) = \varepsilon_{\sigma_\alpha(i,j)} \otimes \eta_{\tau_\alpha(i,j)}. \tag{$*$}$$

For a finite set of rational primes π, we denote by $\mathbb{Z}_\pi$ the semi-localization of $\mathbb{Z}$ at the primes in π. Let π_Λ and π_Γ be the prime divisors of $C(\Lambda)$ and $C(\Gamma)$ respectively. Then – $(C(\Lambda), C(\Gamma)) = 1$ – we have

$$\mathbb{Z}_{\pi_\Lambda} \otimes_{\mathbb{Z}} \Lambda \otimes_{\mathbb{Z}} \Gamma \simeq \Pi_{j=1}^{t}(\mathbb{Z}_{\pi_\Lambda} \otimes_{\mathbb{Z}} \Lambda) \cdot \eta_j.$$

Now Λ was indecomposable, and since we do not invert any prime dividing $C(\Lambda)$, $\mathbb{Z}_{\pi_\Lambda} \otimes_{\mathbb{Z}} \Lambda$ is still indecomposable, and thus α induces an automorphism α_{π_Λ} on

$$\mathbb{Z}_{\pi_\Lambda} \otimes_{\mathbb{Z}} \Lambda \otimes_{\mathbb{Z}} \Gamma \simeq \Pi_{j=1}^{t}(\mathbb{Z}_{\pi_\Lambda} \otimes_{\mathbb{Z}} \Lambda) \cdot \eta_j$$

which maps

$$(\mathbb{Z}_{\pi_\Lambda} \otimes_{\mathbb{Z}} \Lambda) \cdot \eta_j \text{ to } (\mathbb{Z}_{\pi_\Lambda} \otimes_{\mathbb{Z}} \Lambda) \cdot \eta_{\mu_\alpha(j)}$$

for a permutation $\mu_\alpha \in S_t$. Identify $(\mathbb{Z}_{\pi_\Lambda} \otimes_{\mathbb{Z}} \Lambda) \cdot \eta_j$ with $(\mathbb{Z}_{\pi_\Lambda} \otimes_{\mathbb{Z}} \Lambda) \cdot \eta_{\mu_\alpha(j)}$ by sending η_j to $\eta_{\mu_\alpha(j)}$. Then α_{π_Λ} induces an automorphism on $(\mathbb{Z}_{\pi_\Lambda} \otimes_{\mathbb{Z}} \Lambda)$, which is given by a permutation $\nu_{\alpha,j} \in S_s$. Thus

$$\alpha_{\pi_\Lambda}(\varepsilon_i \otimes \eta_j) = \varepsilon_{\nu_{\alpha,j}(i)} \otimes \eta_{\mu_\alpha}(j). \tag{$\dagger$}$$

Similar considerations for π_Γ instead of π_Λ show

$$\alpha_{\pi_\Gamma}(\varepsilon_i \otimes \eta_j) = \varepsilon_{\nu_\alpha}(i) \otimes \eta_{\mu_{\alpha,i}}(j) \qquad (\ddagger)$$

for permutations $\nu_\alpha \in S_s$ and $\mu_{\alpha,i} \in S_t$. Since all the maps α, α_{π_Λ} and α_{π_Γ} must induce the same map on $\mathbb{Q}(\Lambda \otimes_{\mathbb{Z}} \Gamma)$, we conclude - comparing $(*)$, $(\dagger)$ and $(\ddagger)$- σ_α depends only on i and τ_α depends only on j; i.e.

$$\alpha(\varepsilon_i \otimes \eta_j) = \varepsilon_{\beta_\alpha(i)} \otimes \eta_{\gamma_\alpha(j)} \text{ for } \beta_\alpha \in S_s \text{ and } \gamma_\alpha \in S_t. \qquad (\S)$$

If we can show that the permutations β_α and γ_α induce automorphisms of Λ and Γ respectively, the lemma will be proved. However, α induces a homomorphism

$$\Gamma \simeq \varepsilon_1 \otimes \Gamma \longrightarrow \varepsilon_{\beta_\alpha(1)} \otimes \Gamma \simeq \Gamma.$$

With this identification, α induces on Γ the permutation γ_α. Similarly one shows that β_α induces an automorphism on Λ. □

We can use (2.6) to describe the automorphisms of $\Omega(G \times H)$ for finite groups of relatively prime order.

Proposition 2.7. *Let G and H be groups of coprime order.*

a) $\mathrm{Aut}(\Omega(G)) \times \mathrm{Aut}(\Omega(H))$ *is isomorphic to a subgroup of* $\mathrm{Aut}(\Omega(G \times H))$.

b) Let $\Omega(G) = \Pi_{i=1}^s \Omega_i(G)^{m_i}$ and $\Omega(H) = \Pi_{i=1}^t \Omega_i(H)^{n_i}$ be the decomposition of $\Omega(G)$ and $\Omega(H)$ respectively into indecomposable rings, where m_i and n_i denote the multiplicities. Assume that $q_{G/[1]} \in \Omega_1(G)$ and $q_{H/[1]} \in \Omega_1(H)$.

W. l. o. g. we may assume that H has odd order. By the odd order theorem of Feit and Thompson H is soluble. Therefore $\Omega(H)$ is indecomposable, i.e. $t = 1$ and $n_1 = 1$. Finally we assume that $m_i = 1$ for all i, [3] *then*

$$\mathrm{Aut}(\Omega(G \times H)) = \Pi_{i=1}^s \mathrm{Aut}(\Omega_i(G)) \times \mathrm{Aut}(\Omega(H)),$$

and

$$\mathrm{Aut}_n(\Omega(G \times H))$$
$$= \mathrm{Aut}_n(\Omega_1(G)) \times \mathrm{Aut}_n(\Omega_1(H)) \times \Pi_{i=2}^s \mathrm{Aut}(\Omega_i(G)) \times \mathrm{Aut}(\Omega_1(H)).$$

c) Assume that $G \times H$ is soluble. Then

(i) $\mathrm{Aut}(\Omega(G \times H)) \cong \mathrm{Aut}(\Omega(G)) \times \mathrm{Aut}(\Omega(H))$,

[3]The general case with no restrictions on the multiplicities may be described similarly using additionally actions of appropriate symmetric groups on isomorphic indecomposable factors of $\Omega(G) \otimes \Omega(H)$.

(ii) $\mathrm{Aut}_n(\Omega(G \times H)) \cong \mathrm{Aut}_n(\Omega(G)) \times \mathrm{Aut}_n(\Omega(H))$.

PROOF. The natural map

$$\begin{gathered}\Omega(G) \otimes_{\mathbb{Z}} \Omega(H) \to \Omega(G \times H), \\ \text{induced by } G/[U] \otimes H/[V] \to (G \times H)/[U \times V]\end{gathered} \qquad (*)$$

is a ring isomorphism, provided G and H have relatively prime order. In fact every subgroup of $G \times H$ is of the form $U \times V$, $|G|$ and $|H|$ being relatively prime. Because of the natural map (*), we have an injection $\mathrm{Aut}(\Omega(G)) \times \mathrm{Aut}(\Omega(H)) \longrightarrow Aut(\Omega(G \times H))$ and a similar one for the normalized automorphisms. So a) holds. For b) we get

$$\begin{aligned}\mathrm{Aut}(\Omega(G \times H)) &= \mathrm{Aut}(\Omega(G) \otimes \Omega(H)) = \mathrm{Aut}(\Pi_{i=1}^{s} \Omega_i(G) \otimes \Omega(H)) \\ &= \Pi_{i=1}^{s} \mathrm{Aut}(\Omega_i(G) \otimes \Omega(H)).\end{aligned}$$

The last equality follows from the fact that $\Omega_i(G) \otimes \Omega(H)$ and $\Omega_j(G) \otimes \Omega(H)$ are not isomorphic, if i differs from j, because by assumption $\Omega_i(G)$ and $\Omega_j(G)$ are not isomorphic, if $i \neq j$.

Since $(|G|, |H|) = 1$, we get that

$$(C(\Omega(G)), C(\Omega(H))) = 1$$

, and so $(C(\Omega_i(G)), C(\Omega(H))) = 1$; thus we can apply (2.6), to deduce b), if we note that the product decomposition preserves augmentations.

As before we note that for a soluble group, $\Omega(G)$ is indecomposable, and so c) follows from b) (cf. 1.8, 1.9). □

We are now in the position to turn to the

PROOF OF THEOREM 2.4. Nicolson [8] has computed the orbit O of the trivial class [1] of $V(G)$ under the action of $\mathrm{Aut}(\Omega(G))$ (cf. 2.1). An element $[V] \in \Omega(G)$ belongs to this orbit if and only if V is the unique subgroup of a square free order, or V is of even square free order and its Sylow 2-subgroup is contained in each subgroup of G of order 4.

Assume that G is not isomorphic to $C_2 \times C_2 \times \mathcal{O}$ for some group $\mathcal{O}$ of odd order. By Lemma 2.5 we know that in the orbit O there is at most one class of a subgroup of order 2. Then the non-normalized automorphisms described in 2.3 (i) and (ii) commute, and are all of order 2. Consequently they generate an elementary abelian 2-subgroup T of $\mathrm{Aut}(\Omega(G))$ and T acts transitively on O. Clearly $\mathrm{Aut}_n(\Omega(G))$ is a normal subgroup.

By assumption G is soluble. So we can use Proposition 2.2 without any restriction. Now let σ be a normalized automorphism such that σ_* preserves inclusions then σ commutes with each generator of T. In fact, Proposition 2.2 shows that corresponding classes under a normalized automorphism represent subgroups of the same order and the description of the non-normalized automorphisms depends only on the knowledge of these orders and the knowledge whether the unique normal subgroup of order p is contained in a class $[W]$ or not.

Clearly $T \cap \mathrm{Aut}_n(G) = 1$ and part b) of Theorem 2.4 follows.

Assume now that $G \cong C_2 \times C_2 \times \mathcal{O}$ for some group $\mathcal{O}$ of odd order. By 2.3 (iii) and Proposition 2.7 c) we get immediately that $S_4 \times \mathrm{Aut}(\Omega(\mathcal{O}))$ is isomorphic to $\mathrm{Aut}(\Omega(G))$. □

Remarks. a) 2.7 extends a result of Krämer [6, Theorem 1] from nilpotent to soluble groups.

b) Let M be a minimal non-abelian simple group,and let H be a group of order relatively prime to $|M|$. Note that $\Omega(M) \cong \Lambda \times \mathbf{Z}$ and thus $\mathrm{Aut}(\Omega(M)) \cong \mathrm{Aut}(\Lambda)$. By Proposition 2.7 b) we obtain that

$$\mathrm{Aut}(\Omega(M \times H)) \cong \mathrm{Aut}(\Omega(M)) \times \mathrm{Aut}(\Omega(H)) \times \mathrm{Aut}(\Omega(H)).$$

This shows that the formula of Proposition 2.7c) fails in general for non-soluble groups.

c) Proposition 2.2 fails for groups with different conjugacy classes of self-normalizing perfect subgroups, e.g. A_6. $\Omega(A_6)$ is isomorphic to $\mathbb{Z} \times \mathbb{Z} \times \Lambda \subset gh(A_6)$, where the $\mathbb{Z}$-order Λ corresponds to the classes of soluble subgroups, and the two copies of $\mathbb{Z}$ to the idempotents to $[A_5]$ and $[A_6]$. This follows immediately from the congruences of Dress. Consequently there exists an automorphism of $\Omega(A_6)$ permuting the idempotents corresponding to $[A_5]$ and $[A_6]$ and fixing all other elements of $V(A_6)$.

d) For the proof of Theorem 2.4 a) we used the odd order theorem of Feit and Thompson. Note, if one could prove that $S_4 \times \mathrm{Aut}(\Omega(\mathcal{O}))$ is isomorphic to $\mathrm{Aut}(\Omega(C_2 \times C_2 \times \mathcal{O}))$, without its use, the odd order theorem would follow from Proposition 2.7 b).

3. Calculation of $\mathbf{Aut}_n\Omega(\mathbf{G})$

Recall from 2.1 that any $\sigma \in \mathrm{Aut}_n(\Omega(G))$ induces $\sigma_* \in \mathrm{Sym}(V(G))$ stabilizing $G/[G]$ and $G/[1]$. Define a partial ordering[4] on $V(G)$ as follows. Let V be a maximal subgroup of U, then $[V] \preceq [U]$, if $(\#([V],[U]), |U : V|) = 1$. For arbitrary subclasses we define $[V] \preceq [U]$, if there is a chain $[V] = [U_0], \ldots, [U] = [U_n]$ such that $[U_i]$ is maximal in $[U_{i+1}]$ and $[U_i] \preceq [U_{i+1}]$. Note that for soluble groups for each subgroup U there is at least one maximal subgroup V with $[V] \preceq [U]$.

[4] We thank M.Wursthorn for critical questions about the poset structure of $V(G)$ used in an earlier version of this paper. It turned out that the natural poset structure induced by inclusion is not the correct one.

Proposition 3.1. *Assume that G is soluble and let $\sigma \in \mathrm{Aut}_n(\Omega(G))$. Then σ_* is a poset automorphism of $V(G)$ with respect to the special partial ordering of $V(G)$ defined above. Moreover, for $U_* \in \sigma_*([U])$ we have:*

a) $|U_*| = |U|$,

b) $|N_G(U_*)| = |N_G(U)|$,

c) $\prod_{q\in\pi(U_*)} q = \prod_{q\in\pi(U)} q$.

PROOF. By Proposition 2.2 we know that

$$\sigma(G/[W]) = G/[W_*] + \sum_{[V],|V|<|W|} a_{[V]} \cdot G/[V].$$

Let V be a maximal subgroup of W and $|W : V| = p$. We know that $G/[W]^{[V]} = \sigma(G/[W])^{[V_*]} = G/[W_*]^{[V_*]} + a_{[V]} \cdot G/[V_*]$. If $[V_*] \not\leq [W_*]$, then

$$\frac{\#([V],[W]) \cdot |N_G(V)|}{|W|} = \frac{a_{[V_*]} \cdot p \cdot |N_G(V_*)|}{p \cdot |V_*|}.$$

Properties a) and b) have already been proved in Proposition 2.2. Thus $|W| = p \cdot |V_*|$ and $|N_G(V)| = |N_G(V_*)|$. Consequently p divides $\#([V,W])$ and we get that $[V] \preceq [U]$ if and only if $[V_*] \preceq [U_*]$. Therefore σ_* is a poset automorphism. Finally property c) follows immediately from a), b) and Proposition 2.1 (iii). □

$V(G)$ has a natural poset structure induced by inclusion of the subgroups. The following example shows that in general this natural poset structure is not preserved by the bijections of $V(G)$ induced from normalized automorphisms of $\Omega(G)$.

Let G be the dihedral group of order 8. Denote by $[N_1]$ and $[N_2]$ the elements of $V(G)$ consisting of Klein four groups. Let $[U_1]$ and $[U_2]$ be the classes consisting of non-central subgroups of order 2 enumerated such that $[U_i] \leq [N_i]$. Define $\sigma \in \mathrm{Aut}_n(\Omega(G))$ by

$$\sigma(G/[N_1]) = G/[N_2] - G/[V_2] + G/[V_1],$$
$$\sigma(G/[N_2]) = G/[N_1] + G/[V_2] - G/[V_1],$$
$$\sigma(G/[X]) = G/[X] \text{ for all other classes } [X].$$

Then $\sigma_*([U_1]) = [U_1] \not\leq \sigma_*([N_1]) = [N_2]$. Moreover σ does not map transitive G - classes to transitive G - classes.

From Proposition 3.1 two questions arise. Let τ be a poset automorphism with the properties a), b) and c) described in Proposition 3.1.

Question 1. When does τ induce an automorphism σ of $\Omega(G)$ with $\sigma_* = \tau$; i.e. is the image of $-_* : \mathrm{Aut}_n(\Omega(G)) \longrightarrow \mathrm{Sym}(V(G))$ characterized by the properties a), b) and c) ?

Question 2. Define a $\mathbb{Z}$-linear map $\tau_\Omega : \Omega(G) \longrightarrow \Omega(G)$, by $G/[U] \longrightarrow G/\tau([U])$. When is $\tau_\Omega \in \mathrm{Aut}_n(\Omega(G))$?

First we turn to the second question: To simplify the notation, let us denote by $\mathrm{Aut}_n(V(G))$ the subgroup of $\mathrm{Sym}(V(G))$ consisting of all poset automorphisms of $V(G)$ satisfying the conditions 3.1 a), b) and c). Moreover $\tau \in \mathrm{Aut}_n(V(G))$ is said to preserve inclusions, if additionally $[V] \le [U]$ implies $\tau([V]) \le \tau([U])$ for each pair $([V],[U])$.

Theorem 3.2. *a) Let $\sigma \in \mathrm{Aut}(\Omega(G))$. Then σ maps transitive G-classes to transitive G-classes if and only if $\sigma_* \in \mathrm{Aut}_n(V(G))$ and $\#([V],[U]) = \#(\sigma_*([V],\sigma_*([U]))$.*
b) Assume that G is soluble and let $\sigma \in \mathrm{Aut}_n(\Omega(G))$. Then $\sigma(G/[U]) = G/\sigma_*([U])$ for some $[U] \in V(G)$ if and only if $\#([V],[U]) = \#(\sigma_*([V]),\sigma_*([U]))$ for each $[V] \le [U]$.

PROOF. Assume that σ maps transitive G-sets to transitive G-sets, say $\sigma(G/[1]) = G/[U]$. Then $|G|\cdot G/[U] = G/[U]^2$, and so U must be normal and hence $G/[U]^2 = |G/U|\cdot G/[U]$. Thus $U = 1$ and σ must be normalized.

The condition $G/[U]^{[V]} = \sigma(G/[U])^{\sigma_*([V])}$ implies

$$\#([V],[U])\cdot |N_G(V)|/|U| = \#(\sigma_*([V]),\sigma_*([U]))\cdot |N_G(V_*)|/|U_*|, \qquad (*)$$

where we have written V_* and U_* for representatives of $\sigma_*([V])$ and $\sigma_*([U])$ respectively. The number of classes $[W]$ with $G/[U]^{[W]} \neq 0$ coincides with the corresponding number with respect to $\sigma(G/[U]) := G/\sigma_x([U])$ and is just the number of different subclasses of $[U]$ and $\sigma_x([U])$ respectively. By induction on the order of subgroups it follows that the representatives of $[U]$ and $\sigma_x([U])$ have the same order, that σ_x is a poset automorphism of $V(G)$ and that σ_x coincides with σ_*. Moreover $G/[U]^{[U]} = G/\sigma_*([U])^{\sigma_*([U])}$ implies $|N_G(U)/U| = |N_G(U_*)/U_*|$. Therefore properties a) and b) of Proposition 3.1 are satisfied and, since σ is an automorphism, also property c) by Proposition 2.1 (iii). Therefore $\sigma \in \mathrm{Aut}_n(V(G))$ and by (*) it follows that $\#([V],[U]) = \#(\sigma_*([V]),\sigma_*([U]))$.

Conversely, assume that $\tau \in \mathrm{Aut}_n(V(G))$. Then τ induces a unique automorphism τ_x of $gh(G)$. Looking at the table of marks, cf. 1.1, one sees immediately that $\tau_x(G/[U]) = G/\tau([U])$. Hence $\sigma(\tau) := \tau_{\Omega(G)}$ is an automorphism of $\Omega(G)$ with $\tau = \sigma(\tau)_*$. This proves part a).

If G is soluble, we obtain from Proposition 3.1 and equation (*) that

$$\#([V],[U]) = \#(\sigma_*([V]),\sigma_*([U])).$$

For the converse observe that by Proposition 2.2

$$\sigma(G/[U]) \doteq G/\sigma_*([U]) + \sum_{[W],|W|<|U|} a_{[W]} \cdot G/[W].$$

Put $w = \sum_{[W],|W|<|U|} a_{[W]} \cdot G/[W]$.

If $\#([V],[U]) = \#(\sigma_*([V]),\sigma_*([U]))$ for each $[V] \leq [U]$, then $w^{[V]} = 0$ for each $[V] \leq [U]$. W.l.o.g. we may assume that the number of subclasses of $[U]$ is greater than or equal to that of $\sigma_*([U])$. Since $G/[U]^{[W]} = 0$ for each $[W]$ not containing $[U]$ and since σ_* is a bijection, it follows that $w^{[W]} = 0$. Hence $w = 0$. □

Remark 3.3. The proof of Theorem 3.2 b) actually shows that for G soluble and $\sigma \in \mathrm{Aut}_n(\Omega(G))$ we always have

$$\sigma(G/[U]) - G/\sigma_*([U])$$
$$= \sum_{[V],|V|<|U|} \frac{|N_G(V)|}{|U|} \cdot (\#([V],[U]) - \#(\sigma_*[V],\sigma_*[U])) \cdot e_{\sigma_*([V])}.$$

Corollary 3.4. *a) Assume that G is soluble, let N be a normal subgroup of G and let $\sigma \in \mathrm{Aut}_n(\Omega(G))$. Then $\sigma_*([N])$ is normal. Moreover $\sigma(G/[N]) = G/[\sigma_*([N])$, if σ_* preserves inclusions.*
b) If G is hamiltonian (i.e. each subgroup of G is normal), then $\mathrm{Aut}_n(\Omega(G))$ is isomorphic to the order preserving lattice isomorphisms of the subgroup lattice of G.

PROOF. Let $N_* \in \sigma_*([N])$. Then by Proposition 3.1, N_* is normal in G with the same order as N. If $[V] \leq [M]$, then $\#([V],[M]) = |G|/|N_G(V)|$, if M is normal in G. Therefore for each $[V] \leq [N]$ we have $\#([V],[N]) = \#(\sigma_*([V],\sigma_*([N]))$. Now the Corollary follows from Theorem 3.2. □

Remark 3.5. Combining Theorems 2.4 and 3.2 we obtain in particular a description of $\mathrm{Aut}(\Omega(G))$ for G abelian; a result proved by Krämer [6, Theorem 2].

Next we give an answer to Question 1, which facilitates the calculation of $\mathrm{Aut}(\Omega(G))$ with a computer algebra system.

Theorem 3.6. *Assume that G is soluble. Let $\tau \in \mathrm{Aut}_n(V(G))$. There exists $\sigma \in \mathrm{Aut}_n(\Omega(G))$ with $\sigma_* = \tau$ if and only if*

$$\sum_{[V]\leq[U]} (|W|/|U|) \cdot \#([V],[U]) \cdot (\sum_{W\in[W]} \mu(W,V_*)) \qquad (**)$$

is a rational integer for each pair $([U],[W])$ with $[W] \leq [V_*]$ for some $[V] < [U]$. Here μ denotes the Moebius function of the subgroup lattice of G and U_* and V_* are representatives of $\tau[U]$ and $\tau[V]$ respectively.

PROOF. As already noted above, τ induces an automorphism τ_x of $gh(G)$ which permutes the quasi-idempotents of $\Omega(G)$. We have to check whether $\sigma := \tau_{x|\Omega(G)}$ maps $\Omega(G)$ into $\Omega(G)$. (Note that $\sigma(G/[1]) = G/[1]$.) Now, of course this will be the case if and only if $\sigma(G/[U] \in \Omega(G)$ for all $[U] \in V(G)$. By 1.7

$$G/[U] = \sum_{[V]\leq[U]} \#([V],[U]) \cdot (|N_G(V)|/|U|) \cdot e_{[V]},$$

which can be written in terms of quasi-idempotents as

$$G/[U] = \sum_{[V]\leq[U]} \#([V],[U]) \cdot (|V|/(|U| \cdot \prod_{q\in\pi(\bar{V})} q)) \cdot q_{[V]}.$$

Now $\sigma(q_{[V]}) = q_{[V_*]}$ and by 1.7

$$q_{[V_*]} = ((\prod_{q\in\pi(\bar{V}_*)} q)/|V_*|) \cdot \sum_{W\leq V_*} \mu(W,V_*) \cdot |W| \cdot (G/[W]).$$

Hence one computes – using $\prod_{q\in\pi(\bar{V}_*)} q = \prod_{q\in\pi(\bar{V})} q$ and $|V| = |V_*|$ – that

$$\sigma(G/[U]) = \sum_{[V]\leq[U]} (\#([V],[U])/|U|) \cdot \sum_{W\leq V_*} \mu(W,V_*) \cdot |W| \cdot (G/[W]).$$

The coefficient of $G/[W]$ in the expression for $\sigma(G/[U])$ is

$$\sum_{[V]\leq[U]} (\#([V],[U])/|U|) \cdot (\sum_{W\in[W]} \mu(W,V_*) \cdot |W|).$$

If $[W] = [U_*]$, then this coefficient is 1. Hence the result follows. □

Remark 3.7. a) It is not necessary to check all pairs $([U],[W])$. If U is normal, then for each W the expression (**) is a rational integer. The same holds, if $[U]$ represents minimal subgroups or if $[U]$ represents cyclic subgroups, c.f. Proposition 4.1.

b) With any knowledge of the Moebius function the expression (**) may be modified. For example if V is a p-group, then by [9] the Moebius function $\mu(W,V)$ is given by the formula $\mu(W,V) = (-1)^r \cdot p^{r\cdot(r-1)/2}$, if W is normal in V and $V/W \cong (C_p)^r$, $\mu(W,V) = 0$ otherwise.

4. Special groups

Proposition 4.1. *a) Let C be a cyclic subgroup of G and let $\sigma \in \mathrm{Aut}_n(\Omega(G))$. Then $\sigma(G/[C]) = G/\sigma_*([C])$ and $C_* \in \sigma_*([C])$ is cyclic.*

b) $[C]$ consists of a conjugacy class of a nilpotent group if and only if $\sigma_*([C])$ does.

PROOF. If we can show that together with C, C_* is also cyclic, then

$$\#([V],[C]) = (\#(\sigma_*([V]),\sigma_*([U])) = 1$$

for each subgroup V of C, and part a) follows immediately from Section 3.

If C is a minimal subgroup of G, then C_* is a minimal subgroup, since σ_* is a poset automorphism. In order to prove a) and b) we use induction.

First we establish b). For this suppose that C is nilpotent. Taking fixed points with respect to $[C]$ and $[C_*]$, Proposition 2.2 gives $|N_G(C)/C| = |N_G(C_*)/C_*|$. By Proposition 2.1 we know then that the set of primes dividing $|\bar{C}|$ where $\bar{C}$ denotes the commutator quotient of C, coincides with the set of primes dividing $\bar{C}_*$. If C_* is not a p-group we obtain by induction that C_* has normal Sylow subgroups, assuming that each proper subgroup has this property and using the fact that each prime dividing $|C_*|$ divides $|\bar{C}_*|$. Note that by Proposition 2.2 C is a p-group if and only if C_* is a p-group, and thus b) follows.

In order to establish a), it suffices now to show that a) holds for cyclic p-subgroups. Assume that C_* is not a cyclic p-group, but that all its proper subgroups are cyclic. It follows that C_* has precisely $p+1$ maximal subgroups. Let $[V]$ be the unique maximal subclass of $[C]$. By Proposition 3.1 we know that $[C_*]$ has the same property except possibly in the case that $\#([V_*],[C_*]) = 1$ and there is a second subclass $[W]$ with $\#([W],[C_*]) = p$. Taking fixpoints with respect to $[W]$ and $[V_*]$ shows that

$$\sigma(G/[C]) = G/[C_*] - G/[W] + \sum_{|X|<|W|} a_{[X]}G/[X].$$

Now taking fixed points with respect to $[1]$ we get the contradiction $0 = 1 + m \cdot p$ with $m \in \mathbb{Z}$. So we know that $[C_*]$ has the unique subclass $[V_*]$. Taking fixed points with respect to M shows that

$$\sigma(G/[C]) = G/[C_*] - G/[V_*] + \sum_{|X|<|V_*|} a_{[X]}G/[X],$$

if $\#([V_*],[C_*]) = p+1$. Again taking fixed points with respect to the class $[1]$ yields the desired contradiction. □

Proposition 4.2. *Let G be a p-group and let $\sigma \in \mathrm{Aut}_n(\Omega(G))$. Assume that for each proper subclass $[V]$ of $[U]$ the automorphism σ maps $G/[V]$ into $G/[V_*]$. Then $\sigma(G/[U]) = G/[U_*]$, if U and U_* are generated by at most two elements and σ_* preserves inclusions.*

PROOF. By Proposition 4.1 we may assume that the minimum number of generators of U is 2. Then the quasi-idempotent $q_{[U]}$ has by 1.7 and Weisner's formula [9] for the Moebius function μ of the subgroup lattice, the form

$$\begin{aligned} q_{[U]} &= p \cdot \frac{|N_G(U)/U|}{|N_G(U)|} \cdot \sum_{V \leq U} \mu(V,U) \cdot |V| \cdot G/V \\ &= p \cdot G/[U] - (\sum_{[V],|U:V|=p} \#([V],[U]) \cdot G/[V]) + G/[Fr(U)], \end{aligned}$$

where $Fr(U)$ denotes the Frattini subgroup of U– this is the place where we have used that U is generated by 2 elements. Look at the analogous formula for $q_{\sigma_*([U])}$ and take into account that $\sigma(q_{[U]}) = q_{\sigma_*([U])}$. Assume that $\sigma(G/[U]) = \Sigma\, a_{[V]} G/[V]$, and note that the transitive G-classes form a $\mathbb{Z}$-basis of $\Omega(G)$. The only such class in the formula of $q_{\sigma_*([U])}$ with coefficient 1 is $G/[Fr(U_*)]$. Observe that the addition of $\sigma(G/[U]) \cdot p$ changes the coefficients of transitive G-classes only mod p, thus $\sigma(G/[Fr(U)]) = G/[Fr(U_*)]$.

Taking fixed points with respect to maximal subclasses $[M]$ of $[U]$ we see that $\#([M],[U]) \equiv \#([M_*],[U_*])$ mod p. Now U and U_* have precisely $p+1$ maximal subgroups. By assumption the number of maximal subclasses of $[U]$ and $[U_*]$ must coincide. Therefore there is no room left for $\#([M],[U])$ not being equal to $\#([M_*],[U_*])$; hence $\sigma(G/[U]) = G/[U_*]$. □

Corollary 4.3. *Let G be a p-group of order $\leq p^5$. Then a normalized automorphism σ of $\Omega(G)$ maps transitive G-classes to transitive G-classes, if σ_* preserves inclusions.*

PROOF. If U is a subgroup of order p or p^2, then $\sigma(G/[U]) = G/[U_*]$ by Propositions 4.1 and 4.2. If U has order p^4 or p^5, then U is normal. Hence by 3.1 the result follows.

Assume that U has order p^3. Then U has at most p conjugates. By Propositions 4.1 and 4.2 we may assume that U has 3 as minimum number of generators. Moreover, if U is not normal, then U_* is not normal and so we may assume that U and U_* have precisely p conjugates. If V is a maximal subgroup of U, then the number of G-conjugates of V in U is not divisible by p, if $|N_G(V)| = p^4$ or p^5; and is p, if $|N_G(V)| = p^3$. The same holds with respect to U_*. Consequently it follows that $\#([V],[U]) = \#([V_*],[U_*])$, if V is a maximal subgroup of U. Now we know that

$$\sigma(G/[U]) = G/[U_*] + \sum_{|V|=p} \alpha_{[V]} \cdot G/[V_*] + \alpha_{[1]} \cdot G/[1].$$

Taking fixed points with respect to $[V]$ with $|V| = p$ we see that $\#([V],[U]) - \#([V_*],[U_*]) \equiv 0$ mod p^2. Now U has only p^2+p+1 subgroups of order p. Moreover a subgroup of order p of U has at most p^2 G - conjugates. By assumption the numbers of subclasses in $[U]$ and $[U_*]$ coincide, it follows that $\#([V],[U]) = \#([V_*],[U_*])$. □

The following is now immediate from Proposition 4.2.

Corollary 4.4. *Let G be a metacyclic p-group. Then a normalized automorphism σ of $\Omega(G)$ maps transitive G-classes to transitive G-classes provided σ_* preserves inclusions.*

Remarks 4.5. a) The results on dihedral groups and hamiltonian groups, proved in [7], may be derived easily from 3.1, 3.4 and 4.4. The other groups considered in [7] are groups having a cyclic normal subgroup and a cyclic complement of coprime order. They clearly have the property that the normalized automorphisms of $\Omega(G)$ are just given by $\mathrm{Aut}_n(V(G))$.

b) Finally we note that the obvious question whether for p-groups normalized automorphisms of $\Omega(G)$ preserving inclusions on $V(G)$ are precisely the automorphisms of the table of marks has at least for p-groups of small order, a positive answer. At the moment we do not want to commit ourselves as to whether this might be true in general. At least it seems to be worth investigating this question using the formula of Section 3 for p-groups of small order, with a computer algebra system.

References

[1] A. Blass, Natural endomorphisms of Burnside rings, *Trans.Amer.Math.Soc.* **253**(1979), 121–137.

[2] C.W. Curtis and I. Reiner, *Methods of representation theory, Vol.II* (John Wiley and Sons, 1987).

[3] A. Dress, A characterization of solvable groups, *Math.Z.* **110**(1969), 213–217.

[4] A. Dress and C. Siebeneicher, The Burnside ring of profinite groups and the Witt vector construction, *Adv. Math.* **70**(1988), 87–132.

[5] D. Gluck, Idempotent formula for the Burnside algebra with applications to the p-subgroup simplicial complex, *Illinois J. Math.* **25**(1981), 63–67.

[6] H. Krämer, Über die Automorphismengruppe des Burnside Ringes endlicher abelscher Gruppen, *J.Algebra* **30**(1974), 279–293.

[7] G. Ochoa Lezaun, Los grupos de automorfismos de los anillos de Burnside de los grupos finitos hamiltonianos y diedricos, Publ. del Seminario Mat. Garcia de Galdeano Serie II, Secc.2, No.1, 1984, 116 pages.

[8] D.M. Nicolson, The orbit of the regular G-set under the full automorphism group of the Burnside ring of a finite group G, *J.Algebra* **51**(1978), 288–299.

[9] L. Weisner, Some properties of prime-power groups, *Trans. Amer. Math. Soc.* **38**(1935), 485–492.

[10] T. Yoshida, Idempotents of Burnside rings and Dress induction theorem, *J.Algebra* **80**(1983), 90–105.

ON FINITE GENERATION OF UNIT GROUPS FOR GROUP RINGS

JAN KREMPA

Institute of Mathematics, Warsaw University, ul. Banacha 2, 02-097 Warszawa, Poland

Abstract

Let R represent an associative, but nonnecessarily commutative ring with $1 \neq 0$, G a nontrivial group, and RG the group ring of G over R.

Let us consider the following problem: *Find the necessary and sufficient conditions under which the unit group of RG, or the group of normalized units of RG, is finitely generated.*

We are going to survey and extend some known results about this problem. We also formulate several more detailed questions suggested by this survey.

1. Preliminaries

In this paper we assume that all rings are associative with $1 \neq 0$. Subrings with the same unities will be called unital. For convenience of readers we recall some notation and terminology from ring theory.

$\mathrm{U}(A)$ will always denote the unit group of the ring A, A^+ the additive group of A, and $1 + B$ – the set $\{1 + b : b \in B\}$ for any subset $B \subset A$. Let us also agree that $\mathrm{J}(A)$ will stand for the Jacobson radical of the ring A, and $\mathrm{N}(A)$ for the set of all nilpotent elements of A. Further we will say that a ring A is *semisimple* if $\mathrm{J}(A) = 0$, and *reduced* if $\mathrm{N}(A) = 0$. Clearly if A is commutative then $\mathrm{N}(A)$ is an ideal contained in $\mathrm{J}(A)$ and the factor ring $A/\mathrm{N}(A)$ is reduced. Rings having no proper central idempotents will be called here *indecomposable*.

In the sequel we will also use the following notation:

$\boldsymbol{Z}$ the ring of rational integers;

$\boldsymbol{F}_q$ the finite field with q elements;

$A[t]$ the polynomial ring of a variable t over A;

$N \rtimes H$ a semidirect product of a normal subgroup N and a subgroup H.

From now on R will always represent a ring of coefficients, G a nontrivial group, and $R[G]$, or simply RG, the group ring of G over R. Further let $\mathrm{V}(RG)$ be the group of normalized units of RG. We will also use some other standard notation, terminology, and results on group rings used, for example, in [12, 23, 29, 3]. In this paper we are going to discuss the following question:

Question 1. Find the necessary and sufficient conditions under which the group $\mathrm{U}(RG)$ or $\mathrm{V}(RG)$ is finitely generated.

If R and G are commutative then the above question coincides with Problem 7 from [9]. It is also closely related to Problem 98 from [4].

It is clear that $\mathrm{U}(RG) = \mathrm{V}(RG) \rtimes \mathrm{U}(R)$, hence whenever $\mathrm{V}(RG)$ is finitely generated, then $\mathrm{U}(RG)$ is finitely generated if and only if $\mathrm{U}(R)$ is finitely generated. On the other hand, from Example 3.7 we will have that finite generation of $\mathrm{U}(RG)$ does not imply that $\mathrm{V}(RG)$ is finitely generated. However from now on we will concentrate only on groups $\mathrm{U}(RG)$ because proofs in the case of $\mathrm{V}(RG)$ are similar.

Before presenting some results about Problem 1 we recall some examples of units in group rings.

In [12], (see page 10), elements of the form rg where $r \in R$ and $g \in G$ are called *monomials*. Extending this natural term, used also outside of the theory of group rings, a unit which is a monomial in RG will be called here a *monomial unit*. In the literature monomial units are usually called trivial, but there exist other sorts of units which are well known, not necessarily monomial, but still trivial. Below we recall two such families of such units.

If $e \in R$ is an idempotent and $g \in G$, then $v = eg + 1 - e$ is a rather trivial unit from $\mathrm{V}(RG)$. Moreover v is nonmonomial if and only if e is proper, and $g \neq 1$.

If $a \in \mathrm{N}(R)$ then $1 + ag \in \mathrm{U}(RG)$, and $1 - a + ag = 1 + a(g-1) \in \mathrm{V}(RG)$ for every $g \in G$. Such an unit is not monomial if and only if $a \neq 0$ and $g \neq 1$.

Using the units just described, simple calculations and some elementary facts about abelian groups, one can prove the following useful observation (cf. [12] Lemma 3.4.4):

Proposition 1.1. *Let $\mathrm{U}(RG)$ be finitely generated and let $A = \mathrm{A}(RG)$ be the subring of R generated by all coefficients of all elements from $\mathrm{U}(RG)$. Then:*

1. *$\mathrm{U}(R)$ and G are finitely generated;*
2. *A is a finitely generated unital subring of R, and $\mathrm{U}(RG) = \mathrm{U}(AG)$;*
3. *A contains all nilpotent and idempotent elements of R;*
4. *if RG is commutative then either R is reduced, or G is finite.*

The conditions of the above proposition are in general not sufficient for finite generation of $\mathrm{U}(RG)$ even if R and G are rather small. As the first illustration let us quote a result of M. Mirowicz from [20] concerning the case of the infinite dihedral group

$$D_\infty = \langle x, y | x^2 = 1, y^x = y^{-1} \rangle.$$

Theorem 1.2. *Let R be either $\boldsymbol{Z}$, or $\boldsymbol{F}_2$ or $\boldsymbol{F}_3$. Then $\mathrm{U}(R[D_\infty])$ is not finitely generated.*

In [20] Mirowicz gave a full description of the structure of the unit groups which he considered, but the following questions still seem to be open:

Question 2. Let $q > 3$ be a prime or a power of a prime number. Is $\mathrm{U}(F_q[D_\infty])$ finitely generated? Does there exist a ring R such that $\mathrm{U}(R[D_\infty])$ is finitely generated?

As an immediate consequence of Proposition 1.1 we have that, if $\mathrm{U}(RG)$ is finitely generated and G is abelian, then $G = F \otimes T$ where F is free abelian of finite rank and T is finite. Hence with this notation we have:

$$(R[F])[T] \cong R[F \otimes T] \cong RG \cong R[T \otimes F] \cong (R[T])[F].$$

So one can see that, at least in the case when G is abelian, our problem splits into the following cases:

1. G is torsion free;
2. G is finite.

In this survey we will follow this splitting because the arguments used in both cases are rather different.

As a first theorem in case 1, let us quote the following result:

Theorem 1.3. (Karpilovsky [9]) *Let R be a commutative ring and let G be a torsion free abelian group. Then* $\mathrm{U}(RG)$ *is finitely generated if and only if*

1. $\mathrm{U}(R)$ *and G are finitely generated;*
2. *R is a finite product of indecomposable rings;*
3. *R is reduced.*

The proof follows from Proposition 1.1 and some results from [18, 22, 14], or as in [9].

2. Finite groups

In this section we present several facts about Question 1 in case 2. Some of them depend heavily on strong theorems about units of arbitrary rings. Probably the most famous result of this type is the Dirichlet unit theorem. As nontrivial generalizations of this classical result we have:

Theorem 2.1. (Roquette, Samuel) *Let A be a commutative domain. If A is finitely generated then* $\mathrm{U}(A)$ *is finitely generated.*

Theorem 2.2. (Bass [1]) *Let A be a finitely generated commutative ring. Then* $\mathrm{U}(A)$ *is finitely generated if and only if* $(\mathrm{J}(A))^+$ *is finitely generated.*

Corollary 2.3. *Let A be a commutative ring. Then* $\mathrm{U}(A)$ *is finitely generated if and only if* $(\mathrm{J}(A))^+$ *is finitely generated and* $\mathrm{U}(A/\mathrm{J}(A))$ *is finitely generated.*

You can find proofs for example in [1, 11, 12] or [17].

Theorem 2.2 leads to a full solution of Problem 1 for "small" commutative rings. To formulate it, and some further results, let us agree for every natural n that $\mathrm{J}_n(R)$ will denote the set of all $r \in R$ such that $nr \in \mathrm{J}(R)$. It is easy to see that $\mathrm{J}_n(R)$ is an ideal of R containing $\mathrm{J}(R)$. Such ideals are involved in the description of $\mathrm{J}(RG)$ for finite G.

Theorem 2.4. (Karpilovsky [10]) *Let R be a finitely generated commutative ring and G an abelian group of order $n < \infty$. Then* $\mathrm{U}(RG)$ *is finitely generated if and only if* $(\mathrm{J}_n(R))^+$ *is finitely generated.*

Further let C_m denotes the cyclic group of order m. In [16] (see [12, 17]) the following consequences of Corollary 2.3 were proved:

Lemma 2.5. *Let R be commutative and G be a finite abelian group of exponent m. Then* $\mathrm{U}(RG)$ *is finitely generated if and only if* $\mathrm{U}(R[C_m])$ *is finitely generated.*

Theorem 2.6. *Let R be commutative. Then the group* $\mathrm{U}(R[C_2])$ *is finitely generated if and only if* $\mathrm{U}(R)$ *and* $(\mathrm{J}_2(R))^+$ *are finitely generated.*

The two last results suggest induction on the order of G as a method of solving Problem 1 for commutative R and abelian G. Unfortunately, from [16] (see [12, 17]), we have the following examples:

- For every $k \geq 1$ there exists a ring A_{2^k} such that $\mathrm{U}(A_{2^k}[C_{2^k}])$ is finitely generated but $\mathrm{U}(A_{2^k}[C_{2^{k+1}}])$ is not finitely generated.
- For every odd prime p and $k \geq 0$ there exists a ring A_{p^k} such that $\mathrm{U}(A_{p^k}[C_{p^k}])$ is finitely generated but $\mathrm{U}(A_{p^k}[C_{p^{k+1}}])$ is not finitely generated.
- For every relatively prime $k, l \geq 3$ there exists a ring $A_{k,l}$ such that $\mathrm{U}(A_{k,l}[C_k])$ and $\mathrm{U}(A_{k,l}[C_l])$ are finitely generated but $\mathrm{U}(A_{k,l}[C_{kl}])$ is not finitely generated.

All the rings just listed can be choosen as commutative Krull domains of characteristic 0 with invertible orders of suitable groups. It means that rather strong assumptions on R should be supposed in further consideration of a commutative version of Problem 1. So the following question (see Problem 8 from [9]) seems to be natural:

Question 3. Find the solution of Problem 1 for R being a commutative Noetherian domain and G cyclic of finite order $n > 2$. Even the case $n = 3$ seems to be still unsolved.

In the case when either G is nonabelian or R is noncommutative, the situation becomes more complicated. We will illustrate it in the next section. Now let us utilize the following nontrivial generalization of the Dirichlet unit theorem:

Theorem 2.7. (see [30, 26]) *Let A be a ring. If the group A^+ is free abelian of finite rank, then $\mathrm{U}(A)$ is finitely generated, and even finitely presented.*

As a consequence we have:

Corollary 2.8. *Let A be a ring. If A^+ is finitely generated then $\mathrm{U}(A)$ is finitely generated.*

PROOF. Let T be the torsion part of A^+. By assumption T is a finite ideal of A. If $T = 0$ then $\mathrm{U}(A)$ is finitely generated by the above theorem. If $T \neq 0$, but $\mathrm{J}(T) = 0$, then T is a ring with 1 which is finite and semisimple, so there exists an ideal $I \subset A$ which is a ring with 1, such that $A = I \oplus T$. It means that $\mathrm{U}(A) \cong \mathrm{U}(I) \otimes \mathrm{U}(T)$. Now both components of the above direct product are finitely generated, hence $\mathrm{U}(A)$ is finitely generated too.

If $\mathrm{J}(T) \neq 0$ then it is an ideal of A, and we can go to one of previous cases by taking the factor ring $A/\mathrm{J}(T)$. This easily completes the proof. □

Now as an immediate consequence we have the full solution of Problem 1 for rings which are small in an additive sense.

Theorem 2.9. *Let R be a ring such that R^+ is finitely generated and let G be an arbitrary finite group. Then $\mathrm{U}(RG)$ is finitely generated.*

3. Some examples

Now we will show that in the noncommutative case it is very difficult to extend Theorems 2.4 and 2.9 because strong results about units of finitely generated rings are not always valid in this case.

Group rings distinguished by Theorem 1.2 are semisimple and finitely generated. This means that in the noncommutative version of Theorem 2.2 one implication is not true. Now, after a short preparation, we will show that the other implication, and hence also Corollary 2.3, need not be true. Let us start from the following observation:

Lemma 3.1. *A group $X = N \rtimes H$ is finitely generated if and only if H is finitely generated and there exists a finite subset $S \subset N$ such that $N = \langle S^H \rangle$.*

From now on we will start to use typical notation for sets of matrices. Moreover for a ring A we denote by A_u the subring of A generated by $\mathrm{U}(A)$. With this notation we have:

Proposition 3.2. *Let B; C be rings, let M be a B,C-bimodule, and let* $A = \begin{bmatrix} B & M \\ 0 & C \end{bmatrix}$. *Then:*

1. *$\mathrm{U}(A)$ is finitely generated if and only if $\mathrm{U}(B)$ and $\mathrm{U}(C)$ are finitely generated, and M is finitely generated as a B_u, C_u-bimodule;*
2. *A is finitely generated if and only if B and C are finitely generated as rings and M is a finitely generated $B,C-$bimodule.*

PROOF. 1. We have $\mathrm{U}(A) = \begin{bmatrix} \mathrm{U}(B) & M \\ 0 & \mathrm{U}(C) \end{bmatrix}$. Let us take

$$H = \begin{bmatrix} \mathrm{U}(B) & 0 \\ 0 & \mathrm{U}(C) \end{bmatrix}, \text{ and } N = \begin{bmatrix} 1 & M \\ 0 & 1 \end{bmatrix}.$$

Then it is easy to see that $\mathrm{U}(A)$ is a subdirect product of H and N, and $H \cong \mathrm{U}(B) \otimes \mathrm{U}(C)$.

Now if $X \subset M$ is any subset and $S = \begin{bmatrix} 1 & X \\ 0 & 1 \end{bmatrix}$, then it is easy to see that $\langle S^H \rangle = \begin{bmatrix} 1 & B_u X C_u \\ 0 & 1 \end{bmatrix}$. From these facts and Lemma 3.1, the claim 1 follows immediately.

The claim 2 is even easier to prove than 1. □

Example 3.3. Let us take $n \geq 2$, $B = M = C = \boldsymbol{Z}[\frac{1}{n}]$, and $A = \begin{bmatrix} B & M \\ 0 & C \end{bmatrix}$. Then by the above proposition, A is finitely generated as a ring and the group $\mathrm{U}(A)$ is finitely generated, while $(\mathrm{J}(A))^+ \cong M^+$ is not finitely generated.

The examples used in the above arguments were not reduced. However we can show that in the noncommutative case even Theorem 2.1 becomes false. For this let F be the free abelian group with the set of free generators $\{t_n;\ n \in \boldsymbol{Z}\}$, H the free abelian group with free generators x, y, and let $G = F \rtimes H$, where H acts on F by the rules: $(t_n)^x = (t_n)^y = t_{n+1}$ for all $n \in \boldsymbol{Z}$. From Lemma 3.1 it follows that G is finitely generated, for example $G = \langle t_0, x, y \rangle$. Moreover let R be either $\boldsymbol{Z}$ or $\boldsymbol{F}_q[t]$. Then RG is a domain, and from direct calculations or Theorem 4.1 below we have that $\mathrm{U}(RG) = \mathrm{U}(R) \otimes G$.

Example 3.4. With the above notation let us take

$$A = R[t_0, t_0^{-1}, x, y, ax^{-1}, by^{-1}],$$

where $a, b \in R$ are nonunits such that $a + b = 1$. Clearly A is a finitely generated domain, and simple calculation shows, that $\mathrm{U}(A) = \mathrm{U}(R) \otimes F$ is abelian, but certainly not finitely generated.

Now we will show that in Theorem 2.4 the assumption that G is abelian is essential even if R is a finitely generated commutative principal ideal domain.

Example 3.5. Let F be a finite field of characteristic different from 2 and 3, $R = F[t]$, and let $G = S_3$ be the symmetric group on 3 elements. Then it is easy to see that $FG \cong F \oplus F \oplus \begin{bmatrix} F & F \\ F & F \end{bmatrix}$, hence $RG \cong R \otimes_F (FG) \cong R \oplus R \oplus \begin{bmatrix} R & R \\ R & R \end{bmatrix}$. Thus $\mathrm{U}(RG)$ is not finitely generated by an old theorem of Nagao, saying that $\mathrm{GL}_2(R)$ is not finitely generated (see [1, 12]).

If R and G are as above then we know that $\mathrm{U}(RG)$ is not finitely generated, but from Theorem 2.4 $\mathrm{U}(RH)$ is finitely generated for every proper subgroup $H \subset G$. So let us ask the following question:

Question 4. Let G be a finite abelian group and R a ring such that $\mathrm{U}(RG)$ is finitely generated, and let $H \subset G$ be a subgroup. Is $\mathrm{U}(RH)$ finitely generated?

Clearly the answer is 'yes' if either R is commutative or H is a direct summand of G.

Now we will show that in Theorem 2.4 the assumption that R is commutative is essential even for abelian G.

Example 3.6. Let us take $B = C = M = \boldsymbol{Z}[t, t^{-1}]$ and let $R = \begin{bmatrix} B & M \\ 0 & C \end{bmatrix}$. Now let G be any finite abelian group. Then it is easy to see that

$$RG = \begin{bmatrix} B & B \\ 0 & B \end{bmatrix} G \cong \begin{bmatrix} BG & BG \\ 0 & BG \end{bmatrix} \cong \begin{bmatrix} \boldsymbol{Z}G[t, t^{-1}] & \boldsymbol{Z}G[t, t^{-1}] \\ 0 & \boldsymbol{Z}G[t, t^{-1}] \end{bmatrix}.$$

From Proposition 3.2 we have that $\mathrm{U}(RG)$ is finitely generated, while $(\mathrm{J}(R))^+$ is of infinite rank.

Now we have the following illustration of connections between $\mathrm{U}(RG)$ and $\mathrm{V}(RG)$.

Example 3.7. Let R and G be as in the above example. Because $\boldsymbol{Z}G$ is reduced and has no proper idempotents then from [18], or Corollary 4.5 below, it follows that every unit from $\boldsymbol{Z}G[t, t^{-1}]$ is of the form ut^k where

$u \in \mathrm{U}(\boldsymbol{Z}G)$ and $k \in \boldsymbol{Z}$. This gives that $\mathrm{V}(BG) = \mathrm{V}(\boldsymbol{Z}G)$ and hence $\mathrm{V}(RG) = \begin{bmatrix} \mathrm{V}(\boldsymbol{Z}G) & M \\ 0 & \mathrm{V}(\boldsymbol{Z}G) \end{bmatrix}$, where

$$M = \oplus_{i \in \boldsymbol{Z}} (\oplus_{g \in G} \boldsymbol{Z}(g-1)) t^i.$$

Now by Lemma 3.1 we have that $\mathrm{V}(RG)$ is not finitely generated, while $\mathrm{U}(RG)$ is.

The next question is connected with attempts to extend abelian groups when noncommutative rings are considered.

Question 5. Can Lemma 2.5 be extended to the case where R is noncommutative but G is finite abelian?

4. Some subgroups of units

For discussion of Question 1 for torsion free groups we need more information about some subgroups of units of group rings. Let $\mathrm{M}(RG)$ denotes the set of all monomial units from RG. It is clear that $\mathrm{M}(RG)$ is a subgroup of $\mathrm{U}(RG)$ such that $\mathrm{M}(RG) = \mathrm{U}(R) \otimes G$, and $\mathrm{M}(RG) \cap \mathrm{V}(RG) = G$.

The question of finding conditions under which all units of RG are monomial is old, hard, important, and still far from being solved (see [5, 23, 12]). To formulate a partial solution of this question we will use the notion of u.p.-group and t.u.p.-group originally introduced by Higman in [5]. In [31], (see also page 271 of [11], or page 65 of [12]), A. Strojnowski showed that every u.p.-group is a t.u.p.-group. In this way he solved a problem of Higman from [5] and proved the following result:

Theorem 4.1. *Let R be a domain and G a u.p.-group. Then* $\mathrm{U}(RG) = \mathrm{M}(RG)$. *Even more: If $R{*}G$ is an arbitrary crossed product of R and G then all units of this ring are also monomial.*

Remark. The notions of u.p.-group and t.u.p.-group can be immediately extended to semigroups, but they become different in this context. For a suitable example see page 125 of [21]. However it is still not known if the semigroup version of Theorem 4.1 is true.

It is well known, (see [5, 23, 12, 19]), that the class of u.p.-groups is a quite large class of torsion free groups containing all locally indicable groups, and even all right ordered groups. G. M. Bergman in [2] proved that not every right ordered group must be locally indicable, but the following question seems to be still open:

Question 6. Is every u.p.-group right orderable?

The first example of a torsion free group which is not a u.p.-group was constructed in [27], but later a much simpler one was exhibited by Promislow in [25]. He showed that the well known group, say H, with presentation

$$H = \langle x, y \,|\, y^{-1}x^2y = x^{-2},\ x^{-1}y^2x = y^{-2}\rangle$$

is not a u.p.-group being 'almost abelian' and torsion free (cf. [24] proposition 37.1). Due to known results (see [24] Section 37) for this group H and any field K, the group ring KH is a domain. The following question seems to be well motivated:

Question 7. Is $\mathrm{U}(KH) = \mathrm{M}(KH)$ for every field K?

In connection with the second part of Theorem 4.1 one can ask the following question:

Question 8. Let G be a group such that for any domain D any crossed product $D{*}G$ has only monomial units. Is G a u.p.-group?

For questions of this type it is reasonable to introduce the following classes of groups:

- $\mathcal{M}$ The class of all groups G such that $\mathrm{U}(RG) = \mathrm{M}(RG)$ for every domain R;
- $\mathcal{M}_c$ The class of all groups G such that $\mathrm{U}(RG) = \mathrm{M}(RG)$ for every commutative domain R;
- $\mathcal{M}_*$ The class of all groups G such that $\mathrm{U}(R{*}G) = \mathrm{M}(R{*}G)$ for every domain R and any crossed product $R{*}G$.

From the above definition we have that $\mathcal{M}_* \subset \mathcal{M} \subset \mathcal{M}_c$, and by Theorem 4.1 every u.p.-group belonges to $\mathcal{M}_*$. The following questions should be asked immediately:

Question 9. Are the above defined classes distinct? Which group operations can be applied inside these classes?

Remark. Everything which is later shown for u.p.-groups is also true for groups from the class $\mathcal{M}$, but we will not mention this fact again.

Instead of immediately extending Theorem 4.1 to a larger class of coefficients, it is better to show first a more general fact involving reduced rings. So let us recall a result about them, which is still not well known enough.

Lemma 4.2. *Let R be a reduced ring. Then:*

1. *Every idempotent in R is central.*
2. *If $P \subset R$ is a minimal prime ideal then the factor ring R/P is a domain.*

The proof is not very complicated. You can consult Exercises 4, 5 and 6 on page 285 of [24] for details.

For convenience of further notation let $\mathrm{B}(R)$ denote the set of all central idempotents of R. Clearly $0, 1 \in \mathrm{B}(R)$. It is well known that the set $\mathrm{B}(R)$ is a Boolean algebra under standard operations on idempotents, or equivalently under the order given by $e \leq f$ if and only if $ef = e$.

Now let $\mathrm{T}(RG)$, (usually T in [12]), be the subgroup of $\mathrm{U}(RG)$ generated by all elements $eg + 1 - e$ for $e \in \mathrm{B}(R)$ and $g \in G$. Below we give a more convenient description of elements of $\mathrm{T}(RG)$.

Lemma 4.3. (cf. [15, 16, 12]) *Let $v \in RG$. Then $v \in \mathrm{T}(RG)$ if and only if there exists $n \geq 1$ such that $v = \sum_{i=1}^{n} e_i g_i$ where $1 = \sum_{i=1}^{n} e_i$ is a decomposition of $1 \in R$ into a sum of orthogonal central idempotents, and all $g_i \in G$.*

The above lemma allows us to see that properties of $\mathrm{T}(RG)$ are strongly connected with properties of G and $\mathrm{B}(R)$ (see [18, 15, 12]).

The result below, in some cases, was established in [18, 22] and [14].

Theorem 4.4. ([15]) *Let R be reduced and G a u.p.-group. Then $\mathrm{V}(RG) = \mathrm{T}(RG)$. In particular $\mathrm{U}(RG) = \mathrm{U}(R) \otimes \mathrm{T}(RG)$.*

PROOF. Use Lemma 4.2 and follow the proof of Theorem 3.3.11 in [12]. □

As an immediate consequence of the above theorem we have a desired generalization of the first statement of Theorem 4.1.

Corollary 4.5. ([15]) *Let G be a u.p.-group. Then $\mathrm{U}(RG) = \mathrm{M}(RG)$ if and only if R is reduced and indecomposable.*

5. U.P.-groups

In this section we are going to discuss Problem 1 in the case where G is a u.p.-group. As an easy consequence of Theorem 4.4 we have:

Theorem 5.1. ([16]) *Let R be a reduced ring and G a u.p.-group. Then $\mathrm{U}(RG)$ is finitely generated if and only if $\mathrm{U}(R)$ and G are finitely generated, and R is a finite product of indecomposable rings.*

For further use let us introduce the following notion: A group G will be called a *CF-group* if it has only finitely many conjugacy classes. Such groups were considered earlier, but probably without a special name (see [28] page 129). Clearly every finite group is a CF-group, every CF-group which is a FC-group is finite and, for example, the infinite cyclic group is not a CF-group. With this notion we can formulate the following theorem:

Theorem 5.2. ([16]) *Let R be a commutative ring and G a u.p.-group. Then $\mathrm{U}(RG)$ is finitely generated if and only if:*

1. *$\mathrm{U}(R)$ and G are finitely generated.*
2. *R is a finite product of indecomposable rings.*
3. *Either R is reduced or G is a CF-group.*

For the proof see [12] (proof of Theorem 3.4.13), or simplify the proof of Theorem 6.4 below.

The above theorem immediately leads to the following question:

Question 10. Does there exist a finitely generated u.p.-group which is a CF-group?

Clearly not every u.p.-group is a CF-group. The following result from [6] is more interesting:

Theorem 5.3. *If G is a torsion free group then there exists a torsion free CF-group $\widehat{G}$ containing G as a subgroup.*

This theorem shows that there exist quite a lot of torsion free CF-groups. If, in particular, we take the group from [25] or [27] then we will see that a torsion free CF-group need not be a u.p.-group. Now as a special case of Question 10 one can ask the following:

Question 11. Does there exist a CF-group which is a u.p.-group?

CF-groups constructed with the help of Theorem 5.3 are certainly infinitely generated. Several years ago S. V. Ivanov, (see [7, 8]), constructed finitely generated, torsion free CF-groups. Methods used by him are rather complicated, and it is still not known if his groups are u.p.-groups.

Now let us look for classes of groups certainly containing no examples of CF-groups which are u.p.-groups. The following observations will be helpful:

Proposition 5.4. *Let G be a CF-group. Then:*

1. *G has only finitely many normal subgroups.*
2. *There exists a smallest normal subgroup $N \subset G$ such that G/N is finite.*

3. *If H is either a homomorphic image or a subgroup of finite index in G, then H is a CF-group.*

The proof of this proposition is straightforward. As a consequence we obtain by induction on the class of solvability of the group under consideration, the following result:

Proposition 5.5. *Let G be a CF-group. If G is a finite extension of a solvable group then G is finite.*

The above proposition covers an important class of polycyclic-by-finite groups. Another important class is formed by linear groups. For such groups the following result is known:

Theorem 5.6. *Let G be a linear group over a field. If G is a CF-group then G is finite.*

The proof can be similar to that of local finiteness of periodic linear groups (as is proposed in [3], Exercise 1 on page 361).

For a generalization of the above result the following observation is helpful:

Lemma 5.7. *Let R be a ring, and $G \subset \mathrm{U}(R)$ an infinite CF-group. Then there exists a prime ideal P of R such that the natural image of G in $\mathrm{U}(R/P)$ is also infinite.*

PROOF. By Proposition 5.4 we can choose a normal subgroup $N \subset G$ of finite index that has no finite proper images. Now, by the same proposition, there exists an ideal $P \subset R$ which is a maximal element of the family of ideals $I \subset R$ such that $N \cap (1 + I) \neq N$. Further let $J \subset R$ be the ideal generated by $P \cup \{1 - n : n \in N\}$.

Now if $I \supset P$ is an ideal of R then $\{1 - n : n \in N\} \subset I$, which means that $J \subset I$. This says that the ring R/P is subdirectly irreducible.

Now if $J^2 \subset P$ then the group $N/(N \cap (1 + P))$ is a nontrivial abelian CF-group and so is finite, which is impossible by the choice of N. In this way we prove that P is semiprime, and hence a prime ideal of R, which completes the proof. □

Now applying the famous Posner theorem about the structure of prime PI-rings, (see [23]), Theorem 5.6 and the above lemma we obtain:

Theorem 5.8. *Let R be a ring satisfying a nontrivial polynomial identity. If $G \subset \mathrm{U}(R)$ is a CF-group then G is finite.*

For contrast, if G is a torsion free CF-group, and K is a field, then by the theorem of Connell (see [23]), KG is a prime ring with $G \subset \mathrm{U}(KG.)$ This means that while trying to generalize the above theorem some strong conditions on R will have to be assumed.

In connection with Proposition 5.4 the following questions are natural:

Question 12. Is the class of all CF-groups closed under extensions? In particular, is any finite extension of a CF-group again a CF-group?

6. Almost reduced coefficients

In this section we will try to extend results about finite generation of $\mathrm{U}(RG)$ to the case when G is still a u.p.-group but R is not necessarily commutative. Let us look first at the following example:

Example 6.1. Let G be the infinite cyclic group and $R = \begin{bmatrix} Z & Z \\ 0 & Z \end{bmatrix}$. Then $\mathrm{U}(RG)$ is finitely generated by Proposition 3.2 but R is a nonreduced ring satisfying many polynomial identities.

This example shows that one can expect some complications when extending results from Section 5 to noncommutatve coefficients even for abelian groups. However below we propose a common generalization of Theorems 5.1 and 5.2 to the class of almost reduced coefficients, where a ring R will be called *almost reduced* if $\mathrm{N}(R)$ is contained in the center of R.

Clearly reduced rings and commutative rings are almost reduced. The class of almost reduced rings is closed under taking direct products and subrings. From these observations we have that every subdirect product of a reduced ring and a commutative ring is almost reduced; however the converse is not true. Before showing a suitable example let us agree that $[R, R]$ denotes the ideal of R generated by all commutators $[r, s] = rs - sr$, for all $r, s \in R$.

Example 6.2. Let K be a field, $n \geq 2$, and

$$A_n = K\langle x, y\rangle/([x, y]^n, [[x, y], x], [[x, y], y])$$

be the image of the free algebra $K\langle x, y\rangle$ with unity, subject to relations $[x, y]^n = [[x, y], x] = [[x, y], y] = 0$. It is easy to calculate that $[A_n, A_n] = \mathrm{N}(A_n)$ is the ideal of A_n generated by $[x, y]$. Thus A_n is almost reduced, $\mathrm{N}(A_n)$ is a nilpotent ideal of index n and A_n cannot be represented as a subdirect product of a reduced ring and a commutative ring.

Lemma 6.3. *Let A be an almost reduced ring. Then:*

1. *Any idempotent of A lies in* $\mathrm{B}(A)$.
2. *If A is finitely generated as a ring then* $\mathrm{N}(A)$ *is nilpotent.*

PROOF. 1. Let $e \in A$ be an idempotent. Since for every $a \in A$ the product $ea(1-e) \in \mathrm{N}(A)$, then

$$0 = ea(1-e)e = eea(1-e) = ea(1-e).$$

This means that $ea = eae$. Similarly one can check that $ae = eae$, hence $ae = ea$ and $e \in \mathrm{B}(R)$.

2. Let A be finitely generated, $a \in \mathrm{N}(A)$, and $x, y \in A$. Then a and ax are central, so we have $0 = [ax, y] = a[x, y]$. This means that $\mathrm{N}(A))[A, A] = 0$. Because A is finitely generated then it is well known that $\mathrm{N}(A/[A, A])$ is nilpotent. Hence there exists $n \geq 1$ such that $(\mathrm{N}(A))^n \subset [A, A]$ and therefore $(\mathrm{N}(A))^{n+1} = 0$. □

Now we are ready to prove the main result of this section.

Theorem 6.4. *Let R be an almost reduced ring and G a u.p.-group. Then* $\mathrm{U}(RG)$ *is finitely generated if and only if the following conditions are satisfied:*

1. $\mathrm{U}(R)$ *and G are finitely generated.*
2. *R is a finite product of indecomposable rings.*
3. *Either R is reduced, or* $\mathrm{N}(R)^+$ *is finitely generated and G is a CF-group.*

PROOF. In this proof, let $\mathrm{N}(R) = I$ and $\bar{R} = R/I$.

($\Rightarrow$) 1. follows from Proposition 1.1.

From assumptions on R and G it is easy to see that $1 + IG \subset \mathrm{U}(RG)$, hence $\mathrm{U}(\bar{R}G)$ is finitely generated and, by Theorem 5.1, $\bar{R}$ is a finite product of indecomposable rings. It is well known that every idempotent from $\bar{R}$ can be lifted to R, and in fact, by Lemma 6.3, to $\mathrm{B}(R)$. This gives 2.

3. Due to Proposition 1.1 we can assume that R is finitely generated without any change of I. By the just proved condition 2, we can also assume that R is indecomposable. Now by Lemma 6.3 I is nilpotent. The case $I = 0$ trivially gives 3.

Let us assume now that $I \neq 0$ but $I^2 = 0$. In this case let $H = \mathrm{M}(RG)$ and $N = 1 + \oplus_{g \neq 1} Ig$. From Corollary 4.5 one can calculate that $HN = \mathrm{U}(RG)$. Clearly $H \cap N = 1$ and N is a normal subgroup of $\mathrm{U}(RG)$ because R is almost reduced and $I^2 = 0$.

By Lemma 3.1 we have elements $x_1, x_2, \ldots, x_r \in N$ such that N is generated as a group by $\{(x_i)^h : h \in H, i = 1, 2, \ldots, r\}$. Without loss of generality we can assume that $x_i = 1 + b_i y_i$, where $b_i \in I$ and $y_i \in G \setminus \{1\}$. Now if $ug \in H$

then $ugx_i(ug)^{-1} = 1 + b_i gy_i g^{-1}$. This form of the generators of N implies that G is covered by conjugacy classes of elements $1, y_1, \ldots, y_s$ and I^+ is generated by $b_1, \ldots, b_r$. This completes the proof in the case $I^2 = 0$. One can easily go to the general case because I is nilpotent by Lemma 6.3.

($\Leftarrow$) Let $A \subset R$ be the unital subring generated by I. By the assumed 3. and Theorem 2.2 applied to A, we have that $I^{s+1} = 0$ for some $s \geq 0$. If $s = 0$ then $\mathrm{U}(RG)$ is finitely generated by Theorem 5.1. Now let us assume that $s > 0$ and let $(I^s)^+$ be generated by $b_1, b_2, \ldots, b_r$. Moreover let G be covered by conjugacy classes of elements $1, y_1, \ldots, y_t$. By the induction hypothesis we have elements $h_1, \ldots, h_q$ such that $\mathrm{U}((R/I^s)G)$ is finitely generated. We can also assume, that $G \subset \langle h_1, \ldots, h_q \rangle$. Now it is easy to see that $\mathrm{U}(RG) = \langle h_1, \ldots, h_q, \{1 + b_i y_j\} \rangle$ where $i = 1, \ldots, r$, $j = 1, \ldots, t$. This ends the proof.□

From Theorem 1.3 it follows that if R is commutative, G is torsion free abelian and $\mathrm{U}(RG)$ is finitely generated, then R has to be reduced. This result was later extended in [16], (see also [12]), to torsion free nilpotent groups. Now applying the above theorem, Theorem 5.8 and Proposition 5.5 we have a further extension of the just mentioned result.

Proposition 6.5. *Let R be an almost reduced ring and G be a u.p.-group such that* $\mathrm{U}(RG)$ *is finitely generated. If G is either of the form* $\mathrm{U}(A)$ *where A is a PI-ring, or G is solvable-by-finite, then R is reduced.*

Clearly under the above conditions we can apply Theorem 5.1. Finally let us ask the following question connected with possible generalizations of results for finite G.

Question 13. Can Corollary 2.3 be extended to almost reduced rings?

References

[1] H. Bass, *Introcuction to some methods of algebraic K-theory* (Regional Conference Series in Mathematics **20**, 1974).

[2] G.M. Bergman, Right orderable groups that are not locally Indicable, *Pacific J. Math.* **147**(1991), 243–248.

[3] C.W. Curtis and I. Reiner, *Representation theory of finite groups and associative algebras* (Wiley Interscience, New York, 1962).

[4] L. Fuchs, *Infinite abelian groups vol. 2* (Academic Press, New York, 1973).

[5] G. Higman, The units of group rings, *Proc. London Math. Soc.* **46**(1940), 231–248.

[6] G. Higman, B.H. Neumann and H. Neumann, Embedding theorems for groups, *J. London Math. Soc.* **24**(1949), 247–254.

[7] S.V. Ivanov, On some finiteness conditions in semigroup and group theory, *Semigroup Forum*, to appear.

[8] S. V. Ivanov and A.Yu. Ol'shanskii, Some applications of graded diagrams in combinatorial group theory (Lecture Notes of London Math. Soc. **160**, 1991), 258–308.

[9] G. Karpilovsky, *Commutative group algebras* (Marcel Dekker, New York, 1983).

[10] G. Karpilovsky, On finite generation of unit groups of commutative group rings, *Arch. Math. (Basel)* **40**(1983), 503–508.

[11] G. Karpilovsky, *Unit groups of classical rings* (Clarendon Press, Oxford, 1988).

[12] G. Karpilovsky, *Unit groups of group rings* (Longman Group UK limited 1989).

[13] G. Karpilovsky, Units of commutative group algebras, *Exposition Math.* **8** (1990), 247–287.

[14] J. Krempa, On semigroup rings, *Bull. Acad. Polon. Sci.* **25**(1977), 225–231.

[15] J. Krempa, Homomorphisms of group rings, in: Banach Center Publication vol. 9 (PWN Warsaw 1982), 233–255.

[16] J. Krempa, Finitely generated groups of units in group rings, preprint of the Institute of Mathematics, Warsaw University, Warsaw, 1985.

[17] J. Krempa, Unit groups and commutative ring extensions, *Comm. Algebra* **16**(1988), 2349–2361.

[18] D.C. Lantz, R-automorphisms of $R[G]$ for G abelian torsion-free, *Proc. Amer. Math. Soc.* **61**(1976), 1–6.

[19] A.I. Lichtman, On unique product groups, *Comm. Algebra* **9**(1981), 533–551.

[20] M. Mirowicz, Units in group rings of the infinite dihedral group, *Canad. Math. Bull.* **34**(1991), 83–89.

[21] J. Okniński, *Semigroup algebras* (Marcel Dekker Inc., New York, 1991).

[22] M.M. Parmenter, Isomorphic group rings, *Canad. Math. Bull.* **18**(1975), 567–576.

[23] D.S. Passman, *The algebraic structure of group rings* (Wiley-Interscience Publications, New York, 1977).

[24] D.S. Passman, *Infinite crossed products* (Academic Press, New York, 1989).

[25] S.D. Promislow, A simple example of a torsion-free non unique product group, *Bull. London Math. Soc.* **20**(1988), 302–304.

[26] M.S. Raghunathan, *Discrete subgroups of Lie groups* (Springer Verlag, 1972).

[27] E. Rips and Y. Segev, Torsion-free groups without unique product property, *J. Algebra* **108**(1987), 116–126.

[28] D.J.S. Robinson, *Finiteness conditions and generalized soluble groups* (Springer Verlag, Berlin, 1972).

[29] S. K. Sehgal, *Topics in group rings* (Marcel Dekker Inc., New York, 1978).

[30] J.-P. Serre, Arithmetic groups, in *Homological group theory* (C.T.C. Wall (ed.), Cambridge University Press, 1979), 105–136.

[31] A. Strojnowski, A note on u.p.-groups, *Comm. Algebra* **8**(1980), 231–234.

COUNTING FINITE INDEX SUBGROUPS

ALEXANDER LUBOTZKY

Institute of Mathematics, Hebrew University, Jerusalem 91904, Israel

For Karl Gruenberg on his 65th birthday.

Introduction

Let Γ be a finitely generated group. Denote by $a_n(\Gamma)$ (resp.$\sigma_n(\Gamma)$) the number of subgroups of Γ of index n (resp. of index at most n). This paper deals with the connection between the algebraic structure of the group Γ and the arithmetic properties of the sequence $a_n(\Gamma)$, $n = 1, 2, 3, \ldots$, e.g., the growth of the sequence $a_n(\Gamma)$ ("the subgroup growth") or the properties of the function $\zeta_\Gamma(s) = \sum_{n=1}^{\infty} a_n(\Gamma)n^{-s}$ which encodes the sequence $a_n(\Gamma)$.

These studies have two sources of inspiration. The first is the notion of word growth of groups; namely, denote by $b_n^\Sigma(\Gamma)$ the number of elements of Γ whose length is n with respect to a fixed finite set Σ of generators of Γ. Much work has been done on $b_n^\Sigma(\Gamma)$ and its connection with Γ– see [Ba], [Mi], [Wol], [Gro], [Gri] and the references therein. To some extent $b_n^\Sigma(\Gamma)$ measure the growth of Γ from below, while $a_n(\Gamma)$ express its growth from the top. The two types of growth have some vague connection (cf. [LM3]), but the word growth is used here only as a model for the kind of problems we want to face: groups of (subgroup) polynomial growth, intermediate growth etc. It should be noticed however that while the numbers $b_n^\Sigma(\Gamma)$ (but not their growth) depend on a choice of generators, $a_n(\Gamma)$ depend only on Γ. Thus the numbers $a_n(\Gamma)$ are of inherent interest and not merely their growth. This brings us to the second source of inspiration: the theory of rings of algebraic integers and their zeta functions. Here if O is the ring of algebraic integers in a number field k, one writes $\zeta_k(s) = \Sigma r_n n^{-s}$ where r_n is the number of ideals of O of index n. The function $\zeta_k(s)$ is called the Dedekind ζ-function of k and expresses much of the arithmetic of k. Similarly, we will study $\zeta_\Gamma(s) = \Sigma a_n(\Gamma)n^{-s}$ and describe the first steps of an analogous theory for non-commutative groups.

The reader might wonder whether $a_n(\Gamma)$ is the right analogue to r_n. One might suggest other possibilities, for example looking at the number of normal subgroups of index n. At this point it is unclear which definition would lead to a richer theory. We, however, have limited ourselves in this survey to the counting of all finite index subgroups.

Denote by $R(\Gamma)$ the intersection of all finite index subgroups of Γ. Obviously, $a_n(\Gamma/R(\Gamma)) = a_n(\Gamma)$. So, there is no harm in assuming $R(\Gamma) = \{1\}$, i.e., Γ is a residually finite group. In this respect, the subgroup growth is more restricted than the word growth. Anyway, the class of residually finite groups is rich enough, containing, for example, all the finitely generated linear groups. Closely connected with a residually finite group Γ are its pro-finite completion $\hat{\Gamma}$ and its pro-p completion $\Gamma_{\hat{p}}$, p a prime. So our study will lead to the territory of pro-finite groups.

The earliest paper in the mathematical literature which considered systematically counting finite index subgroups is, as far as we can tell, the paper of Marshal Hall [Ha1] in 1949. (So, subgroup growth is an older subject than word growth! In fact, Hurwitz in 1902 had already studied a question which is essentially counting finite index subgroups of surface groups– see [Me4] and the references therein.) In that paper, Hall gave a recursive formula for the number of subgroups of index n in the free group on r generators. Hall's method is based on associating with every subgroup H the permutational representation of Γ on the coset space Γ/H. His method was considerably simplified by various authors who also extended it to other groups which are somehow close to free groups. Most significantly is the work of T. Müller who developed an elaborate theory for the subgroup growth of virtually free groups. This direction is described in Section 1.

A completely new direction was started approximately ten years ago by D. Segal, G. Smith and F. Grunewald ([Sm1], [Se] and [GSS]). They looked at $a_n(\Gamma)$ for a nilpotent group Γ and in particular defined the zeta function $\zeta_\Gamma(s) = \Sigma a_n(\Gamma)n^{-s}$. It is particularly natural to do so for nilpotent groups since:

(a) for such groups $a_n(\Gamma)$ grows polynomially, thus $\zeta_\Gamma(s)$ has a non-empty domain of congruence.

(b) $\zeta_\Gamma(s)$ has an "Euler factorizaton" $\zeta_\Gamma(s) = \prod_p \zeta_{\Gamma,p}(s)$.

By applying the work of Denef [De1] on the rationality of some p-adic integrals they showed that the local factors $\zeta_{\Gamma,p}(s)$ are rational. Just as important, they computed many examples suggesting some very attractive conjectures. These important developments are described in Section 2. This work accentuated the importance of pro-p groups to the topic of counting subgroups and led M. du Sautoy [dS3] to prove that the zeta function of a compact p-adic analytic group is rational. His work in turn opens up the question of explicitly calculating these functions for semi-simple groups. Very little is known in this direction (with the exception of some examples computed by Ilani [Il3]). Simultaneously, it became evident that the subgroup growth is a very useful invariant for pro-p groups: A. Lubotzky and A. Mann proved that a pro-p group G is p-adic analytic if and only if $a_n(G)$ grows polynomially. A. Shalev [Sh1] showed that for non p-adic analytic groups the growth is at least $n^{C\log n}$. Section 3 describes the current situation in this sub-area.

Section 4 considers the question: For which groups Γ, $a_n(\Gamma)$ grows polynomially? A complete answer was given by A. Lubotzky, A. Mann and D. Segal ([LMS], [MS], [LM3], [Se]): This happens if and only if Γ is virtually solvable of finite rank. The proof of this theorem required an ensemble of tools such as the classification of finite simple groups, number theory and the theories of p-adic Lie groups, algebraic groups and arithmetic groups. In particular, it was shown that the growth of congruence subgroups of arithmetic groups (with non-solvable zariski closures) is not polynomially bounded. A more detailed study was done by A. Lubotzky [Lu4] where it is shown that for arithmetic groups in characteristic zero (e.g., $\Gamma = SL_r(\mathbb{Z})$) the growth of the congruence subgroups is $n^{C \log n / \log \log n}$.

Moreover, this type of growth characterizes the congruence subgroup property (CSP). Namely, if Γ fails to have CSP then the growth of $\sigma_n(\Gamma)$ is strictly larger– which means that Γ has "many more" non-congruence subgroups than congruence ones. On the other hand if Γ has the congruence subgroup property (e.g. $\Gamma = SL_r(\mathbb{Z})$, $r \geq 3$), $\sigma_n(\Gamma)$ grows as $n^{C \log n / \log \log n}$ so it has "intermediate subgroup growth" between polynomial and exponential. It should, however, be mentioned that free groups have super-exponential subgroup growth ($\sim e^{Cn \log n}$) and it is not difficult to give examples of solvable groups of exponential subgroup growth. Recently D. Segal and A. Shalev [SS] gave examples of solvable groups with fractionally exponential subgroup growth– thus adding a completely new source of groups of intermediate subgroup growth. The results on congruence subgroups are described in Section 5. They also highlight the connection between counting finite index subgroups and various counting problems in finite groups. The last mentioned area has been developed dramatically in recent years– e.g., the work of Pyber [Py1]– and it gives fruits to our topic as well.

As the reader may have sensed already from this introduction– the topic of "Subgroup Growth" is still in its infancy level. Extensive progress has been made in recent years and more development is anticipated. This makes it a wonderful topic for a series of talks in a conference– but it is an almost impossible task to accomplish a complete survey. This survey should be considered as a temporary report of the state of the art– calling attention to this beautiful chapter of asymptotic group theory. This paper is a short version of notes [Lu6] titled "Subgroup Growth" distributed at the Galway/St Andrews conference on group theory 1993. It was however updated to cover some work which was done in the last months of 1993.

This paper was written while the author was visiting the University of Chicago whose warm hospitality and support are gratefully acknowledged. We are also grateful to A. Mann for some helpful remarks.

1. Counting subgroups and permutational representations

The first paper in the literature in which the question of counting subgroups of a given index was considered is the 1949 paper of Marshal Hall [Ha1] in which a recursive formula was given for the number of subgroup of index n in the free group on r generators. Hall's method was extended and simplified by Dey [De2] and Wolfhart [Wo] to get the following form: Let Γ be a finitely generated group and H a subgroup of index n. There is an action of Γ on the set Γ/H of left cosets of H, which defines a permutational representation of Γ on a set of n elements. Identify Γ/H with the set $\{1,2,\ldots,n\}$ such that H is corresponding to 1. There are $(n-1)!$ ways to make this identification. Thus H defines $(n-1)!$ homomorphisms from Γ to S_n. Every such homomorphism $\varphi:\Gamma\to S_n$ satisfies (i) $\varphi(\Gamma)$ is transitive on $\{1,2,\ldots,n\}$ and (ii) $\operatorname{Stab}_{\Gamma,\varphi}(1)=\{\gamma\in\Gamma|\varphi(\gamma)(1)=1\}=H$. Conversely, every transitive permutational representation of degree n (i.e. $\varphi:\Gamma\to S_n$ satisfying (i)) defines an index n subgroup $H=\operatorname{Stab}_{\Gamma,\varphi}(1)$. Hence:

Proposition 1.1. *Let $t_n(\Gamma)$ be the number of transitive permutational representations of Γ on the set $\{1,2,\ldots,n\}$. Then $a_n(\Gamma)=t_n(\Gamma)/(n-1)!$ where $a_n(\Gamma)$ is the number of subgroups of Γ of index n.*

Example 1.2. $\Gamma=\mathbb{Z}$, $a_n(\Gamma)=1$ for every n, while $t_n(\Gamma)$ is equal to the number of n-cycles in S_n which is $(n-1)!$.

It remains to count the number of transitive actions. Let the number of all homomorphisms from Γ to S_n be $h_n(\Gamma)=|\operatorname{Hom}(\Gamma,S_n)|$. We have:

Lemma 1.3. *Let Γ be a group. Then:*

$$h_n(\Gamma)=\sum_{k=1}^{n}\binom{n-1}{k-1}t_k(\Gamma)\,h_{n-k}(\Gamma)$$

PROOF. Indeed, for every $1\le k\le n$ there are $\binom{n-1}{k-1}$ ways to choose the orbit of 1, $t_k(\Gamma)$ ways to act on this orbit and $h_{n-k}(\Gamma)$ ways to act on its complement in $\{1,2,\ldots,n\}$. □

(1.1) and (1.3) imply:

Corollary 1.4. *Let Γ be any group. Then:*

$$a_n(\Gamma)=\frac{1}{(n-1)!}h_n(\Gamma)-\sum_{k=1}^{n-1}\frac{1}{(n-k)!}h_{n-k}(\Gamma)\,a_k(\Gamma).$$

For some groups, $h_n(\Gamma)$ are easy to compute, e.g., for the free group on r generators $h_n(F_r) = (n!)^r$. Hence:

Corollary 1.5. (M. Hall [Hal]) *Let F_r be the free group on r generators. Then:*

$$a_n(F_r) = n(n!)^{r-1} - \sum_{k=1}^{n-1}(n-k)!^{r-1}a_k(F_r).$$

To estimate the growth of $a_n(\Gamma)$ for $\Gamma = F_r\,(r \geq 2)$ we notice that "most" r-tuples of permutations in S_n acts transitively on $\{1,2,\ldots,n\}$, i.e., $\frac{t_n(F_r)}{h_n(F_r)} \to 1$ as $n \to \infty$. Indeed $h_n(F_r) = (n!)^r$ while the number of r-tuple which are *not* transitive is bounded by $P = \sum_{k=1}^{n-1}\binom{n-1}{k-1}h_k(F_r)h_{n-k}(F_r) = \sum_{k=1}^{n-1}\binom{n-1}{k-1}(k!)^r((n-k)!)^r$ as the proof of Lemma 1.3 shows. Now, it is easy to see that $\lim\limits_{n\to\infty}\frac{P}{(n!)^r} = 0$.

We mention in passing the result of Dixon [Di] that most r-tuples $(r \geq 2)$ of permutations of S_n not merely act transitively but actually generate either S_n or A_n. But the transitivity suffices to deduce:

Proposition 1.6. (Newman [Ne2])

$$a_n(F_r) \sim n\cdot(n!)^{r-1}.$$

PROOF. By (1.1), $a_n(F_r) = t_n(F_r)/(n-1)! \sim \frac{h_n(F_r)}{(n-1)!} = n(n!)^{r-1}$. □

The next case which was considered in the literature is the case of a free product $\Gamma = *_{i=1}^r A_i$. Clearly $h_n(\Gamma) = \Pi_{i=1}^r h_n(A_i)$ and hence (1.4) implies:

Corollary 1.7. (Dey [De2]) *Let $\Gamma = *_{i=1}^r A_i$ and let $h_n^i = h_n(A_i) = |\,Hom(A_i, S_n)|$. Then*

$$a_n(\Gamma) = \frac{1}{(n-1)!}(\Pi_{i=1}^r h_n^i) - \sum_{k=1}^{n-1}\frac{1}{(n-k)!}a_k(\Gamma)(\Pi_{i=1}^r h_{n-k}^i).$$

Of course (1.5) is a special case of (1.7) when $A_i \simeq \mathbb{Z}$ and $h_n^i = n!$ for every i and n. However, in general it is not an easy task to compute $h_n(A)$ even if A is a finite group. If $A = \mathbb{Z}/d\mathbb{Z}$ then $h_n(\mathbb{Z}/d\mathbb{Z})$ is the number of degree n permutations of order dividing d. This function has received a considerable amount of attention (see [MW], [Wi], and the references therein).

For example Moser and Wyman [MW] proved:

Proposition 1.8. *Let p be a prime. Then*

$$h_n(\mathbb{Z}/p\mathbb{Z}) \sim K_p\,exp\left(\frac{p-1}{p}n\log n - \frac{p-1}{p}n + n^{1/P}\right)$$

where $K_p = p^{-1/2}$ for $p > 2$ and $K_2 = 2^{-1/2}e^{-1/4}$.

Newman [Ne2] showed that also for a free product of finite cyclic groups "most" permutational actions are transitive (provided this is not the infinite dihedral group) and hence $t_n \sim h_n$. A case of particular interest is $\mathrm{PSL}_2(\mathbb{Z}) \simeq \mathbb{Z}/2\mathbb{Z} * \mathbb{Z}/3\mathbb{Z}$. Hence

$$a_n(\mathrm{PSL}_2(\mathbb{Z})) \sim h_n(\mathbb{Z}/2\mathbb{Z})\, h_n(\mathbb{Z}/3\mathbb{Z})/(n-1)!$$

and therefore one can deduce from (1.8) that:

Proposition 1.9. (Newman [Ne2])

$$a_n(\ PSL_2(\mathbb{Z})) \sim (12\pi e^{1/2})^{-1/2}\ exp\left(\frac{n\log n}{6} - \frac{n}{6} + n^{1/2} + n^{1/3} + \frac{\log n}{2}\right)$$

He also computed $a_n(\mathrm{PSL}_2(\mathbb{Z}))$ for many $n-s$. For example:

$$a_{100}(\mathrm{PSL}_2(\mathbb{Z})) = 159299552010504751878902805384624$$

We will come back to this in Section 5 when we will show that $\mathrm{PSL}_2(\mathbb{Z})$ has far fewer congruence subgroups. Thus the congruence subgroup property fails in a very strong sense.

A different approach to computing $a_n(\mathrm{PSL}_2(\mathbb{Z}))$ is given by Stothers [St]. A recursive formula for this sequence was given by Godsil, Imrich, and Razen in [GIR]. In a series of papers Gardy and Newman ([GN1], [GN2], [GN3]) established some linear recurrences when $a_n(\Gamma)$ are considered modulo a fixed integer m when Γ is either a free group or a free product of cyclic groups.

Let us now look again at (1.4) for a general group Γ. With the notation introduced there, write $A(X) = A_\Gamma(X) = \sum_{n=1}^{\infty} a_n(\Gamma)X^n$ and $B(X) = B_\Gamma(X) = \sum_{n=0}^{\infty} b_n(\Gamma)X^n$ where $b_0(\Gamma) = 1$ and $b_n(\Gamma) = h_n(\Gamma)/n!$. Now, (1.4) means

$$n\, b_n(\Gamma) = \sum_{k=1}^{n} a_k(\Gamma) b_{n-k}(\Gamma)$$

which formally means

$$X\, B'(X) = A(X)\, B(X)$$

i.e.,

$$\frac{A(X)}{X} = \frac{B'(X)}{B(X)} = \log(B(X))'$$

and hence:

Proposition 1.10. $B(X) = \ exp\left(\int \frac{A(X)}{X}\, dX\right).$

Note that $\int \frac{A(X)}{X} dX = \sum_{n=1}^{\infty} a_n(X) \frac{X^n}{n}$.

The last proposition has some non-trivial applications which are outside the main theme of this paper– but just to mention in brief: For a prime p let $\overline{\tau}_p(n) = \frac{\tau_p(n)}{n!}$ where $\tau_p(n)$ is the number of elements of order dividing p in S_n. Then:

$$1 + \sum_{n=1}^{\infty} \overline{\tau}_p(n) X^n = \exp\left(X + \frac{X^p}{p}\right)$$

This is deduced from (1.10) by considering the finite (!) group $\Gamma = \mathbb{Z}/p\mathbb{Z}$. So in some cases (1.10) can be useful to get information on $\mathrm{Hom}(\Gamma, S_n)$ from $a_n(\Gamma)$, rather than the opposite direction which will be our more common use of (1.10). More general results of this kind can be deduced very quickly from (1.10) using various finite groups. Special cases of it were studied over forty years ago (by more direct methods– see [MW] and [Wi] for history and references).

Far reaching generalizations of most of the above mentioned results were obtained recently by T. Müller [Mu5]. According to (1.10), $\Sigma b_n(X) X^n = \exp\left(\int \frac{A(X)}{X} dX\right)$, hence if G is a finite group of order m,

$$\sum \frac{|\,\mathrm{Hom}(G, S_n)|}{n!} X^n = \exp\left(\sum_{d|m} \frac{a_d(G)}{d} X^d\right)$$

Denote $P(X) = P_G(X) = \sum_{d|m} \frac{a_d(G)}{d} X^d = \sum_{i=1}^{m} C_i X^i$, then $P(X)$ is a real polynomial with non-negative coefficients, $C_1 \neq 0$, and $C_i = 0$ for $\frac{m}{2} < i < m$. Müller developed a machinery which gives a detailed asymptotic expansion for the coefficients of $\exp(P(X))$ for such $P(X)$. This way he obtained asymptotic expansion of $\mathrm{Hom}(G, S_n)$ for every finite group G. The precise result is too long to be mentioned here, but here is a corollary.

Theorem 1.11. (T. Müller [Mu5]) *Let G be a finite group of order m. Then*

$$|\,Hom(G, S_n)| \sim K_G\, n^{(1-1/m)n}\ exp\left(-(1-1/m)n + \sum_{\substack{d|m \\ d<m}} \frac{a_d(G)}{d} n^{d/m}\right)$$

where

$$K_G = \begin{cases} m^{-1/2} & if\ 2 \nmid m \\ m^{-1/2}\ exp\left(-\frac{(a_{m/2}(G))^2}{2m}\right) & if\ 2|m \end{cases}$$

Theorem 1.11 is an impressive generalization of Proposition 1.8 and [Wi], which proved a similar result for cyclic groups (but Müller's result is stronger even in the cyclic case as he gives the full expansion). More important for our context is that the Theorem can be used to handle $a_n(\Gamma)$ for Γ which is a free product of finite groups in a way generalizing the deduction of (1.9) from (1.8). Here also Müller was able to give a detailed asymptotic expansion, but we bring only the asymptotic values:

Theorem 1.12. ([Mu5]) *Let $\Gamma = *_{i=1}^{s} G_i$ be a free product of $2 \le s < \infty$ non-trivial finite groups of orders $m_1, \ldots, m_s$ respectively. If $s = 2$ assume not both G_1 and G_2 are cyclic of order 2. Then*

$$a_n(\Gamma) \sim L_\Gamma \cdot \Phi_\Gamma(n) \text{ as } n \to \infty$$

where

$$L_\Gamma = (2\pi m_1 \cdot \ldots \cdot m_s)^{-1/2}\ exp\left(-\textstyle\sum_{\{i|\,2|m_i\}} \frac{(a_{m_{i_k}}(G_i))^2}{2m_i}\right)$$

$$\Phi_\Gamma(n) = n^{-h(\Gamma)n}\ exp\left(h(\Gamma)n + \textstyle\sum_{i=1}^{s} \sum_{\substack{d_i < m_i \\ d_i | m_i}} \frac{a_{d_i}(G_i)}{d_i} n^{d_i/m_i} + \frac{1}{2}\log n\right)$$

and

$$h(\Gamma) = \text{ Euler characteristic of } \Gamma = \frac{1 - (m_1 - 1) \cdot \ldots \cdot (m_s - 1)}{m_1 \cdot \ldots \cdot m_s}$$

Note that Proposition 1.9 is a very special case of 1.12. As mentioned, the results of Müller are even stronger for the previously known special cases. For example for $\Gamma = \mathrm{PSL}_2(\mathbb{Z})$ he shows:

$$\begin{aligned} a_n(\Gamma) &= (12\pi e^{1/2})^{-1/2} n^{n/6} \exp\left(-\frac{n}{6} + n^{1/2} + n^{1/3} + \frac{1}{2}\log n\right) \\ &\cdot\left\{1 - n^{-1/6} - \frac{1}{6}n^{-1/3} - \frac{13}{24}n^{-1/2} - \frac{7}{36}n^{-2/3}\right. \\ &\left.+ \frac{253}{240}n^{-5/6} - \frac{67963}{51840}n^{-1} - \frac{2449841}{362880}n^{-7/6} + 0\left(n^{-4/3}\right)\right\} \end{aligned}$$

We mention that along the way Müller shows that if Γ is as in Theorem 1.12, then as for the free group, $t_n(\Gamma) \sim h_n(\Gamma)$, i.e., "with probability one" the actions of Γ on $\{1, \ldots, n\}$ are transitive. The following generalization of Dixon's theorem mentioned above was conjectured in [Lu6] and was proved by Pyber [Py2].

Theorem 1.13. *Let G_1 and G_2 be two fixed non-trivial finite groups, not both of order 2. Then with probability 1 as n going to infinity, the images of G_1 and G_2 generate either A_n or S_n when we run over all possible homomorphisms from G_1 and G_2 (i.e., from $G_1 * G_2$) into S_n.*

Remark 1.14. We restrict ourselves to the problem of counting *all* subgroups. Much work has been done on counting free subgroups of virtually free groups. This is sometimes an easier problem as free subgroups correspond to some kind of fixed point free actions which are somewhat easier to be counted. The reader is referred to [Mu1], [Mu2], [Mu3], [St2] and the references therein.

2. Nilpotent groups and zeta functions

As mentioned in the first section, the subject of counting finite index subgroups started with the paper of M. Hall [Ha1] in 1949 which dealt with free groups. Over the next thirty-five years all papers on the topic elaborated on this and studied mainly groups which are virtually free. Approximately ten years ago, Dan Segal, Geff Smith and Fritz Grunewald ([Sm], [Se] and [GSS]- the last one appeared only in 1988 but was circulated around a few years earlier) initiated the study of the subject in "small" groups; solvable and especially nilpotent. If Γ is a finitely generated nilpotent group then it is particularly convenient to encode $a_n(\Gamma)$ (= the number of subgroups of index n in Γ) via a Dirichlet series $\zeta_\Gamma(s) = \Sigma a_n(\Gamma)n^{-s} = \Sigma[\Gamma : H]^{-s}$ where H runs over all finite index subgroups of Γ. The function $\zeta_\Gamma(s)$ is called the *Zeta function of* Γ. It has two pleasant properties:

(a) If Γ is nilpotent then $a_n(\Gamma)$ grows polynomially with n and $\zeta_\Gamma(s)$ is therefore not merely a formal series but actually converges for $Re(s) > \alpha$ where $\alpha = \alpha(\Gamma)$ is some real number.

(b) For a nilpotent Γ every subgroup H of index $n = p_1^{\alpha_1} \cdot \ldots \cdot p_r^{\alpha_r}$ is an intersection in a unique way of subgroups H_i $(1 \leq i \leq r)$ of index $p_i^{\alpha_i}$. Thus $a_n(\Gamma) = \Pi_{i=1}^r a_{p_i^{\alpha_i}}(\Gamma)$ and hence $\zeta_\Gamma(s)$ has Euler product decomposition $\zeta_\Gamma(s) = \prod_p \zeta_{\Gamma,p}(s)$ where the product runs over all primes and $\zeta_{\Gamma,p}(s) = \sum_{i=0}^\infty a_{p^i}(\Gamma)p^{-is}$.

So the zeta function $\zeta_\Gamma(s)$ of a nilpotent group share some of the features of Dedekind zeta function of a number field K, $\zeta_K(s) = \sum_M [O : M]^{-s}$ where O is the ring of integers of K and M runs over all finite index ideals of O. Many of the zeta functions $\zeta_\Gamma(s)$ which were computed in [Sm] and [GSS] are expressed via such $\zeta_K(s)$ and especially via $\zeta(s) = \zeta_{\mathbb{Q}}(s) = \Sigma n^{-s} = \prod_p (1 - p^{-s})^{-1}$.

It is very tempting to believe that the other properties of the classical zeta functions are also shared by $\zeta_\Gamma(s)$, e.g., the existence of a functional equation. But as of now very little is known. The reader might also suggest that the appropriate analoge of Dedekind zeta function for groups should be $\zeta_\Gamma^\Delta(s) = \sum_N [\Gamma : N]^{-s}$ where N runs only over the *normal* subgroups of finite index. Indeed, in [Sm] and [GSS], $\zeta_\Gamma^\Delta(s)$ was also studied beside $\zeta_\Gamma(s)$, as well as two other related functions. Only the future will tell which one is more suitable for group theoretic use.

We start with the free abelian groups:

Theorem 2.1. *Let $\Gamma = \mathbb{Z}^r$ be the free abelian group of rank r. Then*

$$\zeta_{\mathbb{Z}^r}(s) = \zeta(s) \cdot \zeta(s-1) \cdot \ldots \cdot \zeta(s-r+1)$$

where ζ is the classical Riemann ζ-function.

There are five (!) different proofs in the literature for this not too difficult, yet not completely trivial, result. (See [BR2], [Sm], [Il1], [GSS], and [Man3]. In [Lu6], the first four are described in detail.) We sketch here the proof from [GSS], which while applied to a general nilpotent group gives the important Theorem 2.13 below.

We start with a simple Lemma:

Lemma 2.2. *Let Γ be a group and $\hat{\Gamma}$ its pro-finite completion. Then*

(a) *For every n, $a_n(\Gamma) = a_n(\hat{\Gamma})$ where for a pro-finite group G, by $a_n(G)$ we mean the number of* closed *subgroups of index n.*

(b) *If $\hat{\Gamma}$ is pro-nilpotent (i.e., if every finite quotient of Γ is nilpotent or equivalently $\hat{\Gamma} = \prod_p \Gamma_{\hat{p}}$ where p runs over all primes and $\Gamma_{\hat{p}}$ denotes the pro-p completion of Γ) then:*

(i) $\zeta_{\Gamma,p}(s) = \zeta_{\Gamma_{\hat{p}}}(s)$

ii) $\zeta_\Gamma(s) = \prod_p \zeta_{\Gamma,p}(s)$

This lemma is very simple and we omit the proof. But one warning should be made: $\zeta_{\Gamma,p}(s)$ is defined as $\sum_{i=0}^{\infty} a_{p^i}(\Gamma)p^{-is}$, i.e., encoding all subgroups of p- power index. For general groups Γ this is *not* the same as $\zeta_{\Gamma_{\hat{p}}}(s)$ which captures only the *sub-normal* subgroups of Γ of p-power index.

Anyway for $\Gamma = \mathbb{Z}^r$ or more generally for Γ nilpotent, we can compute $\zeta_\Gamma(s)$ via $\zeta_{\Gamma_{\hat{p}}}(s)$. Theorem 2.1 is therefore equivalent to the assertion:

$$\zeta_{\mathbb{Z}_p^r}(s) = \prod_{i=0}^{r-1} (1 - p^i p^{-s})^{-1}$$

Let $G = \mathbb{Z}_p^r$ with the standard basis $\{e_1, e_2, \ldots, e_r\}$. A finite index subgroup H of G has a basis of the following form:

$$h_1 = (\lambda_{11}, \ldots, \lambda_{1n})$$
$$h_2 = (0, \lambda_{22}, \ldots, \lambda_{2n})$$

$$h_i = (0, \ldots 0, \lambda_{ii}, \ldots, \lambda_{in})$$
$$\vdots$$
$$h_n = (0, \ldots, 0, \lambda_{rr})$$

obtained in the following way: $H \cap \mathbb{Z}_p e_r = \lambda_{rr} e_r$, and

$$(H \cap \ \mathrm{Span}_{\mathbb{Z}_p}\{e_i, e_{i+1}, \ldots, e_r\}) \equiv \lambda_{ii}\, e_i(\text{modulo } \ \mathrm{Span}_{\mathbb{Z}_p}\{e_{i+1}, \ldots, e_r\}).$$

A basis of H of this form will be called a good basis. It is easy to see that $[G : H] = |\lambda_{11}|^{-1} \cdot \ldots \cdot |\lambda_{nn}|^{-1} = p^{\alpha_1} \cdot \ldots \cdot p^{\alpha_r}$ (where $\lambda_{ii} = p^{\alpha_i} u_i$ and u_i is a unit of $\mathbb{Z}_p$). Let $M(H)$ denote the subset of the upper triangular matrices M obtained by taking bases for H of the above form. Let μ be the normalized Haar measure of the additive group of the upper triangular matrices over $\mathbb{Z}_p$.

Lemma 2.3. $\mu(M(H)) = (1-p^{-1})^r p^{-\alpha_1} p^{-2\alpha_2} \cdot \ldots \cdot p^{-r\alpha_r} = (1-p^{-1})^r \prod_{i=1}^r |\lambda_{ii}|^i$.

PROOF. Note first that $M(H)$ is an open set. If $\{h_1, \ldots, h_r\}$ is a good basis as above, then any other good basis $\{h'_1, \ldots, h'_r\}$ can be written as $h'_i = \lambda_{ii}\, u_i + \nu_{i+1}$ where ν_{i+1} is in the $\mathbb{Z}_p$-span of $\{h_{i+1}, \ldots, h_r\}$ and $u_i \in \mathbb{Z}_p^*$- the group of units of $\mathbb{Z}_p$. Thus h_r can be "moved" in a subset of $\mathbb{Z}_p$ of measure $(1-p^{-1})|\lambda_{rr}| = (1-p^{-1})p^{-\alpha_r}$, h_{r-1} can be moved by multiplication of $\lambda_{r-1,\, r-1}$ by a unit and by adding $p^{\alpha_r}\, \mathbb{Z}_p\, e_r$, so as an element of $\mathbb{Z}_p \times \mathbb{Z}_p$, h_{r-1} can vary along a subset of measure $(1-p^{-1})p^{-\alpha_{r-1}} \cdot p^{-\alpha_r}$. Similarly h_{r-2} can be multiplied by $\mathbb{Z}_p^*$, i.e., $\lambda_{r-2,r-2}$ can be changed within a subset of $\mathbb{Z}_p$ of measure $(1-p^{-1})p^{-\alpha_{r-2}}$ and the pair $(\lambda_{r-2,r-1}, \lambda_{r-2,r})$ can be changed by addition of elements from the set $\mathbb{Z}_p\, \lambda_{r-1,\, r-1}\, e_{r-1} + \mathbb{Z}_p\, \lambda_{r,r} e_r$. This shows that h_{r-2} can be "moved" within a subset of $\mathbb{Z}_p^3$ of measure $(1-p^{-1})p^{-\alpha_{r-2}} \cdot p^{-\alpha_{r-1}} \cdot p^{-\alpha_r}$. In a similar way h_i can be a vector from a subset of $\mathbb{Z}_p^{r-i+1}$ of measure $(1-p^{-1})p^{-\alpha_{i+1}} \cdot \ldots \cdot p^{-\alpha_r}$. Now, μ is the product measure of all these. Hence:

$$\mu(M(H)) = (1-p^{-1})^r p^{-r\alpha_r}\, p^{-(r-1)\alpha_{r-1}} \cdot \ldots \cdot p^{2\alpha_2} \cdot p^{-\alpha_1}$$

as claimed. □

As said before $[G : H] = |\lambda_{11}|^{-1} \cdot \ldots \cdot |\lambda_{rr}|^{-1}$. Thus:

Corollary 2.4. $[G : H]^{-s} = \frac{1}{(1-p^{-1})^r} \int_{M(H)} |\lambda_{11}|^s \cdot \ldots |\lambda_{rr}|^s \cdot |\lambda_{11}|^{-1} |\lambda_{22}|^{-2} \cdot \ldots \cdot |\lambda_{rr}|^{-r} d\mu$.

So we can replace the sum $\zeta_{\mathbb{Z}_p^r}(s) = \sum_{H \leq_f G} [G : H]^{-s}$ by an integral

$$\zeta_{\mathbb{Z}_p^r}(s) = \frac{1}{(1-p^{-1})^r} \int_{\bigcup_H M(H)} |\lambda_{11}|^{s-1} \cdot \ldots \cdot |\lambda_{rr}|^{s-r}\, d\mu.$$

To evaluate this integral note that $\cup_H M(H)$ is equal to the set all upper triangular matrices over $\mathbb{Z}_p$ with non-zero entries along the diagonal. Those with determinant zero form a set of measure zero and therefore can be ignored. Thus:

$$\begin{aligned}\zeta_{\mathbb{Z}_p^r}(s) &= \frac{1}{(1-p^{-1})^r}\int_T |\lambda_{11}|^{s-1}\cdot\ldots\cdot|\lambda_{rr}|^{s-r} \\ &= \frac{1}{(1-p^{-1})^r}\Pi_{i=1}^r\int_{\mathbb{Z}_p}|\lambda_{ii}|^{s-i}d\nu\end{aligned}$$

where $d\nu$ is the normalized Haar measure of $\mathbb{Z}_p$. A simple computation (which will be used often) shows:

Lemma 2.5. $\int_{\mathbb{Z}_p}|\lambda|^s d\nu = \sum_{i=0}^{\infty}(1-p^{-1})p^{-i}p^{-is} = \frac{1-p^{-1}}{1-p^{-s-1}}$

Thus $\zeta_{\mathbb{Z}_p^r} = (1-p^{-s})^{-1}(1-p^{-(s-1)})^{-1}\cdot\ldots\cdot(1-p^{-(s-(r-1))})^{-1}$ which proves (2.1). □

This proof can be carried a long way for an arbitrary finitely generated, torsion free, nilpotent group Γ. For such Γ there is a series of normal subgroups $\Gamma=\Gamma_1\geq\Gamma_2\geq\ldots\geq\Gamma_r\geq\Gamma_{r+1}=\{1\}$ such that $\Gamma_i/\Gamma_{i+1}\simeq\mathbb{Z}$ for $i=1,\ldots,r$, where r is the Hirsch length of Γ. For $i=1,\ldots,r$ choose $x_i\in\Gamma_i$ such that $x_i\Gamma_{i+1}$ generate Γ_i/Γ_{i+1}. Every element x of Γ can be represented *uniquely* as $x=x_1^{a_1}\cdot\ldots\cdot x_r^{a_r}$ with $a_i\in\mathbb{Z}$ and $\{x_1,\ldots,x_r\}$ is called a Malćev basis for Γ. P. Hall showed that by considering $(a_1,\ldots,a_r)$ as the cordinates of x, the group operation in Γ are given by polynomial functions whose coefficients are in $\mathbb{Q}$ (See [*Ha*]).

We can define now $G=\Gamma^{\mathbb{Z}_p}$ to be the space $\mathbb{Z}_p^r$ where the group operation are given by the same polynomials expressing the group operations of Γ. It is easy to see that $\Gamma^{\mathbb{Z}_p}$ is a pro-p group and in fact isomorphic to the pro-p completion of Γ. Thus $\zeta_{\Gamma,p}=\zeta_G$. The groups $G_i=\Gamma_i^{\mathbb{Z}_p}$ define a filtration $G=G_1\geq G_2\geq\ldots\geq G_r\geq G_{r+1}=\{1\}$ of G.

Definition 2.6. Let H be a finite index subgroup of G. A subset $\{h_1,\ldots,h_r\}\leq H$ is called a *good basis* for H if for every $i=1,\ldots,r$, h_iG_{i+1} generate $(H\cap G_i)G_{i+1}/G_{i+1}$. This is indeed a basis for H and every element x of H can be represented uniquely as $x=h_1^{\lambda_1}\cdot\ldots\cdot h_r^{\lambda_r}$ with $\lambda_i\in\mathbb{Z}_p$. (h^λ for $h\in H$ and $\lambda\in\mathbb{Z}_p$ is well defined– see [DDMS, Chapter 1]). We consider $(\lambda_1,\ldots,\lambda_2)$ as the coordinates of x in $\mathbb{Z}_p^r$.

The coordinates of a good basis of H have the form:

$$h_i=(0,\ldots,0,\lambda_{ii},\ldots,\lambda_{ir}),\ \ i=1,\ldots,r$$

(By an abuse of the language we will identify an element and its vector of coordinates.) So as before we can associate with a good basis an upper triangular matrix. Again, $[G : H] = \Pi_{i=1}^{r}|\lambda_{ii}|^{-1}$. Let $M(H)$ be the set of all upper triangular $r \times r$ matrices over $\mathbb{Z}_p$ obtained from good bases of H.

Lemma 2.7. *$M(H)$ is open and $\mu(M(H)) = (1-p^{-1})^r \, \Pi_{i=1}^{r}|\lambda_{ii}|^i$.*

(Note the $|\lambda_{ii}|$ are determined by H and not by the given basis as $|\lambda_{ii}|^{-1} = |(H \cap G_i)G_{i+1}/G_{i+1}|$).

The proof of (2.7) is identical to that of (2.3). The commutativity of $\mathbb{Z}_p^r$ did not play any role there. Just as for $\mathbb{Z}_p^r$ we can deduce:

Proposition 2.8. *$\zeta_{\Gamma,p}(s) = \zeta_G(s) = (1-p^{-1})^{-r} \int_M \Pi_{i=1}^{r}|\lambda_{ii}|^{s-i} d\mu$ where M is the union of all $M(H)$ where H runs over all finite index subgroups of G.*

Unlike the case $G = \mathbb{Z}_p^r$ it is not so easy to describe M for general groups. Still when this can be done (2.8) can give a complete answer. We illustrate this by:

Theorem 2.9. *Let Γ be the discrete Heizenberg group, i.e.*

$$\Gamma = \left\{ \begin{pmatrix} 1 & a & b \\ 0 & 1 & c \\ 0 & 0 & 1 \end{pmatrix} \middle| \, a,b,c \in \mathbb{Z} \right\}$$

Then:

$$\zeta_\Gamma(s) = \frac{\zeta(s)\,\zeta(s-1)\,\zeta(2s-2)\,\zeta(2s-3)}{\zeta(3s-3)}$$

PROOF. The Heizenberg group is $P = \langle x,y,z \,|\, (x,y) = z,\ (x,z) = (y,z) = 1\rangle$ and its pro-p completion $G = \Gamma_{\hat{p}}$ has, of course, the same presentation, just being considered as a presentation within the category of pro-p groups. Let $G_3($ resp : $G_2)$ be the closed subgroup generated by $z($ resp : z and $y)$ and $G_1 = G$. Proposition 2.8 gives a formula for $\zeta_{\Gamma,p}(s) = \zeta_G(s)$, but we have to recognize M– the set of upper triangular matrices which represent good bases of finite index subgroups of G. If $A = \begin{pmatrix} \lambda_{11} & \lambda_{12} & \lambda_{13} \\ 0 & \lambda_{22} & \lambda_{23} \\ 0 & 0 & \lambda_{33} \end{pmatrix}$ is such a matrix then by Definition 2.6 it represents a good basis for an open subgroup H if and only if the following three conditions are satisfied:

(i) H is generated (as a closed subgroup) by $x^{\lambda_{11}}\,y^{\lambda_{12}}\,z^{\lambda_{13}}$, $y^{\lambda_{22}}\,z^{\lambda_{23}}$ and $z^{\lambda_{33}}$

(ii) $H \cap G_2$ is generated by $y^{\lambda_{22}}\,z^{\lambda_{23}}$ and $z^{\lambda_{33}}$ and

(iii) $H \cap G_3$ is generated by $z^{\lambda_{33}}$

So if we take H to be the subgroup generated by the three elements in (i), then it is open if and only if $\lambda_{11} \cdot \lambda_{22} \cdot \lambda_{33} \neq 0$. Assume this, then (iii) is satisfied if and only if the commutator $(x^{\lambda_{11}} y^{\lambda_{12}} z^{\lambda_{13}}, y^{\lambda_{12}} z^{\lambda_{23}})(= (x^{\lambda_{11}}, y^{\lambda_{22}}) = z^{\lambda_{11}\lambda_{22}})$ is in the subgroup generated by $z^{\lambda_{33}}$. This happens if and only if λ_{33} divides $\lambda_{11} \cdot \lambda_{22}$ in $\mathbb{Z}_p$ i.e., $v(\lambda_{33}) \leq v(\lambda_{11}) + v(\lambda_{22})$, where v is the p-adic valuation. If condition (iii) is satisfied then one easily checks that (ii) also follows. We conclude that M is the set of all upper triangular matrices of type A above for which all $\lambda_{ii} (i = 1, 2, 3)$ are non-zero and $v(\lambda_{33}) \leq v(\lambda_{11}) + v(\lambda_{22})$. Thus, by (2.8),

$$\begin{aligned} \zeta_\Gamma^p(s) &= \zeta_G(s) = (1-p^{-1})^{-3} \int_M |\lambda_{11}|^{s-1} |\lambda_{22}|^{s-2} |\lambda_{33}|^{s-3} d\mu \\ &= (1-p^{-1})^{-3}(1-p^{-1})^3 \sum_{e_1=0}^{\infty} \sum_{e_2=0}^{\infty} \sum_{e_3=0}^{e_1+e_2} p^{-e_1 s} p^{e_2(s-1)} p^{e_3(s-2)}. \end{aligned}$$

In the last equality we are using Lemma 2.5. A simple computation now finishes the proof of 2.9. □

Let's now look again in the general case of torsion free nilpotent groups: Proposition 2.8 gives a quite explicit integral which expresses $\zeta_{\Gamma,p}(s)$. The only difficulty is the range of integration M. This is the set of all upper triangular matrices with row-columns $h_i = (0, \dots, 0, \lambda_{ii}, \dots, \lambda_{ir}), i = 1, \dots, r$, which represent good bases for open subgroups of $G = \Gamma_{\hat{p}}$.

Lemma 2.10 *An ordered set of rows $\{h_1, \dots, h_r\}$ represents a good basis of some finite index subgroup H of G if and only if*

(i) $\Pi \lambda_{ii} \neq 0$ *and*

(ii) If $i \geq j$ then the commutator (h_i, h_j) is in the subgroup generated by $h_{j+1}, \dots, h_r$, i.e., there exist $\beta_{j+1}, \dots, \beta_r$ in $\mathbb{Z}_p$ such that $(h_i, h_j) = h_{j+1}^{\beta_{j+1}} \cdot \ldots \cdot h_r^{\beta_r}$.

PROOF. Clearly a good basis, i.e., a basis $\{h_1, \dots, h_r\}$ of H for which $H \cap G_i = \langle h_i, h_{i+1}, \dots, h_r \rangle$ should satisfy the conditions. Conversely, assume (i) and (ii) are satisfied and let H be the subgroup generated by $\{h_1, \dots, h_r\}$. Condition (i) implies that H is of finite index. Denote $H_i = \langle h_i, \dots, h_r \rangle$ and assume by induction that $H_j = H \cap G_j$ for $j < i$ and let's prove it for $j = i$: Condition (ii) implies that H_i is normal in H. As $H_{i-1} = H \cap G_{i-1}$ is generated by $h_{i-1}, h_i, \dots, h_r$ and H_i is normal in H_{i-1} we get: $H_{i-1} = H_i \cdot \langle \overline{h_{i-1}} \rangle$. So: $H \cap G_i = H_{i-1} \cap G_i = H_i \cdot \langle \overline{h_{i-1}} \rangle \cap G_i = H_i \cap G_i = H_i$ and the lemma is proven. □

Now comes a crucial observation: Conditions (i) and (ii) of Lemma 2.10 show that M is a *definable* subset of the upper triangular matrices over $\mathbb{Z}_p$. This means that M can be described by first order statements. This is clear for condition (i). But some explanation is needed regarding condition (ii):

We mentioned earlier that P. Hall showed that by choosing a Malćev basis for Γ (and thus identifying Γ with $\mathbb{Z}^r$ as a set) the group operation of Γ are given by rational polynomials. In fact more is true: The k-power operation in Γ is polynomial in Γ *and* k, i.e., the map $\psi : \Gamma \times \mathbb{Z} = \mathbb{Z}^r \times \mathbb{Z} \to \Gamma = \mathbb{Z}^r$ given by $\psi(\gamma, k) = \gamma^k$ for $\gamma \in \Gamma$ and $k \in \mathbb{Z}$ is a polynomial map from $\mathbb{Z}^{r+1}$ to $\mathbb{Z}^r$ with rational coefficients. Therefore this map can also be extended to $\Gamma_{\hat{p}} \times \mathbb{Z}_p \to \Gamma_{\hat{p}}$.

Thus if we look at condition (ii) it says that for the commutator of h_i and h_j there exist $\beta_{j+1}, \ldots, \beta_n$ such that $\psi(h_{j+1}, \beta_{j+1}) \cdot \ldots \cdot \psi(h_r, \beta_r) = (h_i, h_j)$. As the commutator and the product are polynomials this is a statement in the first order language. We have:

Proposition 2.11. *The domain of integration M in Proposition 2.8 is a definable set.*

Everything is now ready to apply the following theorem:

Theorem 2.12. (Denef [Del]) *Let M be a definable subset of $\mathbb{Z}_p^m$ and $h : \mathbb{Z}_p^m \to \mathbb{Z}_p$ be a definable function. Then $Z_M(s) = \int_M |h(x)|^s d\,\mu$, where $d\,\mu$ is the Haar measure of $\mathbb{Z}_p^m$, is a rational function of p^{-s}.*

A definable function $h : \mathbb{Z}_p^m \to \mathbb{Z}_p$ is a function whose graph $\{(y, h(y)) \in \mathbb{Z}_p^{m+1} | y \in \mathbb{Z}_p^m\}$ is a definable subset.

While we will not go here into the proof of this Theorem, it is worth mentioning that it relies on work of Macintyre [Ma1] in logic. Macintyre studied the first order theory of the p-adic numbers: He looked at the language of valued fields (i.e., the language of fields plus one unary predicate saying whether an element of the field is in the valued ring or not). Then added to this language a sequence of unary predicate symbols $P_1, P_2, P_3, \ldots$ whose interpretation is $x \in P_n$ if X is an n power in the field. Macintyre proved that the theory of p-adic numbers admits an elimination of quantifiers in the extended language. Now, each formula $\varphi(X_1, X_2, \ldots, X_m)$ of the language defines a set $A = \{(\underline{a}) \in \mathbb{Q}_p^m | \mathbb{Q}_p \models \varphi(\underline{a})\}$ (these are the "definable sets"). The resutl of Macintyre means that every definable set has a simpler form: It is a Boolean combination of sets of the forms $\{x \in \mathbb{Q}_p^m | \exists\, y \in \mathbb{Q}_p \text{ s.t.} f(x) = y^n\}$ with $f \in \mathbb{Z}_p[x_1, \ldots, x_m]$ and $n \in \mathbb{N}$. Thus to prove Denef's Theorem one can assume that the domain of integration has such a form. This was the starting point of the beautiful work of Denef who used it to prove some conjectures of Serre and Igusa on the rationality of the generating function of the number of solutions mod p^n of some diophantine equations.

Theorem 2.12, Proposition 2.11 and Proposition 2.8 give us now the main theorem of this chapter:

Theorem 2.13. (Grunewald-Segal-Smith [GSS]) *The p-th Euler factor of the zeta function of a finitely generated torsion free nilpotent group is a rational function of* p^{-s}

Rationality means in particular that the coefficients $a_{p^n}(\Gamma)$ satisfie a linear recurrence relation. So:

Corollary 2.14. *There exist positive integers* l *and* k *such that the sequence* $(a_{p^i}(\Gamma))_{i>l}$ *satisfies a linear recurrence relation over* $\mathbb{Z}$ *of length at most* k.

Remark 2.15 (Uniformity in p) The set M of Propositions 2.8 and 2.11 was actually definable in a way which is independent of the prime p. The results of Denef [De2] and Macintyre [Ma2] show that in such a case the rational functions of (2.12) and (2.13) may be taken to have numerators and denominators of bounded degrees– independent of p. This also implies that l and k in (2.14) can be chosen to work for all primes p. Grunewald, Segal and Smith [GSS] suggest even a stronger possibility: Given Γ, a torsion free finitely generated nilpotent group, then there exist finitely many rational functions $W_1(X,Y),\ldots,W_n(X,Y)$ of two variables over $\mathbb{Q}$ such that for each prime p there is an i for which $\zeta_{\Gamma,p}(s) = W_i(p,p^{-s})$. The many computations and results in [GSS] give quite strong support to conjecture that this is indeed the case.

Remark 2.16. (Uniformity for groups of the same Hirsch length)) Recently M. du Sautoy proved: For a given r, there exists a polynomial $f(Y,X)$ in $\mathbb{Q}[Y,X]$ such that if G is a finitely generated torsion free nilpotent group Γ of Hirsch length r and p a prime, then there exists a polynomial $Q(X) \in \mathbb{Q}[X]$, depending on Γ and p, such $\zeta_{\Gamma,p}(s) = \frac{Q(p^{-s})}{f(p,p^{-s})}$.

We end this section mentioning some results about a different zeta function associated with a finitely generated group. Namely, let $\hat{a}_n(\Gamma)$ be the number of subgroups of Γ of index n whose profinite completion is isomorphic to $\hat{\Gamma}$. In [GSS], is was shown that $\hat{\xi}_\Gamma(s) = \sum \hat{a}_n(\Gamma)n^{-s}$ has Euler product decomposition $\hat{\xi}_\Gamma(s) = \prod_p \hat{\xi}_{\Gamma,p}(s)$ and Theorem 2.13 is also valid for it. Moreover, there exists a $\mathbb{Z}$-Lie ring L, with $\hat{\xi}_{\Gamma,p}(s) = \hat{\xi}_{L,p}(s)$ for almost every prime, where $\hat{\xi}_{L,p}(s)$ is defined in the clear analogous way. It is also shown there that for almost all primes p,

$$\hat{\xi}_{\Gamma,p}(s) = \int_{G_p^+} |\det g|_p^s d\mu_p, \qquad (*)$$

where $G \leq \mathrm{GL}(L)$ is the algebraic group of automorphisms of L, $G_p^+ = G(\mathbb{Q}_p) \cap \mathrm{End}(L \otimes \mathbb{Z}_p)$ and μ_p is the Haar integral of $G(\mathbb{Q}_p)$ normalized so that $\mu(G(\mathbb{Z}_p)) = 1$. The integral in $(*)$ was computed by Igusa [Ig] for a reductive group G (under some assumptions on the representation of G). Igusa's work generalized earlier results of Satake and MacDonald for some classical groups. In our context, G is typically not reductive. In [dSL], duSautoy and Lubotzky show how to reduce the computation of (*), under some assumptions, to the reductive case. This way they can use Igusa's work to get: (a) explicit computation of $\hat{\xi}_{\Gamma,p}(s)$ for some interesting examples, (b) uniformity results of the type conjectured in [GSS]– see Remark 2.15 above– and (c) a functional equation for $\hat{\xi}_{\Gamma,p}$ (but not for $\hat{\xi}_{\Gamma}$).

Incidentally, the functional equation expresses the symmetry in a root system of a reductive group G between the positive and negative roots– see [Ig] and [dSL].

While the method applies only to $\hat{\xi}_{\Gamma,p}$ and not to $\xi_{\Gamma,p}$, it supports similar conjectures for $\xi_{\Gamma,p}$.

For a comprehensive survey of various zeta functions associated with groups, the reader is referred to [dS6].

3. Pro-p groups

The subject of counting finite index subgroups is intimately connected with pro-finite groups. This is of no surprise and we have already made use of it. This section will be devoted to counting questions for the pro-finite groups themselves. Beside the intrinsic interest, some of the results on pro-finite groups are useful for applications to discrete groups. In the context of pro-finite groups G, $a_n(G)$ denotes of course the number of open subgroups of index n.

For the free pro-finite group $\hat{F}_2$, there is nothing new to say: $a_n(\hat{F}_r) = a_n(F_r)$ and $a_n(F_r)$ was discussed in length in Section 1. More interesting is the case of a free pro-p group on r generators denoted $F_{r,\hat{p}}$.

For such a free pro-p group $F = F_{r,\hat{p}}$, $a_n(F) = 0$ unless $n = p^k$ and $a_{p^k}(F)$ is equal to the number of subnormal subgroups of index p^k in F_r.

The number $a_{p^k}(F_{r,\hat{p}})$ can be calculated recursively using P. Hall's enumeration principle.

Proposition 3.1. (Ilani [Il1]) *For $k \geq 1$,*

$$a_{p^k}(F_{r,\hat{p}}) = \sum_{t=1}^{r} (-1)^{t+1} \begin{bmatrix} r \\ t \end{bmatrix} p^{t(t-1)/2}\, a_{p^{k-t}}(F_{p^t(r-1),\hat{p}})$$

where $\begin{bmatrix} r \\ t \end{bmatrix}$ is the number of subspaces of codimension t in the r-dimensional vector space $\mathbb{F}_p$.

The above proposition gives a legitimate recursive formula, but it uses $a_{p^l}(F_{s,\hat{p}})$ to express $a_{p^k}(F_{r,\hat{p}})$ with $s \neq r$ ($s > r$ but $l < k$). Ilani (loc. cit.)

was able to deduce from (3.1) a recursion relation which expresses $a_{p^k}(F_{r,\hat{p}})$ using only $a_{p^t}(F_{r,\hat{p}})$ for $t < k$.

Proposition 3.2. *For $k \geq 1$,*

$$a_{p^k}(F_{r,\hat{p}}) = \sum_{l=1}^{k}(-1)^{t+1}\, p^{t(t-1)/2} \left[\begin{array}{c} p^{k-t}(r-1)+1 \\ t \end{array} \right] a_{p^{k-t}}(F_{r,\hat{p}}).$$

Proposition 3.3. *Let $G = F_{r,\hat{p}}$ be the free pro-p group on $r \geq 2$ generators. Then:*

$$p^{\frac{r-1}{p-1}(p^n-1)-\frac{n(n-1)}{2}} \leq a_{p^n}(G) \leq p^{\frac{r-1}{p-1}(p^n-1)}$$

and hence:

$$\lim(a_{p^n}(G))^{p^{-n}} = p^{(r-1)/(p-1)}.$$

In particular $a_{p^n}(G)$ grows exponentially as a function of p^n.

This last result was extended significantly by Mann [Man3] and Pyber-Shalev [PS1]:

Theorem 3.4. *Let G be a finitely generated pro-finite group. Then*

(a) (Mann) *If G is pro-solvable then $a_n(G)$ grows at most exponentially.*

(b) (Pyber-Shalev) *If $a_n(G)$ grows super exponentially (i.e., $\frac{\log a_n(G)}{n}$ is unbounded), then every finite groups is a quotient of some finite index subgroup of G.*

Theorem 3.4(a) implies in particular that for a finitely generated solvable group Γ, $a_n(\Gamma)$ grows at most exponentially (since $a_n(\Gamma) = a_n(\hat{\Gamma})$). There are finitely generated (pro-p and discrete) solvable groups of exponential growth, e.g., $\Gamma = C_p \wr \mathbb{Z}$. It is however interesting to observe that finitely *presented* groups behave differently.

Proposition 3.5. *Let G be a finitely presented solvable pro-p group. Then $a_n(G) \leq C^{\sqrt{n}}$ for some constant C.*

PROOF. Wilson [Wn1] showed that G satisfies the Golod-Shafarevitz inequality. Namely, for arbitrary finite presentation $\langle X; R\rangle$ of G, one has

$$|R| - (|X| - d(G)) \geq \frac{d(G)^2}{4}, \qquad (*)$$

where $d(G)$ denotes the minimal number of generators of G. Moreover, Wilson deduced that this implies that there exists a constant c such that $d(H) \leq c[G : H]^{1/2}$ for every open subgroup H of G. Indeed, if G has a presentation with d generators and r relations and $[G : H] = h$, then H has a presentation with at most hd generators and hr relations. Hence, by (*) applied to H, $d(H)^2 \leq c[G:H]$.

Now, a subgroup H of index $n = p^l$ in a pro-p group is contained in a subgroup K of index p^{l-1} and given K there are at most $\frac{p^{d(K)}-1}{p-1}$ possibilities for H. From this one can easily deduce the proposition. □

Segal and Shalev [SS] constructed for every $d \geq 2$, finitely presented (pro-p and discrete) metabelian groups G with $a_n(G)$ growth like $C^{n^{1/d}}$.

Let's pass now to groups of "slow growth". The next theorem characterizes pro-p groups of polynomial subgroup growth (PSG) as the p-adic analytic pro-p groups. It plays an important role in the characterization of discrete groups of polynomial subgroup growth– to be described in the next chapter.

The equivalence of (a) and (b) is due to Lubotzky and Mann ([LM2], cf. [DDMS]) while the equivalence of (a) and (c) was shown by Shalev [Sh1]:

Theorem 3.6. *Let G be a pro-p group. The following three conditions are equivalent:*

(a) G is a p-adic analytic group.

(b) G has polynomial subgroup growth (PSG), i.e., $a_n(G) \leq n^c$ for some constant c and every n.

(c) $a_n(G) \leq C\, n^{(\frac{1}{8}-\epsilon)\log n}$ for some $C, \epsilon > 0$ and every n.

The theorem shows that there is a gap in the possible growths: if a pro-p group has growth $O(n^{(\frac{1}{8}-\epsilon)\log n})$ for some $\epsilon > 0$, then it actually has polynomial growth. The following result of Shalev [Sh1] (see also [LS]) shows that this is essentially best possible:

Theorem 3.7. *Let $G = Ker(SL_2(\mathbb{F}_p[t]) \to SL_2(\mathbb{F}_p))$. Then $a_n(G) = O(n^{(2+\epsilon)\log n})$ for every $\epsilon > 0$.*

On the other hand, Shalev [Sh2] showed that in the category of pro-finite groups there is no gap between polynomial and non-polynomial subgroup growth. Namely:

Theorem 3.8. *For every function $f\colon \mathbb{N} \longrightarrow \mathbb{N}$ such that $f(1) \geq 1$ and $\frac{\log f(n)}{\log n} \to \infty$, there is a finitely generated non-PSG pro-finite group G satisfying $a_n(G) \leq f(n)$ for every n.*

Lubotzky [Lu4] showed that among the finitely generated linear groups there is a gap between polynomial and non-polynomial groups.

Theorem 3.9. *Let Γ be a finitely generated non-PSG linear group (over some field F). Then there exists a constant C such that $a_n(\Gamma) \geq n^{c\log n/\log\log n}$ for infinitely many n.*

As we will see in Section 5, this result is the best possible. The answer to the following interesting problem however is not known.

Problem 3.10. Is there a gap between PSG and non-PSG in the category of all finitely generated groups? If so, what is the minimum?

We believe that groups with growth $n^{c\log n/(\log\log n)^2}$ exist, and maybe this is the minimal possible non-PSG for general finitely generated groups.

Let's go back to pro-p groups.

The next theorem deals with the regularity behavior of the number of finite index subgroups in a compact p-adic analytic groups rather than the growth. It is a far reaching extension of Theorem 2.13.

Theorem 3.11. (du Sautoy [dS3]) *Let G be a compact p-adic analytic group. Then $\zeta_{G,p}(s) = \sum_{n=1}^{\infty} a_{p^n} p^{-ns}$ is a rational function in p^{-s} with rational coefficients.*

The theorem does not assume that G is pro-p (though it has a pro-p subgroup of finite index).

The proof of 3.11 borrows its main strategy from the proof of (2.13), i.e., to express the zeta function as a p-adic integral over a set representing good bases for finite index subgroups. But (3.11) is not merely much more difficult than (2.13); it is also impractical. The parametrization given for (2.13), enables (at least in principle, and as illustrated in the proof of (2.1) and in the proof of (2.9), also in practice) to calculate the ζ-function explicitly. This however seems impossible by the proof of (3.11). As of now the only non-nilpotent pro-p groups for which the ζ-function was computed explicitly are congruence subgroups of $SL_2(\mathbb{Z}_p)$, $p \geq 3$.

Theorem 3.12. (Ilani [Il3]) *Let p be a prime greater than two and $G = Ker(SL_2(\mathbb{Z}_p) \to SL_2(\mathbb{F}_p))$. Then G is a uniform pro-p group and*

$$a_{p^n}(G) = \begin{cases} \frac{p+1}{2(p-1)}(n-1)p^{n+1} + \frac{p^3-2p-2}{(p-1)^2(p+1)}p^{n+1} + \frac{p^{(n+3)/2}}{(p-1)^2} + \frac{1}{(p-1)^2(p+1)} & n \text{ odd} \\ \frac{p+1}{2(p-1)}n\,p^{n+1} - \frac{2p+1}{(p-1)^2(p+1)}p^{n+1} + \frac{p^{(n+4)/2}}{(p-1)^2} + \frac{1}{(p-1)^2(p+1)} & n \text{ even.} \end{cases}$$

Hence

$$\zeta_G(s) = \zeta_{G,p}(s) = \Sigma a_{p^n}(G)P^{-n}$$
$$= \frac{1}{(1-p^{-s})(1-p^{1-s})(1-p^{2-s})} - \frac{p^7\,p^{-2s}}{(p-1)(p^2-1)(1-p^{2-s})} +$$
$$\frac{p^2\,p^{-2s}}{p-1}\left(\frac{p+1}{(1-p^{1-s})^2(1+p^{1-s})} + \frac{p^{1-s}(p^2-p-1)-1}{(p^2-1)(1-p^{2-2s})} + \frac{p^{-s}+1}{(p-1)(1-p^{1-2s})}\right).$$

The proof of (3.12) is based on another result of Ilani [Il2] who studied the connection between subgroups of G and sub-Lie-algebras of G - when G is a uniform pro-p group. Recall (cf. [DDMS, Chapter 4]) that on such a G, a $\mathbb{Z}_p$-Lie algebra structure is defined.

Theorem 3.13. (Ilani [Il2]) *If G is a uniform pro-p group of dimension d and $p \geq d$, then every (closed) subgroup of G is a $\mathbb{Z}_p$-sub-Lie algebra and every $\mathbb{Z}_p$-sub-Lie algebra is a (closed) subgroup.*

So, by (3.13), the problem of calculating $a_{p^n}(G)$ is equivalent to calculating the number of subalgebras of a $\mathbb{Z}_p$-Lie algebra. The latter is a much easier task as was shown in [GSS]. In fact the method of "good bases" as presented in Section 2, for abelian and nilpotent groups can be adapted for Lie algebras. This can give a much easier proof for du Sautoy's Theorem (3.11), but only when $p \geq \dim(G)$. Anyway, (3.13) is useful for proving (3.12). The $\mathbb{Z}_p$-Lie algebra corresponding to $G = \operatorname{Ker}(SL_2(\mathbb{Z}_p) \to SL_2(\mathbb{F}_p))$ is $sl_2(\mathbb{Z}_p)$. The calculation, however, needed for $sl_2(\mathbb{Z}_p)$ is not that easy and Ilani used a computer to work it out. He also gave a computer-free proof of (3.12) by explicitly analyzing the subalgebras of $sl_2(\mathbb{Z}_p/p^r\mathbb{Z}_p)$, making an essential use of the fact that $sl_2(\mathbb{Z}_p)$ is of a very small dimension, i.e., three. Neither of his methods seems to suggest how to tackle the following interesting problem:

Problem 3.14. Let $G = \operatorname{Ker}(SL_n(\mathbb{Z}_p) \to SL_n(\mathbb{F}_p))$. Calculate $\zeta_G(s)$. More generally, if $\underline{G}$ is a Chevalley group scheme and $G = \operatorname{Ker}(\underline{G}(\mathbb{Z}_p) \to \underline{G}(\mathbb{F}_p))$, calculate $\zeta_G(s)$. It is natural to expect that $\zeta_G(s)$ can be expressed using invariants derived from the root system of $\underline{G}$.

Theorem (3.13) enables one to translate the problem into a problem on the Lie algebra $sl_n(\mathbb{Z}_p)$– at least if p is large enough.

4. Groups of polynomial subgroup growth

A group Γ is said to have *polynomial subgroup growth* (a PSG-group) if there exists c such that $a_n(\Gamma) \leq n^c$ for every $n \in \mathbb{N}$, or equivalently $\sigma_n(\Gamma) \leq n^{c+1}$ where $\sigma_n(\Gamma) = \sum_{i=1}^{n} a_n(\Gamma)$. Denote $\alpha(\Gamma) = \limsup_{n\to\infty} \frac{\log \sigma_n(G)}{\log n}$.

Theorem 4.1. (Lubotzky-Mann-Segal [LMS], [MS], [LM3], [Se]) *Let Γ be a finitely generated residually finite group. Then Γ has polynomial subgroup growth if and only if Γ is virtually solvable of finite rank.*

Recall that a group is virtually solvable if it contains a solvable subgroup of finite index. It is of finite rank if every finitely generated subgroup is generated by a bounded number of generators.

Theorem 4.1 joins a number of theorems which have been proven in recent years showing that some finiteness properties of infinite residually finite groups implies finiteness or virtual solvability. It deserves a notice that various old conjectures which turned out to be false for general groups are true for residually finite groups. The most famous example is the Burnside problem: A finitely generated group of finite exponent can be infinite as was shown by Adian and Novikov, but the recent solution of the restricted Burnside problem by Zel'manov says that a residually finite finitely generated group of finite exponent is finite.

Similarly, the examples of simple infinite groups whose proper subgroup are all cyclic, constructed by Ol'sanski and Rips show that finitely generated groups of finite rank need not be solvable. The story with residually finite groups is however different. Before stating the theorem, let us recall that a group Γ is said to have upper rank $\leq r$ if the rank of every finite quotient of it is at most r or equivalently the rank of $\hat{\Gamma}$ as a pro-finite group is at most r.

Theorem 4.2. *Let Γ be a finitely generated residually finite group. Then the following three conditions are equivalent:*

(1) Γ is of finite rank.

(2) Γ is of finite upper rank.

(3) Γ is virtually solvable of finite rank.

The equivalence of (1) and (3) is due to Lubotzky and Mann [LM2] and the equivalence of (1) and (3) was proved by Mann and Segal [MS] and independently by Wilson.

The proof of Theorem 4.1 is quite involved and uses diverse methods. As it has already received various expositions in the literature (cf. [Man1], [DDMS], and [LMS]), we will be very brief here:

The easier direction is the one saying that solvable groups of finite rank are PSG (see [Se] and [Lu6]). For the other direction: assume first that Γ is a subgroup of $GL_d(\mathbb{Q})$ for some d. It is therefore, as Γ is finitely generated, a subgroup of $\Delta = G(\mathbb{Z}[\frac{1}{p_1}, \ldots, \frac{1}{p_l}])$ where G is some $\mathbb{Q}$-subgroup of GL_d and $S = \{p_1, \ldots, p_l\}$ is a finite set of primes and Γ is Zariski closed in G. As our goal is to prove that Γ is virtually solvable, one can even reduce to the case where G is connected, semi-simple, and simply connected (see [LM3]). Now, the strong approximation theorem for linear groups ([No], [MVW], or [We]) can be applied to conclude that the closure of Γ in the congruence topology of Δ is a finite index subgroup Δ_0 of Δ. It implies that for every n, $\sigma_n(\Gamma) \geq \gamma_n(\Gamma)$, where $\sigma_n(\Gamma) = \sum_{i=1}^{n} a_n(\Gamma)$ and $\gamma_n(\Delta_0)$ is the number of *congruence* subgroups of index at most n in Δ_0. Using the Prime Number Theorem, one can estimate $\gamma_n(\Delta)$ to show that it does not grow polynomially and hence Γ is virtually solvable.

The case of a general linear group over $\mathbb{C}$ is carried out by showing (using results of Jordan, Wehrfritz, and Platonov) that a non-virtually-solvable finitely generated linear group over $\mathbb{C}$ has a representation over $\mathbb{Q}$ whose image is not virtually solvable.

The case of residually-p groups is handled as follows: The pro-p completion $\Gamma_{\hat{p}}$ of Γ is also PSG and hence by Theorem 3.6 above, it is p-adic analytic hence linear over $\mathbb{Q}_p$ and so Γ is linear over $\mathbb{Q}_p$ and over $\mathbb{C}$ as well. Note that by handling the residually-p case we also cover linear groups in positive characteristic. In fact with slightly more care about the counting of congruence subgroups (see Section 5) we have also just proven Theorem 3.9 (which is actually valid for all residually-p groups).

To finish the proof of Theorem 4.1 for general Γ, one analyzes, using the classification of finite simple groups, the possible composition factors of finite quotients of Γ. This (with the aid of the linear case) reduces the proof to the case where every finite quotient of Γ is solvable, i.e., $\hat{\Gamma}$ is pro-solvable PSG group. Such a group is shown to have finite rank, which means that Γ has finite upper rank. We can now apply Theorem 4.2 (whose proof is described in detail in [DDMS]) to finish the proof of 4.1.

It should be emphasized that Theorem 4.1 is valid only for finitely generated groups. There are PSG-countable groups (even linear!) which are not virtually solvable. For example, let $\Gamma = SL_n(D)$, where $D = \mathbb{Q} \cap \hat{\mathbb{Z}}_p$. From the congruence subgroup property, one deduces that $\hat{\Gamma} = SL_n(\hat{\mathbb{Z}}_p)$ which is a p-adic analytic group. Thus, $\hat{\Gamma}$ and Γ are PSG-groups. No characterization of non-finitely generated PSG-groups is known. It is also not known exactly when a general pro-finite group is PSG. Here are the two main results in this direction:

Theorem 4.3. (Mann-Segal [MS], [Man2]) *A pro-solvable group G is PSG if and only if it is of finite rank.*

Theorem 4.4. *Let G be a pro-finite group of polynomial subgroup growth. Then: G has normal subgroups $K \leq H \leq G$ such that*

(i) $(G : H) < \infty$.

(ii) *H/K is a (finite or an infinite) product of finite simple groups $\prod_{i \in I} F_i$ such that each F_i is a simple group of Lie type of the form $L_{n_i}(p_i^{r_i})$ where n_i (= the Lie-rank of F_i) and r_i are bounded.*

(iii) *K is a pro-solvable group.*

Theorem 4.4 is based on [MS] and an argument of Shalev. We finally mention another result of Mann on PSG-pro-finite groups:

Proposition 4.5. *A PSG-pro-finite group G is finitely generated (as a pro-finite group).*

PROOF. Assume $a_n(G) = O(n^r)$ and let k be a positive integer. $G^k = G\times\cdots\times G$ is endowed with a Haar measure μ. A k-tuple $\alpha = (a_1, \ldots, a_k) \in G^k$ does not generate G if and only if it is in some proper open subgroup. This shows that

$$\mu(\{\alpha \in G^k | a_1, \ldots, a_k \text{ do not generate } G\}) \leq \sum_{1<[G:H]<\infty} \frac{1}{|H|^k} = \sum_{n=2}^{\infty} \frac{a_n(G)}{n^k}.$$

Now, since G is PSG, for some k, the left hand sum is strictly less than 1 and so with a positive probability, k elements generate G, and in particular, G is finitely generated. □

Moreover, in [KL] it was shown that for $G = \hat{\mathbb{Z}}^r$ (but in fact the argument is valid for every PSG-group) we have:

(∗) *There exists an integer k such that with probability 1, some k-tuple of elements of G generate an open subgroup of G.*

In particular, this applies for p-adic analytic pro-p groups. This made Mann and Lubotzky ask:

Problem 4.6. Assume G is a pro-p group satisfying (∗). Is G p-adic analytic?

A positive answer will give a nice probabilistic characterization of analytic pro-p groups.

Mann ([Man3]) called a pro-finite group G for which there exists k with $\mu(\{\alpha \in G^k | \alpha \text{ generates } G\}) > 0$, *positively finitely generated* (PFG for short). So PSG is PFG. He also observes that in the proof of 4.5, it suffices to know that $m_n(G)$ grows polynomially, where $m_n(G)$ is the number of maximal subgroups of index n. He went ahead to show that for every pro-solvable group, $m_n(G)$ grows polynomially, and hence,

Theorem 4.7. ([Man3]) *A finitely generated pro-solvable group is positively finitely generated.*

Mann and Shalev [MaSh] showed an equivalence:

Theorem 4.8. *A finitely generated pro-finite group G is positively finitely generated if and only if $m_n(G)$ grows polynomially.*

We will close this section with a few more results and questions about the precise rate of growth of $a_n(G)$ for a PSG group.

It is actually more convenient to talk about $\sigma_n(G) = \sum_{i=1}^{n} a_i(G)$. Let $\alpha(G) = \limsup \frac{\log \sigma_n(G)}{\log n}$. Clearly $\alpha(G) < \infty$ if and only if $a_n(G)$ and $\sigma_n(G)$

grows (at most) polynomially– in which case we say G has polynomial subgroup growth and $\alpha(G)$ is the smallest real number α such that $\sigma_n(G) = 0(n^{\alpha+\epsilon})$ for every $\epsilon > 0$.

If $\zeta_G(s) = \Sigma\, a_n(G)n^{-s}$ then $\alpha(G)$ can be read of $\zeta_G(s)$; by the Tauberian theorem, $\zeta_G(s)$ may be extended analytically to the half space $Re(s) \geq \alpha(G)$ except for a pole at $s = \alpha(G)$. (If this is a simple pole then in fact $\sigma_n(G) = O(n^\alpha)$, but if it is, say, a double pole then we can only deduce $\sigma_n(G) = O(n^\alpha \log n)$). Anyway, when $\zeta_G(s)$ is explicitly known or in those cases in which it was proved to be rational we can derive some conclusions on $\alpha(G)$ and hence on the rate of growth of $\sigma_n(G)$.

For example, let G be a p-adic analytic pro-p group. By du Sautoy's theorem $\zeta_G(s)$ is a rational function of p^{-s} with rational coefficients. Actually, his result is more precise: it also says that the denominator of $\zeta_G(s)$ is of the form $\Pi_{i=1}^{l}(1 - p^{-a_i s - b_i})$ for some $l \in \mathbb{N} \cup \{0\}$, and $a_1, \ldots, a_l, b_1, \ldots, b_l \in \mathbb{Z}$. This shows that for real s the denominator is zero only for $s = -b_i/a_i$, $i = 1, \ldots, l$. In particular we deduce:

Proposition 4.9. (du Sautoy [dS3]) *Let G be a p-adic analytic pro-p group. Then $\alpha(G) = \limsup \frac{\log \sigma_n(G)}{\log n}$ is rational.*

The analogous result for nilpotent groups is not known, but it is quite likely to hold:

Conjecture 4.10. Let Γ be a nilpotent group then $\alpha(\Gamma)$ is rational.

It is not difficult to find examples which show that $\alpha(G)$ can be rational and not necessarily integer. For example, Theorem 2.9 shows that if $G = \Gamma_{\hat{p}}$ is the pro-p completion of the Heizenberg group then $\zeta_G(s) = \frac{\zeta_p(s)\zeta_p(s-1)\zeta_p(2s-2)\zeta_p(2s-3)}{\zeta_p(3s-3)}$, i.e., $\zeta_G(s)$ has poles for $s = 0, 1$ and $\frac{3}{2}$ (with a double pole for $s = 1$). Anyway $\alpha(G) = \frac{3}{2}$. The same remark applies for $\alpha(\Gamma)$ when Γ is the discrete Heizenberg group. In this sense subgroup growth is different from the classical word growth $b_n(\Gamma)$. Recall that $b_n(\Gamma)$ is defined as the number of elements of Γ of length at most n with respect to a fixed finite set of generators Σ. Denote $\beta(\Gamma) = \limsup \frac{\log b_n(\Gamma)}{\log n}$. It is easy to see that $\beta(\Gamma)$ depends only on Γ and not on Σ, and that if Γ is nilpotent then $\beta(\Gamma) < \infty$. In [Ba], Bass gave a formula for $\beta(\Gamma)$, from which one sees that $\beta(\Gamma)$ is always an integer. Moreover there are two constant $0 \leq C_1, C_2 \in \mathbb{R}$ such that for every n, $C_1 n^{\beta(\Gamma)} \leq b_n(\Gamma) \leq C_2 n^{\beta(\Gamma)}$. It is not known if an analogue of this later result holds for $\sigma_n(\Gamma)$, when Γ is a nilpotent group and $\beta(\Gamma)$ is replaced by $\alpha(\Gamma)$. This is not the case for uniform pro-p groups: The computation of Ilani (3.12), shows that for $G = \mathrm{Ker}(SL_2(\mathbb{Z}_p) \to SL_2(\mathbb{F}_p))$, $a_n(G)$ grows like $n \log n$ and so $\sigma_n(\Gamma)$ grows as $n^2 \log n$, so $\sigma_n(G) = O(n^{2+e})$ for every $e > 0$ but not $O(n^2)$.

Anyway, it will be very interesting to answer the following:

Problem 4.11. For a finitely generated nilpotent group and for a uniform pro-p group G, give a formula for $\alpha(G)$.

Another remark is in order here: When Γ is a finitely generated torsion free nilpotent group and H is a finite index subgroup of Γ. Then it is shown in [GSS], that $\alpha(G) = \alpha(H)$. This is not always the case for groups of polynomial subgroup growth. It was observed in [LM3] that if Γ is the infinite dihedral group then $\alpha(\Gamma) = 2$ while for the infinite cyclic group H, which is an index two subgroup in Γ, $\alpha(H) = 1$. This is again inconsistent with word growth: there $\beta(\Gamma) = \beta(H)$ whenever H is a finite index subgroup of Γ. The study of $\alpha(H)$ when H varies over finite index subgroups of a PSG group deserves some more attention.

5. Counting congruence subgroups

This section is devoted to the growth of the number of congruence subgroups in an arithmetic group. Beside its intrinsic interest as a "non-commutative analytic number theory", these examples have produced the first examples of groups of intermediate subgroup growth– i.e., growth which is greater than polynomial and smaller than exponential. (More examples at the higher end of the intermediate growth range were given recently by Dan Segal and Aner Shalev [SS].)

We shall also relate the subgroup growth of arithmetic groups with the congruence subgroup property, showing that the latter can be characterized by means of subgroup growth. This enables one to formulate a "congruence subgroup problem" for groups which do not have an arithmetic structure. In particular, it suggests the study of the subgroup growth of fundamental groups of hyperbolic manifolds. We will present some (very) partial results in this direction.

Let us start with counting the congruence subgroups: It turns out that there is a fundamental difference between arithmetic groups over global fields in characteristic zero and those of positive characteristic. They are different in the results as well as in the methods of proof. We will therefore handle them separately. The notations however will be presented simultaneously.

Let K be either $\mathbb{Q}$– the field of rational numbers or $\mathbb{F}_p(x)$– the field of rational functions over the finite field of order p. Let O be the ring of integers of K, i.e., $O = \mathbb{Z}$ or $\mathbb{F}_p[x]$. Let G be a simple, simply connected, connected algebraic group defined over K. For the simplicity of the exposition we will also assume that G splits over K. (The interested reader is referred to [Lu4] for the general case including S-arithmetic groups etc. Most readers will find it useful to assume $G = SL_r$. All essential ideas appear already in this case.) So G can be thought of as a Chevalley group (defined over $\mathbb{Z}$) and we fix an embedding of G into GL_r. Let $\Gamma = G(O)$ (e.g., $\Gamma = SL_r(\mathbb{Z})$ or $\Gamma = SL_r(\mathbb{F}_p[t])$).

For an ideal $I \neq \{0\}$ of O we denote $\Gamma(I) = \mathrm{Ker}(G(O) \to G(O/I))$. As O/I is finite, $\Gamma(I)$ is always a finite index normal subgroup– called a principal congruence subgroup. A subgroup Δ of Γ is called a congruence subgroup if it contains $\Gamma(I)$ for some $I \neq \{0\}$.

Let $\gamma_n(\Gamma)$ denote the number of congruence subgroups of Γ of index *at most* n.

Theorem 5.1. (Lubotzky [Lu4]) *Assume char$(K) = 0$. Then there exist positive constants C_1 and C_2 such that*

$$n^{C_1 \log n/\log\log n} \leq \gamma_n(\Gamma) \leq n^{C_2 \log n/\log\log n}.$$

Theorem 5.2. (Lubotzky [Lu4]) *Assume char$(K) = p > 0$. Then there exists positive constants C_3 and C_4 such that*

$$n^{C_3 \log n} \leq \gamma_n(\Gamma) \leq n^{C_4 (\log n)^2}$$

The theorems show that in case all finite index subgroups of Γ are congruence subgroups, Γ has intermediate subgroup growth. This is known to be the case for $\Gamma = SL_r(O)$ when $r \geq 3$ (cf. [Rp], [Ra] and the references therein). Hence:

Corollary 5.3.

(a) If $\Gamma = SL_r(\mathbb{Z})$, $r \geq 3$, then $n^{C_1 \log n/\log\log n} \leq \sigma_n(\Gamma) \leq n^{C_2 \log n/\log\log n}$.

(b) If $\Gamma = SL_r(\mathbb{F}_p[t])$, $r \geq 3$, then $n^{C_3 \log n} \leq \sigma_n(\Gamma) \leq n^{C_4(\log n)^2}$.

Here C_1, C_2, C_3 and C_4 are some positive constants, and $\sigma_n(\Gamma)$ is the number of subgroups of Γ of index at most n.

In fact the proof can be used to get some explicit estimate on these constants. For example for $\Gamma = SL_2(\mathbb{Z})$ we can take in Theorem 5.1, $C_1 = \frac{1}{144}$ and $C_2 = 18$.

We now sketch the proofs of Theorem 5.1 and 5.2. In order to visualize better the difference between the zero and positive characteristic we will give first the proofs for the lower bounds:

Proof of lower bound of (5.1)

By the strong approximation theorem (see [Pr]– but think on SL_r), $\Gamma = G(\mathbb{Z})$ is mapped onto $G(\mathbb{Z}/M\mathbb{Z})$ for every $M \in \mathbb{Z}$. If $M = p_1^{\alpha_1} \cdot \ldots \cdot p_t^{\alpha_t}$ where p_i, $i = 1, \ldots, t$, are different primes, then by the Chinese Remainder Theorem $G(\mathbb{Z}/M\mathbb{Z}) = \Pi_{i=1}^t G(\mathbb{Z}/p_i^{\alpha_i}\mathbb{Z})$. For every prime p, $G(\mathbb{F}_p)$ has a cyclic subgroup of order $p-1$ (coming from the torus isomorphic to $(\mathbb{F}_p^*)^{\mathrm{rank}(G)}$ since G splits; but also in general a theorem of Lang ensures that over a finite field, there is at least a one dimensional torus). Thus if $M = p_1 \cdot \ldots \cdot p_t$ a product of different primes chosen so that for $i = 1, \ldots, t$, $p_i \equiv 1(\bmod m)$ for some number m, then $G(\mathbb{Z}/M\mathbb{Z})$ contains a subgroup isomorphic to $(\mathbb{Z}/m\mathbb{Z})^t$. Now if m itself is a product of distinct primes, say, $m = q_1 \cdot \ldots \cdot q_s$ then $(\mathbb{Z}/m\mathbb{Z})^t \simeq \Pi_{i=1}^s(\mathbb{Z}/q_i\mathbb{Z})^t$ contains at least $\Pi_{i=1}^s q_i^{t^2/4} = m^{t^2/4}$ subgroups. As the order of $G(\mathbb{Z}/M\mathbb{Z})$ is at most M^d with $d = \dim G$, we get $m^{t^2/4}$ subgroups of Γ of index $\leq M^{\dim G}$. We show now how the primes p_i and q_j can be chosen to ensure that $\gamma_n \geq n^{C \log n/\log\log n}$ for some constant C:

Denote $\gamma =$ Euler constant $= 0.57721\ldots$ and $\gamma' = e^{-\gamma}$. Let $N = n^{\gamma'/d}$ where $d = \dim(G)$, $\tau = \log(N)$ and $\beta = \log(\tau) = \log\log(N)$. Let $q_1, \ldots, q_s$ be the list of primes smaller than β and $m = \Pi_{j=1}^s q_j$. By the prime number theorem $m \approx e^\beta = \tau$ and by [El, Ex. 1.20, p. 31], $\frac{m}{\varphi(m)} \approx \frac{\log\beta}{\gamma'} = \frac{\log\log\tau}{\gamma'}$. Let $\Pi = \Pi(m\tau/\log\log\tau; m, 1)$ be the set of the t primes less than $m\tau/\log\log\tau$ and congruent to 1 mod m. From the prime number theorem along arithmetic progressions (cf. [El, Theorem 8.8, p. 277]) it follows that

$$t = \frac{m\tau}{\varphi(m)(\log\log\tau)\log(m\tau/\log\log\tau)} \approx \frac{1}{2\gamma'} \cdot \frac{\tau}{\log\tau}$$

(where here and always $f \approx g$ if $\lim_{n\to\infty} \frac{f(n)}{g(n)} = 1$). The same theorem says also that the product M of these t primes satisfies:

$$\log M \approx m/\varphi(m) \cdot \frac{\tau}{\log\log\tau} \approx \frac{\log N}{\gamma'}.$$

The discussion above shows that between Γ and $\Gamma(M)$ there are at least $m^{t^2/4}$ subgroups whose index is at most M^d which is approximately $N^{d/\gamma'} = n$. Thus $\gamma_n \geq m^{t^2/4}$, so $\log\gamma_n \gtrsim \frac{t^2}{4}\log m \approx \frac{1}{16\gamma'^2}\frac{\tau^2}{\log\tau}$. As $\tau = \frac{\gamma'}{d}\log n$ we deduce that $\log\gamma_n \gtrsim \frac{1}{16d^2}\frac{(\log n)^2}{\log\log n}$. This proves the lower bound with $C_1 = \frac{1}{16d^2}$. □

Proof of lower bound of (5.2)

The group $\Gamma = G(\mathbb{F}_p[t])$ is dense in the pro-finite group $H = G(\mathbb{F}_p[[t]])$, where $\mathbb{F}_p[[t]]$ denotes the ring of formal power series over $\mathbb{F}_p$. The group H is virtually pro-p. In fact, it is not difficult to see (cf. [LS]) that $H(1) = \operatorname{Ker}(G(\mathbb{F}_p[[t]] \to G(\mathbb{F}_p))$ is a pro-p group. $H(1)$ has analytic structure over $\mathbb{F}_p[[t]]$, but it is not p-adic analytic group. For example, it has an infinite torsion subgroup while p-adic analytic pro-p group cannot have such a subgroup (cf. [DDMS]. See also [LS] for a more general statement). Thus by Shalev's theorem (3.6), $\sigma_n(H(1))$ and so also $\sigma_n(H)$ grows at least as $n^{C\log n}$ for a suitable constant C. The same applies also for Γ, which proves our claim. □

Proof of the upper bound of (5.1)

A crucial ingredient in the proof of the upper bound is the following:

Proposition 5.4. ("level $\leq$ index") *Let $\Gamma = G(\mathbb{Z})$ be as in 5.1 and H a congruence subgroup of Γ. Then $H \supseteq \Gamma(m)$ for some $m \leq [\Gamma : H]$.*

Corollary 5.5. *Let $\Gamma = G(\mathbb{Z})$ be as in (5.1). Then*

$$\gamma_n(\Gamma) \leq \sum_{m=1}^{n} \|G(\mathbb{Z}/m\mathbb{Z})\|,$$

where for a finite group F we denote by $\|F\|$ the total number of its subgroups.

The problem is therefore transformed now to a problem on finite groups. In the following proposition we collect some useful easy results:

Proposition 5.6. *Let F be a finite group. Then:*

(i) $\text{rank}(F) \leq \log_2 |F|$.

(ii) $\|F\| \leq |F|^{\text{rank}(F)}$.

Proposition 5.7. *There exists a constant $C = C(G)$ such that $\text{rank}(G(\mathbb{F}_p))$ is bounded by C.*

PROOF. By a result of Aschbacher and Guralnick ([AG]) every finite group is generated by a solvable subgroup plus one element. It suffices therefore to bound the number of generators of solvable subgroups of $G(\mathbb{F}_p)$. Let M be a solvable subgroup of $G(\mathbb{F}_p)$. So $M = PQ$ where P is a p-sylow subgroup of M and Q is of order prime to p. As $G \hookrightarrow GL_r$, the order of P is bounded by p^{r^2} and so $d(P) \leq r^2$. Now Q, being of order prime to p, can be lifted to $GL_r(\mathbb{C})$. (This is a classical result. Here is a less classical proof: $Q \leq GL_r(\mathbb{F}_p)$. Look at the preimage R of Q in $GL_r(\mathbb{Z}_p)$. It has a normal pro-p subgroup $N = \text{Ker}(GL_r(\mathbb{Z}_p) \to GL_r(\mathbb{F}_p))$ and $R/N = Q$ is of order prime to p. So by the Schur-Zassenhaus theorem, R is a semi-direct product of N and Q' where Q' is a subgroup of R isomorphic to Q. Thus Q is isomorphic to a subgroup of $GL_r(\mathbb{Z}_p)$ and of $GL_r(\mathbb{C})$). By a classical theorem of Jordan, Q has an abelian subgroup A of bounded index. A is diagonalizable and so $d(A) \leq r$. Altogether $d(Q)$ is bounded as a function of r and so is $d(M)$ and $\text{rank}(G(\mathbb{F}_p))$. □

Corollary 5.8. $\text{rank}(G(\mathbb{Z}/p^\alpha\mathbb{Z}))$ *is bounded independent of p and α.*

PROOF. As all $G(\mathbb{Z}/p^\alpha\mathbb{Z})$ are images of $G(\mathbb{Z}_p)$, it suffices to prove that $G(\mathbb{Z}_p)$ or $GL_r(\mathbb{Z}_p)$ is of bounded rank (depending on r but not on p). Now, by [DDMS, Chapter 5], $N = \text{Ker}(GL_r(\mathbb{Z}_p) \to GL_r(\mathbb{F}_p))$ is a uniform pro-p group of rank r^2. This with (5.7) proves (5.8). □

Now we can complete the proof of the upper bound of (5.1): For $m \in \mathbb{Z}$, write $m = p_1^{\alpha_1} \cdot \ldots \cdot p_l^{\alpha_l}$ where p_i, $i = 1 \ldots, l$ are the distinct prime divisors of m. By the prime number theorem $l \leq \frac{\log m}{\log \log m}$ and so by (5.8):

$$\begin{aligned} \text{rank}\,(G(\mathbb{Z}/m\mathbb{Z})) &= \text{rank}\left(\prod_{i=1}^{l} G(\mathbb{Z}/p_i^{\alpha_i}\mathbb{Z})\right) \\ &\leq \sum_{i=1}^{l} \text{rank}\,(G(\mathbb{Z}/p_i^{\alpha_i}\mathbb{Z})) = O\left(\frac{\log m}{\log \log m}\right). \end{aligned}$$

(5.6 ii) now implies that $\|G(\mathbb{Z}/m\mathbb{Z})\| \leq m^{C\frac{\log m}{\log \log m}}$. This finishes the proof in light of (5.5). □

Remark 5.9. We used along the way the result of Ashbacher and Guralnick ([AG]) which requires the classification of finite simple groups. But, as was noticed by L. Pyber, this can be avoided (see [Lu4]).

Proof of upper bound of (5.2)

Proposition 5.10. *Let $\Gamma = G(\mathbb{F}_p[t])$ be as in (5.2). Then there exists a constant C such that any congruence subgroup of Γ of index n contains a* subnormal *congruence subgroup of index at most n^C.*

The proof is based on Babai-Cameron-Palfy theorem [BCP].

Proposition 5.11. *Every subnormal congruence subgroup of $\Gamma = G(\mathbb{F}_p[t])$ of index n contains a principal congruence subgroup of index at most $n^{C' \log n}$ for some constant C'.*

Proposition 5.12. (L. Pyber [Py2]) *If F is a finite group, then $a_n(F) \leq |F|^{2 \log n}$.*

The last three propositions imply the upper bound of (5.2). □

We turn now back to assume $\text{char}(K) = 0$ so $\Gamma = G(\mathbb{Z})$. The congruence subgroups of Γ form a basis of neighborhoods of the identity of Γ and thus define a topology on Γ– called the congruence topology. The completion of Γ wth respect to this topology is $G(\hat{\mathbb{Z}})$, by the strong approximation theorem. In general the congruence topology is weaker than the pro-finite topology and so the homomorphism $\pi : \hat{\Gamma} = \widehat{G(\mathbb{Z})} \to G(\hat{\mathbb{Z}})$ induced by the indentity map from Γ to Γ is an epimorphism but not a monomorphism. It is a monomorphism if and only if every finite index subgroup of Γ is a congruence subgroup. The original congruence subgroup problem asks whether this is the case, but it turns out that all the important applications of it need only that $\mathrm{Ker}(\pi)$ is finite. So following the ususal tradition we say that Γ has the *congruence subgroup property* (CSP for short) if $\mathrm{Ker}(\pi)$ is finite. It is not difficult to see that in this case $\sigma_n(\Gamma)$ has the same type of growth as $\gamma_n(\Gamma)$. The next result, from Lubotzky [Lu4], actually shows that CSP can be characterized by the property that $\sigma_n(\Gamma)$ has the same type of growth as $\gamma_n(\Gamma)$. The result is even stronger:

Theorem 5.13. *Let $\Gamma = G(\mathbb{Z})$ as in (5.1). Then Γ has the congruence subgroup property if and only if for every $\varepsilon > 0$ there exists a constant C_ε such that $\sigma_n(\Gamma) \leq C_\varepsilon n^{\varepsilon \log n}$ for every n.*

Theorem 5.13 is maybe even more interesting when expressed in the negative form: If Γ does not have CSP then for some $\varepsilon > 0$, $\sigma_n(\Gamma) \geq n^{\varepsilon \log n}$ for infinitely many n's, i.e., the subgroup rate of growth of Γ is strictly bigger than the rate of growth of the congruence subgroups.

Another interesting aspect of Theorem 5.13 is that it gives a purely group theoretical characterization to CSP, which is an arithmetic property. In particular, we can now formulate a congruence subgroup problem for groups without an arithmetic structure. This is especially interesting for non-arithmetic lattices in the simple Lie groups of rank one $SO(n,1)$ and $SU(n,1)$ in which non-arithmetic lattices are known to exist (for every n in the first family and for $n = 2$ and 3 in the second). It is very natural to conjecture (and it is compatible with Serre's conjecture on CSP– see [Sr]) that all lattices in $SO(n,1)$ and $SO(U,1)$ have subgroup growth at least $n^{C \log n}$. We actually believe that they even have exponential or super-exponential subgroup growth. At this point, however, only a very partial result is known:

Proposition 5.14.

(i) Let $H = SO(2,1) \approx PSL_2(\mathbb{R})$ and Γ a lattice (= a discrete subgroup of finite covolume) in H. Then Γ has a super-exponential growth.

(ii) Let $H = SO(3,1) \approx PSL_2(\mathbb{C})$ and Γ a lattice in H. Then $\sigma_n(\Gamma) \geq n^{C \log n}$ for some constant C.

PROOF. (i) The structure of lattices in $PSL_2(\mathbb{R})$ is well known: Γ either has a free non-abelian subgroup of finite index or it contains a finite index surface group of genus $g \geq 2$. Such a surface group is mapped epimorphically onto a free group on g generators. Thus in either case the subgroup growth of Γ is like that of a free group, i.e., super-exponential by (1.6).

(ii) Γ being a lattice in $SO(3,1)$ has a torsion free subgroup Δ which is a fundamental group of a 3-manifold. A 3-manifold group always has a presentation with no more relations than generators (see [Lu1] for details). By choosing Δ in a suitable way we can arrange that for $p = 2$ the pro-p complion $\Delta_{\hat{p}}$ of Δ, satisfies $d(\Delta_{\hat{p}}) = d \geq 5$ and it has a presentation (as a pro-p group) with r relation where $r \leq d$. Thus $r < \frac{d^2}{4}$, i.e., $\Delta_{\hat{p}}$ does not satisfy the Golod-Shafarevitz inequality. This implies ([Lu1, Theorem 1], [DDMS]) that $\Delta_{\hat{p}}$ is not p-adic analytic. Thus by (3.6), $\sigma_n(\Delta_{\hat{p}})$ grows at least at $n^{C \log n}$ and by (2.2) the same applies for Δ and hence for Γ. □

Experience with lattices in semi-simple groups show that discrete groups with Kazdhan property (T) tend to have the CSP. We end this chapter with a conjecture:

Conjecture 5.15. Let Γ be a discrete group with Kazdhan property (T). Then $\sigma_n(\Gamma)$ grows at most exponentially (one can even aspire to more ambitious bounds).

Recall that Γ has property (T) if the trivial representation of Γ is an isolated point in the dual space of the irreducible unitary representations of Γ. Though not apparent from the definition, one can show that property (T) puts severe restrictions on the finite quotients of Γ (cf. [Lu5]) so it is not unrealistic to expect some control on the subgroup growth.

A discrete group is called amenable if $L^\infty(\Gamma)$ carries an invariant mean. Amenable groups are very different from groups with property (T). Still we wonder whether for a finitely generated amenable group Γ, $\sigma_n(\Gamma)$ grows at most exponentially. Solvable groups are amenable and for solvable groups, Mann [Man3] indeed showed that their subgroup growth is at most exponential. (Note however that for some solvable groups it is exponential.)

References

[AG] M. Aschbacher, R. Guralnik, Solvable generation of groups and Sylow subgroups of the lower central series, *J. Algebra* **77**(1982), 189–201.

[BCP] L. Babai, P.J. Cameron and P.P. Palfy , On the orders of primitive groups with restricted non-abelian composition factors , *J. Algebra* **79**(1982), 161–168.

[Ba] H. Bass, The degree of polynomial growth of finitely generated nilpotent groups, *Proc. London Math. Soc.* **25**(1972), 603–614.

[BMS] H. Bass, J. Milnor and J.P. Serre, Solution of the congruence subgroup problem for $SL(n), (n \geq 3)$ and $Sp(2n), (n \geq 2)$, *Publ. Math. IHES* **33**(1967), 59–137.

[BR1] C.J. Bushnell and I. Reiner, Solomon's conjectures and the local functional equation for zeta functions of orders, *Bull. Amer. Math. Soc.* **2**(1980), 306–310.

[BR2] C.J. Bushnell and I. Reiner, Zeta function of arithmetic orders and Solomon's conjecture, *Math. Z.* **173**(1980), 135–161.

[De1] J. Denef, The rationality of the Poincaré series associated to the p-adic points on a variety, *Invent. Math.* **77**(1984), 1–22.

[De2] J. Denef, On the degree of Igusa's local zeta function, *Amer. J. Math.* **109**(1987), 991–1008.

[DvdD] J. Denef, L. van den Dries, p-adic and real subanalytic sets, *Ann. Math.* **128**(1988), 79–138.

[De1] I.M.S. Dey, *Schreier systems in free products*, Ph.D. Thesis, Manchester, 1963.

[De2] I.M.S. Dey, Schreier systems is free products, *Proc. Glasgow Math. Soc.* **7**(1965), 61–79.

[Di] J.D. Dixon, The probability of generating the symmetric group, *Math. Z.* **110**(1969), 199–205.

[DDMS] J.D. Dixon, M.P.F. du Sautoy, A. Mann and D. Segal, *Analytic Pro-p Groups* (LMS Lecture Notes Series **157**, Cambridge Univ. Press, 1991).

[DM] A. Dress and Th. Müller, Logarithm of generating functions and combinatorial decomposition of functions, Universitüt Bielefeld, preprint.

[dS1] M.P.F. du Sautoy, Finitely generated groups, p-adic analytic groups and Poincaré series, *Bull. Amer. Math. Soc.* **23**(1990), 121–126 (also Appendix C of [DDMS]).

[dS2] M.P.F. du Sautoy, Applications of p-adic methods to group theory, in *p-adic Methods and their Applications* (A. Baker and R. Plymen (eds.), Oxford Univ., to appear).

[dS3] M.P.F. du Sautoy, Finitely generated groups, p-adic analytic groups and Poincaré series, *Ann. Math.* **137**(1993), 639–670.

[dS4] M.P.F. du Sautoy, Zeta functions of groups and rings: uniformity, *Israel J. Math.* **86**(1994), 1–23.

[dS5] M.P.F. du Sautoy, Counting congruence subgroups in arithmetic subgroups, preprint.

[dS6] M.P.F. du Sautoy, Zeta functions on groups, preprint.

[dSL] M.P.F. du Sautoy and A. Lubotzky, Functional equations and uniformity for local zeta functions of nilpotent groups, in preparation.

[El] W. and F. Ellison, *Prime Numbers* (A. Wiley & Sons, 1985).

[FS] B. Fine and D. Spellman, Counting subgroups in the Hecke groups, *Internat. J. Algebra Comput.* **3**(1993), 43–49.

[GIR] C. Godsil, W. Imrich and R. Razen, On the number of subgroups of given index in the modular group, *Mh. Math.* **87**(1979), 273–280.

[GN1] M. Grady, M. Newman, Some divisibility properties of the subgroup counting function for free products, *Math. Comp.* **58**(1992), 347–353.

[GN2] M. Grady and M. Newman, Counting subgroups of given index in Hecke groups, *Contemp. Math., Amer. Math. Soc.*, to appear.

[GN3] M. Grady and M. Newman, Residue periodicity in subgroup counting functions, preprint.

[Gri] R.I. Grigorchuck, On growth in group theory, in *Proc. of Inter. Congress of Math.* (Kyoto, Japan, 1990), 325–338.

[Gro] M. Gromov, Groups of polynomial growth and expanding maps, *Publ. Math. IHES* **53**(1981), 53–78.

[GSS] F.J. Grunewald, D. Segal and G.C. Smith, Subgroups of finite index in nilpotent groups, *Invent. Math.* **93** (1988), 185–223.

[Ha1] M. Hall, Subgroups of finite index in free groups, *Canad. J. Math.* **1**(1949), 187–190.

[Ha2] M. Hall, *The Theory of Groups* (Macmillan, New York, 1959).

[Ha] P. Hall, *Nilpotent Groups* (Queen Mary College Math. Notes, 1969).

[Ig] J.I. Igusa, Universal p-adic zeta functions and their functional equations, *Amer. J. Math.* **111**(1989), 671–716.

[Il1] I. Ilani, Counting finite index subgroups and the P. Hall enumeration principle, *Israel J. Math.* **68**(1989), 18–26.

[Il2] I. Ilani Analytic pro-p groups and their Lie algebras, preprint.

[Il3] I. Ilani, Counting finite index subgroups in $SL_2(\mathbb{Z}_p)$, in preparation.

[Im] W. Imrich, On the number of subgroups of given index in $SL_2(\mathbb{Z})$, *Arch. Math.* **31**(1978), 224–231.

[Jo] G.A. Jones, Congruence and non-congruence subgroups of the modular group: a survey, (in *Proc. of Groups-St. Andrews 1985*, E.F. Robertson and C.M. Campbell (eds.), LMS Lecture Notes Series **121**, Cambridge Univ. Press, 1986), 223–234.

[KL] W.A. Kantor and A. Lubotzky, The probability of generating a finite classical group, *Geom. Dedicata* **36**(1990), 67–87.

[Lu1] A. Lubotzky, Group presentation, p-adic analytic groups and lattices in $SL_2(\mathbb{C})$, *Ann. Math.* **118**(1983), 115–130.

[Lu2] A. Lubotzky, Dimension functions for discrete groups, in *Proc. of Groups - St. Andrews 1985* (E.F. Robertson and C.M. Campbell (eds.), London Math. Soc. Lecture Notes **121**, Cambridge University Press 1986), 254–262.

[Lu3] A. Lubotzky, A group theoretic characterization of linear groups, *J. Algebra* **113**(1988), 207–214.

[Lu4] A. Lubotzky, Subgroup growth and congruence subgroups, preprint.

[Lu5] A. Lubotzky, *Discrete Groups, Expanding Graphs and Invariant Measures* (Progress in Math., Birkhauser Verlag), to appear.

[Lu6] A. Lubotzky, Subgroup Growth, Lecture notes for a series of talks in the Conference on Group Theory, Groups Galway/St Andrews 1993, August 1993.

[LM1] A. Lubotzky and A. Mann, Powerful p-groups I, II, *J. Algebra* **105**(1987), 484–515.

[LM2] A. Lubotzky and A. Mann, Residually finite groups of finite rank, *Math. Proc. Cambridge Philos. Soc.* **106**(1989), 385–388.

[LM3] A. Lubotzky and A. Mann, On groups of polynomial subgroup growth, *Invent. Math.* **104**(1991), 521–533.

[LMS] A. Lubotzky, A. Mann and D. Segal, Finitely generated groups of polynomial subgroup growth, *Israel J. Math.* **82**(1993), 363–371.

[LS] A. Lubotzky and A. Shalev, On some Λ-analytic pro-p groups, *Israel J. Math.* **85**(1994), 307–337.

[M] I.G. Macdonald, *Symmetric Functions and Hall Polynomial* (Oxford Univ. Press, 1979).

[Ma1] A.J. Macintyre, On definable subsets of p-adic fields, *J. Symbolic Logic* **41**(1976), 605–610.

[Ma2] A.J. Macintyre, Rationality of p-adic Poincaré series: uniformity in p, *Ann. Pure. Appl. Logic* **49**(1990), 31–74.

[Man1] A. Mann, Some applications of powerful p-groups, in *Proc. of Groups - St. Andrews 1989* (C.M. Campbell and E.F. Robertson (eds.), LMS Lecture Notes Series **160**, Cambridge Univ. Press, 1991), 370–385.

[Man2] A. Mann, Some properties of polynomial subgroup growth groups, *Israel J. Math.* **82**(1993), 373–380.

[Man3] A. Mann, Positively finitely generated groups, preprint.

[MS] A. Mann and D. Segal, Uniform finiteness conditions in residually finite groups, *Proc. London Math. Soc.* **61**(1990), 529–545.

[MaSh] A. Mann and A. Shalev, Maximal subgroups of finite simple groups and positively finitely generated groups, preprint.

[MVW] C.R. Matthews, L.N. Vaserstein and B. Weisfaler, Congruence properties of Zariski dense subgroups I, *Proc. Lond. Math. Soc.* **48**(1984), 514–532.

[Me1] A.D. Mednyh, Determination of the number of nonequivalent coverings over a compact Riemann surface, *Soviet Math. Dokl.* **19**(1978), 318–320.

[Me2] A.D. Mednyh, On unramified coverings of compact Riemann surfaces, *Soviet Math. Dokl.* **20**(1979), 85–88.

[Me3] A.D. Mednyh, On the solution of the Hurwitz problem on the number of nonequivalent coverings over a compact Riemann surface, *Soviet Math. Dokl.* **24** (1981), 541–545.

[Me4] A.D. Mednyh, Hurwitz problem on the number of nonequivalent coverings of a compact Riemann surface, *Sib. Math. J.* **23**(1982), 415–420.

[MP] A.D. Mednyh and G.G. Pozdnyakova, Number of nonequivalent coverings over a nonorientable compact surface, *Sib. J. Math.* **27**(1986), 99–106.

[Mi] J. Milnor, Growth of finitely generated solvable groups, *J. Diff. Geom.*

2(1968), 443–449.

[MW] L. Moser and M. Wyman, On solution of $X^d = 1$ in symmetric groups, *Canad. J. Math.* **7**(1955), 159–168.

[Mu1] T. Müller, *Kombinatorische Aspekte endlich erzeungter virtuell freier Gruppen*, Ph.D. Thesis, Universität Frankfurt am Main, 1989.

[Mu2] T. Müller, Combinatorial aspects of finitely generated virtually free groups (extended abstract), in *Proc. of Groups-St. Andrews 1989, vol. 2* (C.M. Campbell and E.F. Robertson (eds.), London Math. Soc. Lecture Notes **160**, Cambridge Univ. Press, 1990), 386–395.

[Mu3] T. Müller, Combinatorial aspects of finitely generated virtually free groups, *J. London Math. Soc.* **44**(1991), 75–94.

[Mu4] T. Müller, A group-theoretical generalization of Pascal's triangle, *European J. Combin.* **12**(1991), 43–49.

[Mu5] T. Müller, Finite group actions, subgroups of finite index in free products and asymptotic expansion of $e^{p(z)}$, preprint .

[Ne1] M. Newman, The number of subgroups of the classical modular group of index N (tables) (National Bureau of Standards, Washington D.C., 1976).

[Ne2] M. Newman, Asymptotic formulas related to free products of cyclic groups, *Math. Comp.* **30**(1976), 838–846.

[No] M. Nori, On subgroups of $GL_n(F_p)$, *Invent. Math.* **88**(1987), 257–275.

[Pr] G. Prasad, Strong approximation for semi-simple groups over function fields, *Ann. Math.* **105**(1977), 553–572.

[Py1] L. Pyber, Enumerating finite groups of given order, *Ann. Math.* **137** (1993), 203–220.

[Py2] L. Pyber, Dixon-like theorems, in preparation.

[PS1] L. Pyber and A. Shalev, Groups with super exponential subgroup growth, preprint.

[PS2] L. Pyber and A. Shalev, Subgroup growth and finite permutation groups, preprint.

[Ra] M.S. Raghunathan, On the congruence subgroup problem, *Publ. Math. IHES* **46**(1976), 107–161.

[Rp] A.S. Rapinchuk, Congruence Subgroup Problem for algebraic groups: old and new, *Astérisque* **209**(1992), 73–84.

[Se] D. Segal, Subgroups of finite index in solvable groups I, in *Proc. Groups-St. Andrews 1985* (Campbell and Robertson (eds.), Cambridge Univ. Press, 1986), 307–314.

[SS] D. Segal and A. Shalev, Groups with fractionally exponential subgroup growth, *J. Pure Appl. Alg.* **88**(1993), 205–223.

[Sh1] A. Shalev, Growth functions, p-adic analytic groups and groups of finite co-class, *J. London Math. Soc.*, to appear.

[Sh2] A. Shalev, Subgroup growth and sieve methods, preprint.

[Sm1] G.C. Smith, *Zeta functions of torsion free finitely generated nilpotent groups*, Ph.D. Thesis, University of Manchester, 1983.

[Sm2] G.C. Smith, Compressibility in nilpotent groups, *Bull. London Math. Soc.* **17**(1985), 453–457.

[Sr] J.P. Serre, Le problème des groupes de congruences pour SL_2, *Ann. Math.* **92** (1970), 489–527.

[St1] W.W. Stothers, The number of subgroups of given index in the modular groups, *Proc. Roy. Soc. Edinburgh* **78**(1977), 105–112.

[St2] W.W. Stothers, Free subgroups of the free product of cyclic groups, *Math. Comp.* **32**(1978), 1274–1280.

[St3] W.W. Stothers, On a result of Peterson concerning the modular group, *Proc. Roy. Soc. Edinburgh* **87**(1981), 263–270.

[St4] W.W. Stothers, Subgroups of finite index in free product with amalgamated subgroup, *Math. Comp.* **36**(1981), 653–662.

[St5] W.W. Stothers, Level and index in the modular group, *Proc. Royal Soc. Edinburgh* **99**(1984), 115–126.

[We] T.S. Weigel, On the profinite completion of arithmetic groups of split type, preprint.

[We] B. Weisfeiler, Strong approximation for Zariski dense subgroups of semisimple algebraic grups, *Ann. Math.* **120**(1984), 271–315.

[Wi] H.S. Wilf, The asymptotics of $e^{p(z)}$ and the number of elements of each order in S_n, *Bull. Amer. Math. Soc.* **15**(1986), 228–232.

[Wn1] J.S. Wilson, Finite presentations of pro-p groups and discrete groups, *Invent. Math.* **105**(1991), 177–183.

[Wo] K. Wohlfahrt, Uber einen Satz von Dey und die Modulgruppe, *Arch. Math.* **29**(1977), 455–457.

[Wol] J.A. Wolf, Growth of finitely generated solvable groups and curvature of Riemannian manifolds, *J. Diff. Geom.* **2**(1968), 421–446.

THE QUANTUM DOUBLE OF A FINITE GROUP AND ITS ROLE IN CONFORMAL FIELD THEORY

GEOFFREY MASON[1]

Department of Mathematics, University of California Santa Cruz, Santa Cruz, California 95064, U.S.A.

1. Introduction

One can associate to a (finite) group G a certain Hopf algebra $D(G)$, called the *quantum double* of G [Dr]. The purpose of this paper is to bring this construction to the attention of group theorists, and to explain how it arises in conformal field theory: specifically, the theory of holomorphic orbifolds. In Section 4 we couch these ideas in terms of so-called elliptic modules which, as Charles Thomas has suggested [T], may be related to equivariant elliptic cohomology.

The reader may consult [Dr], [DV^3], [BA1], [BA2], [DPR], [L] for more information and background.

2. The algebra D(G)

Throughout, let G be a finite group and F a field. Let $F[G]$ denote the group algebra over F, and $F[G]^*$ the dual space. We may take a basis of $F[G]^*$ to be the Dirac delta functions $e(g), g \in G$, whose value is 1 at g and 0 at $h \in G, h \neq g$. Thus $F[G]^*$ is an algebra with

$$e(g)e(h) = \begin{cases} e(g), & g = h \\ 0, & g \neq h \end{cases} \tag{2.1}$$

$F[G]$ is endowed with its usual comultiplication

$$\Delta : g \mapsto g \otimes g \tag{2.2}$$

which turns $F[G]$ into a bialgebra (the main point being that $\Delta : F[G] \to F[G] \otimes F[G]$ is an algebra morphism), and of course $F[G]$ even becomes a Hopf algebra with the appropriate definitions - which we do not need.

Now $F[G]$ induces algebra morphisms on $F[G]^*$ via

$$g : e(h) \mapsto e(ghg^{-1}), g, h \in G \tag{2.3}$$

[1]Supported by the National Science Foundation.

and as such we can define the *smash product* of $F[G]$ and $F[G]^*$ as in, for example, Section 7.2 of [Sw]. Thus we have an associative algebra

$$D(G) = F[G] \otimes F[G]^* \tag{2.4}$$

with multiplication defined by

$$(a \otimes e(x)) \cdot (b \otimes e(y)) = ab \otimes e(x)e(aya^{-1}). \tag{2.5}$$

Remarks. One does not really need the paraphernalia preceding (2.5): it is easy to verify that with (2.1) and (2.5) $D(G)$ becomes an associative algebra. $D(G)$ is a particularly simple example of the *quantum double construction* of Drinfeld [Dr].

If F has characteristic zero, or more generally if its characteristic does not divide the order $|G|$ of G, then $D(G)$ is semi-simple. In general, let us set for a conjugacy class $K \subseteq G$,

$$D(K) = F < a \otimes e(x) \mid a \in G, x \in K >, \tag{2.6}$$

and for $x \in G$ set

$$D(x) = F < a \otimes e(x) \mid a \in G > . \tag{2.7}$$

Then we have decompositions

$$D(G) = \oplus_K D(K) \tag{2.8}$$
$$D(K) = \oplus_{x \in K} D(x) \tag{2.9}$$

where (2.8) is a direct sum of 2-sided ideals $D(K)$ and K ranges over the conjugacy classes of G; and (2.9) is a direct sum of right ideals. We further set

$$S(x) = F < a \otimes e(x) \mid a \in C_G(x) > \tag{2.10}$$
$$N(x) = F < a \otimes e(x) \mid a \in G \setminus C_G(x) > . \tag{2.11}$$

Then of course $D(x) = S(x) \oplus N(x)$ and $N(x)$ is a 2-sided nilpotent ideal of $D(x)$, indeed $N(x)D(x) = 0$. Note that $S(x) \cong F[C_G(x)]$ as algebras. One can prove:

Theorem 2.1. *Let $J(x) = J(S(x))D(x)$ (where for an associative algebra $A, J(A)$ is the Jacobson radical of A). Then*

$$J(D(K)) = \bigoplus_{x \in K} J(x).$$

In particular, if char F does not divide $|G|$ then $J(S(x)) = 0$ and so $D(K)$ is semi-simple.

We have stated Theorem 2.1 in a general form in anticipation of future applications, but unless stated otherwise we will from now on assume that the field F is the complex numbers.

Again fix a conjugacy class $K \subseteq G$, fix $x \in K$, and let $C_G(x)g_i, 1 \leq i \leq d$, be the distinct left cosets of $C_G(x)$ in G. Thus $K = \{g_i^{-1}xg_i, 1 \leq i \leq d\}$.

Theorem 2.2. *For a conjugacy class L of $C_G(x)$, set*

$$z(L) = \sum_{a \in L} \sum_{i=1}^{d} g_i^{-1} a g_i \otimes e(g_i^{-1} x g_i). \tag{2.12}$$

Then the elements $z(L)$, as L ranges over the classes of $C_G(x)$, are a basis for the center of $D(K)$. In fact, there is an algebra isomorphism $Z(F[C_G(x)]) \xrightarrow{\cong} Z(D(K))$ defined by $\sum_{a \in L} a \mapsto z(L)$.

Corollary 2.3. *Let $P(G) = \{(x,y) \in G \times G \mid xy = yx\}$ and let $G \setminus P(G)$ be the orbit space where G acts by simultaneous conjugation. Then $D(G)$ has precisely $|G \setminus P(G)|$ simple modules.*

Remark. More generally, if F is algebraically closed (say) of characteristic p then $D(G)$ has exactly $|G \setminus P'(G)|$ simple modules, where now $P'(G)$ is the subset of $P(G)$ in which the second element y is required to be a p'-element (i.e., of order coprime to p).

To construct the simple $D(K)$- (or $D(G)$-)modules we proceed as follows. Let M be a right $F[C_G(x)]$-module, which we can also regard as a right $S(x)$-module. As $D(x)$ is an $S(x) - D(K)$ bimodule then $M \otimes_{S(x)} D(x)$ is a right $D(K)$-module, and we have:

Theorem 2.4. *For $M \in \mathrm{Mod}-F[C_G(x)], N \in \mathrm{Mod}-D(K)$, the functions*

$$M \mapsto M \otimes_{S(x)} D(x)$$
$$N \mapsto N(1 \otimes e(x))$$

define an equivalence between the categories of right $C_G(x)$ and $D(K)$-modules. Thus, simple $C_G(x)$-modules are mapped to simple $D(K)$-modules, and conversely.

Remarks. (I) The same functor constructs the simple $D(K)$-modules from the simple $C_G(x)$-modules even if F has characteristic p.

(II) The group of units of $D(K)$ contains a canonical copy G^* of G defined via the map

$$g \mapsto g^* = \sum_{a \in K} g \otimes e(a). \tag{2.13}$$

Then G^* acts on $M \otimes_{S(x)} D(x)$ in the same way that G acts on $\mathrm{Ind}_{C_G(x)}^G(M) = \bigoplus_{i=1}^d M \otimes g_i$.

(III) One can realize Theorem 2.4 as a Morita equivalence.

The picture that emerges is that $D(G)$ is an algebra whose simple modules are precisely the induced modules $\mathrm{Ind}_{C_G(x)}^G(M)$ where M ranges over the simple $C_G(x)$-modules and x ranges over representatives of the conjugacy classes of G.

One can generalize all of this even further: take a left G-set X and in place of $F[G]^*$ take the algebra $F[X]$ of F-valued functions on X. In (2.5), the r.h.s. becomes $ab \otimes e(x)e(a \circ y)$ where $a \circ y$ is the left action of $a \in G$ on $y \in X$. The simple modules are now those induced from the simple modules of the isotropy subgroups of G in its action on X. There is also a version with F of characteristic p: details left to the vicarious reader.

One case of possible interest is that in which we take X to consist of all n-tuples of commuting elements $(x_1, \ldots, x_n), x_i \in G$, with G acting by conjugation simultaneously on the x_i. We may set $D_n(G) = F[G] \otimes F[X]$ in this case, so that our earlier $D(G)$ is now $D_1(G)$. Thus the simple $D_n(G)$-modules are obtained by inducing the simple modules of all centralizers $C_G(< x_1, \ldots, x_n >)$. We will not pursue this more general situation systematically.

3. $D(G)$ as Hopf algebra

It might be worth recalling why we need to be concerned with Hopf algebras. If A is an associative F-algebra and V, W are (right) FA-modules, then $V \otimes W$ is naturally an $A \otimes A$-module, but generally $V \otimes W$ is *not* an A-module in any natural way. However it becomes such if there is an algebra morphism $\Delta : A \to A \otimes A$, for then we can restrict $V \otimes W$ to the image of A under Δ. And the co-associativity of Δ ensures that as A-modules we have an isomorphism $U \otimes (V \otimes W) \cong (U \otimes V) \otimes W$ given by the natural identification.

In the usual representation theory of groups we have $A = F[G]$, and this is all taken for granted - one uses the co-multiplication (2.2) almost unconsciously, so familiar has it become. But there are other co-associative co-multiplications of interest, even on $F(G)$. For example

$$g \mapsto \sum_{\substack{h,k \in G \\ hk=g}} h \otimes k$$

is one such. It turns out that an analogue of this is relevant for $D(G)$. Namely, define $\Delta : D(G) \to D(G) \otimes D(G)$ via

$$\Delta : a \otimes e(x) \mapsto \sum_{\substack{h,k \in G \\ hk=x}} (a \otimes e(h)) \otimes (a \otimes e(k)). \tag{3.1}$$

For the sake of completeness we record also the antipode S and counit ϵ, though we make no use of them here:

$$S : a \otimes e(x) \mapsto a^{-1} \otimes e(a^{-1}x^{-1}a) \tag{3.2}$$

$$\epsilon : a \otimes e(x) \mapsto \begin{cases} 1, & x = 1 \\ 0, & x \neq 1 \end{cases} . \tag{3.3}$$

Thus $S : D(G) \to D(G), \epsilon : D(G) \to F$, and we leave it to the reader to confirm that these do indeed turn $D(G)$ into a Hopf algebra (cf. [Sw, Chapter IV]).

In any case, after these formalities, we have managed to define a (strictly) associative tensor product on the objects of Mod-$D(G)$. Thus one can form the corresponding Grothendieck ring which is generated by the simple $D(G)$-modules. Note that we have said nothing about the *commutivity* of this ring, and because of space limitations we shall not do so.

Group-theorists might be amused to see the interpretation of this tensor product into more classical language. Thus let K, L be conjugacy classes in G, let $x \in K, y \in L$, and let M, N be $C_G(x)$- and $C_G(y)$-modules, respectively. Let $g_1, g_2, \ldots, h_1, h_2, \ldots$ be left coset representatives for $C_G(x)$ and $C_G(y)$, respectively. Then as explained above, $\mathrm{Ind}_{C_G(x)}^G(M)$ and $\mathrm{Ind}_{C_G(y)}^G(N)$ correspond to $D(G)$-modules, and for $z \in G$ and P a simple $C_G(z)$-module we want to know the multiplicity of $\mathrm{Ind}_{C_G(z)}^G(P)$ in $M \hat{\otimes} N$ (the latter tensor product being that in the category of $D(G)$-modules). The answer runs as follows: consider the usual tensor product $\mathrm{Ind}_{C_G(x)}^G(M) \otimes \mathrm{Ind}_{C_G(y)}^G(N)$ of $F[G]$-modules. It contains the subspace

$$X(z) = \bigoplus_{*} (M \otimes g_i) \otimes (N \otimes h_j) \tag{3.4}$$

where in (3.4) the sum is over all indices i, j such that $(g_i^{-1} x g_i)(h_j^{-1} y h_j) = z$. Then $X(z)$ is in fact a $C_G(z)$-module, and the multiplicity of $\mathrm{Ind}_{C_G(z)}^G(P)$ is equal to the multiplicity of P in $X(z)$ in the usual sense.

4. Intermezzo: elliptic characters

Complex representation theory of G amounts to studying the spaces Hom $(G, GL(n, F))$ or, what amounts to the same thing, $\mathrm{Rep}(G, GL(n, F))$. Here, and below, $\mathrm{Rep}(A, B)$ is the space of homomorphisms of the group A into the group B modulo conjugation by elements of B. Thus it is the space of inequivalent representations of A into B.

Rather than mapping G into $GL(n, F)$ (or into any other group, for that matter) one might want to consider the space

$$\mathrm{Rep}(\pi, G) \tag{4.1}$$

of inequivalent representations of π into G for suitable groups π: a sort of co-representation theory if you will. There is a well-known geometrical meaning one can assign to this set: let $E\pi \to B\pi$ be the universal π-bundle. Then any representation of π into G corresponds to a principal G-bundle over $B\pi$ so that (4.1) is in bijection with classes of such G-bundles.

Similarly one has "class-functions", namely Map(Rep(π, G), F), which is perhaps better written as $\text{Map}_G(\text{Hom}(\pi, G), F)$. These then are F-valued functions on principal G-bundles over $B\pi$. But which (if any) of these functions are distinguished in the way that characters are in usual character theory? (This is the case when $\pi = \mathbf{Z}$. Put this way, it is by no means clear without historical hindsight what the distinguished elements of $\text{Map}_G(\text{Hom}(\mathbf{Z}, G), F)$ should be!)

Now 2-dimensional conformal field theory (CFT), at least that part of it concerned with G-orbifolds, suggests that if, in (4.1), we take $\pi = \pi_g$ to be the fundamental group of a compact Riemann surface of genus g, then an interesting theory should ensue. We shall limit ourselves to the first non-trivial case, namely $g = 1$. But note, however, the general result:

Theorem 4.1. (E. Verlinde [V])

$$|\text{Rep}(\pi_g, G)| = \sum_x \sum_{\chi_x} \left(\frac{|C_G(x)|}{\chi_x(1)} \right)^{2(g-1)}$$

where x runs over representatives of the conjugacy classes of G and χ_x over the irreducible characters of $C_G(x)$.

What is remarkable is not only the beautiful formula itself, but the fact that it was found not by a group-theorist but by a theoretical physicist interested in CFT.

Now the case of genus $g = 1$ has the great simplifying property that $\pi_g \cong \mathbf{Z} \oplus \mathbf{Z}$ is abelian. And the formula in Theorem 4.1 is easy in this case: it says precisely that $|\text{Rep}(\mathbf{Z} \oplus \mathbf{Z}, G)| = |G \setminus P(G)|$ in the notation of Cor. 2.3. Thus Cor. 2.3 can be restated as the assertion

$$|\text{Rep}(\mathbf{Z} \oplus \mathbf{Z}, G)| = \#\ \text{simple}\ D(G)\text{-modules}.$$

More generally, we also see from Section 3 that

$$|\text{Rep}(\mathbf{Z}^n, G)| = \#\ \text{simple}\ D_{n-1}(G)\text{-modules}.$$

One might call $\text{Map}_G(\text{Hom}(\mathbf{Z}^2, G), F)$ the space of elliptic class-functions on G. Then $D(G)$-modules have a right to be called "elliptic G-modules," and "elliptic characters" are the characters of elliptic G-modules.

This is all related to the work of Hopkins, Kuhn and Ravenel [HKR] on extraordinary cohomology theories, and in particular to elliptic cohomology, as has been emphasized to me by Charles Thomas (see also [T]). One expects elliptic cohomology to be intimately related to CFT (cf. [W]). In the succeeding sections we shall explain how $D(G)$ intervenes in the theory of G-orbifolds.

5. VOA's

In quantum theory, the quantum states correspond to the non-zero vectors (or 1-dimensional subspaces) of a Hilbert space, while observables correspond to certain operators on the Hilbert space. Very roughly, the notion of vertex operator algebra (VOA) is what emerges along these lines if one quantizes the loop space of a suitable manifold (say). From our point-of-view, though the geometry provides a useful heuristic, it is the rigorous and axiomatic approach of VOA's that will be most useful. As a general reference we suggest the book [FLM].

We will discuss the complicated definition of VOA only partially. One begins with an (infinite-dimensional) complex vector space V. For one thing, there should be a representation of the Virasoro algebra Vir on V (cf. [KR]). Thus there are operators $L_n (n \in \mathbf{Z})$ on V satisfying

$$[L_m, L_n] = (m-n)L_{m+n} + \frac{1}{12}(m^3 - m)\delta_{m+n,0}c \tag{5.1}$$

for a certain rational number c called the *central charge (or rank)* of V. We are really only concerned with the L_0 operator: it is required that the spectrum (i.e., set of eigenvalues) of L_0 consists of a set of integers which is bounded below, and each eigenspace has finite-dimension. Then we can introduce the so-called L_0-grading $V = \coprod_{n \geq n_0} V_n$ of V with $V_n = \{v \in V | L_0 v = nv\}$. Thus L_0 becomes a degree operator, in complete analogy with the degree operator familiar from representation theory of Kac-Moody Lie algebras [K].

Next we require a linear map $V \to (\mathrm{End} V)[[z, z^{-1}]]$ for z indeterminate. For $v \in V$, the image is denoted by $Y(v, z)$. One usually writes

$$Y(v, z) = \sum_{m \in \mathbf{Z}} v_m z^{-m-1}. \tag{5.2}$$

$Y(v, z)$ is a vertex operator, and each $v_m \in \mathrm{End} V$. We also require that there is a "Virasoro element" ω in V (in fact it will lie in V_2). Namely, the vertex operator corresponding to ω realizes Vir in the sense that

$$Y(\omega, z) = \sum_{m \in \mathbf{Z}} L_m z^{-m-2}. \tag{5.3}$$

The remaining axioms are mainly concerned with how the brackets of operators u_m, v_n behave. They are quite crucial, of course, but we will have to pass over them here.

A linear isomorphism f of V is called an *automorphism* in case it satisfies

$$fY(v, z)f^{-1} = Y(fv, z), v \in V \tag{5.4}$$
$$fw = w.$$

We will concern ourselves with a finite group G of automorphisms of V, together with the subspace V^G of G-invariants of V. In fact V^G is itself a VOA, and in some ways the pair $V \supseteq V^G$ is a bit like a pair of fields with relative Galois group G.

Now VOA's have modules, as we will explain, but the more general notion of *g-twisted module* is equally important. So choose $g \in G$ of order N. A *g-twisted V-module* is a complex vector space M together with a linear map $V \rightarrow (\mathrm{End}M)[[z^{1/N}, z^{-1/N}]]$. The image of $v \in V$ is now denoted

$$Y_g(v,z) = \sum_{m \in \frac{1}{N}\mathbf{Z}} v_m z^{-m-1}, v_m \in (\mathrm{End}M). \tag{5.5}$$

Axioms analogous to those for VOA are required, in particular the components of $Y_g(\omega, z)$ generate a representation of Vir on M in which L_0 again is a degree operator with eigenvalues in $\frac{1}{N}\mathbf{Z}$, bounded below, and with finite dimensional eigenspaces. Thus M gets the corresponding grading $M = \coprod_{m \in \frac{1}{N}\mathbf{Z}} M_n$.

If we take $g = 1$ above, but retain the fractional grading, then what obtains is precisely the definition of a *V-module*. In particular, V is itself a V-module. A V-module M is called *simple* (or irreducible) in case no proper subspace of M is invariant under all component operators $v_m, v \in V$. Likewise one can speak of simple g-twisted V-modules, completely reducible g-twisted V-modules, etc. For more details, see for example [DM].

We call the VOA V *holomorphic (or self-dual)* in case each V-module is completely reducible and if also V is itself the unique simple V-module. Thus the category of V-modules is completely straightforward, and in this regard V is rather like a complete matrix algebra over F.

Constructing VOA's is not easy! The reader should look at [FLM] to convince themselves of this. The famous Moonshine Module $V^\natural$ (loc. cit) is an example. One knows that $\mathrm{Aut}(V^\natural) = \mathbb{M}$ (Monster) and that $V^\natural$ is indeed holomorphic [D].

We are going to consider the following:

Conjecture 5.1. Let V be a holomorphic VOA and G a finite group of automorphisms of V. Then the fixed-point VOA V^G is (i) completely reducible, and (ii) has only finitely many (isomorphism classes of) simple modules.

The way in which the simple V^G-modules are expected to arise is related to $D(G)$, as we will now explain.

6. Ordinary orbifold models

Throughout this section V is a holomorphic VOA as defined in Section 5, and G a finite group of automorphisms of V. Thus V is the unique simple

V-module. To proceed one needs

Conjecture 6.1. If $g \in G$ then there is (up to isomorphism) a unique simple g-twisted V-module.

There are no general results asserting the existence of *any* simple g-twisted module, nor any asserting the uniqueness. For positive results in some special cases see [DM]. If Conjecture 6.1 holds we shall call the set $\{V_g, g \in G\}$ an *orbifold model* for G. Here, V_g denotes the unique simple g-twisted V-module.

Now for $g, h \in G$, there is a simple way to *twist* V_g by h in the more classical sense of the word, to obtain a simple $h^{-1}gh$-twisted V-module. (This is easy, but one cannot see it from the bare-bones definitions given in Section 5. See Section 2 of [DM] for details.) Formally, we may write $h \circ V_g$ for the h-twist of V_g. So assuming Conjecture 6.1 we have an isomorphism of twisted V-modules

$$h \circ V_g \cong V_{h^{-1}gh}. \tag{6.1}$$

More precisely, there is for each $h \in G$ a linear map (6.2)

$$\phi(g,h) : V_g \to V_{h^{-1}gh}. \tag{6.2}$$

which intertwines the corresponding Y maps. Regarding this as a right action, we have the compatibility relation

$$\phi(g, h_1h_2) = c_g(h_1,h_2)\phi(g,h_1)\phi(h_1^{-1}gh_1, h_2). \tag{6.3}$$

This follows in the usual way from Schur's lemma since V_g is simple as g-twisted module; of course $c_g(h_1,h_2) \in \mathbf{C}^*$. Define

$$c(h_1,h_2) = \sum_{g \in G} c_g(h_1,h_2)e(g). \tag{6.4}$$

Thus $c : G \times G \to U(F[G]^*)$, the group of units of $F[G]^*$. If we regard this group as a multiplicative right G-module via right conjugation, the associative law in G implies:

Lemma 6.1. $c \in C^2(G, U(F[G]^*))$.

Thus the uniqueness assertion of Conjecture 6.1 implies the existence of the canonical 2-cocycle c. We call the orbifold model $\{V_g\}$ *ordinary* in case c is a 2-coboundary.

One readily calculates that if in (6.3) we take h_1, h_2 to commute with g then the map

$$c_g : (h_1,h_2) \mapsto c_g(h_1,h_2) \tag{6.5}$$

is an element of $C^2(C_G(g), F^*)$, where $F^* = F - \{0\}$. This is not surprising for two reasons: first, in physics the spaces V_g are a priori *projective* vector spaces, so we should expect projective representations of $C_G(g)$ on V_g from (6.2). Secondly, we have $U(F[G]^*) = \prod_K U(F[K])$ where K ranges over the conjugacy classes of G and $F[K] = F$-valued functions on K. Since $F[K]$ is a transitive G-module (that is, G acts transitively on the basis K) then the Eckmann-Shapiro lemma yields

$$H^2(G, U(F[G]^*)) \simeq \prod_x H^2(C_G(x), F^*) \tag{6.6}$$

for a set of representatives $\{x\}$ of the conjugacy classes of G.

In particular, $\{V_g\}$ is ordinary if and only if, each of the elements (6.5) of $C^2(C_G(g), F^*)$ are 2-coboundaries.

It is interesting to ask if *any* element of $H^2(G, U(F[G]^*))$ can be realized as coming from some G-orbifold model. This is answered affirmatively in [DM] for any group G in the case that c is trivial. We also mention the papers [DW], [FQ] where related questions are answered in the context of topological quantum field theory.

For the remainder of this section we assume that $\{V_g\}$ is an ordinary orbifold model. So we may as well assume that each $C_G(g)$ acts on V_g by an *ordinary* (i.e., not properly projective) representation.

Because of the maps $\phi(g, h)$ (6.2), the spaces V_g and $V_{h^{-1}gh} (h \in G)$ are essentially identical as twisted V-modules. More precisely, it follows easily from the definitions that each V_g is an ordinary (i.e., untwisted) V^G-module, and that $\phi(g, h)$ furnishes an isomorphism of V^G-modules. To put these "conjugate" spaces on a level footing, for each conjugacy class K of G we set

$$V_K = \bigoplus_{g \in K} V_g. \tag{6.7}$$

If $g \in K$ has order N, let

$$V_g = \coprod_{n \in \frac{1}{N}\mathbf{Z}} (V_g)_n \tag{6.8}$$

be the L_0-grading of V_g as described in Section 5.

Now G fixes the Virasoro element ω (cf. (5.4)), which implies that G commutes with each of its component operators L_n. In particular as $C_G(g)$ acts on V_g it preserves each of the L_0-eigenspaces $(V_g)_n$, so we can think of (6.8) as a sequence of $C_G(g)$-modules. Furthermore, $\phi(g, h)$ preserves L_0-gradings and V_K inherits a grading

$$V_K = \coprod_{n \in \frac{1}{N}\mathbf{Z}} (V_K)_n, \; (V_K)_n = \bigoplus_{g \in K} (V_g)_n. \tag{6.9}$$

Our first point is that $\bigoplus_{g \in K} (V_g)_n$ may also be written as $\mathrm{Ind}_{C_G(g)}^G((V_g)_n)$ for any (fixed) $g \in K$. But this is just a $D(K)$-module as explained in Section

2. Thus the decomposition (6.9) of V_K is one of $D(K)$-(or $D(G)$-) modules

$$V_K = \bigoplus_{\chi} M(K,\chi) \otimes V(K,\chi) \tag{6.10}$$

where $M(K,\chi)$ ranges over the simple $D(K)$-modules and $V(K,\chi)$ is the corresponding V^G-module. This decomposition holds because of Theorem 2.4 and because it holds for each V_g individually.

Now the map $M(K,\chi) \mapsto V(K,\chi)$ induces a map

$$\text{Mod} - D(G) \to V^G\text{- Mod} \tag{6.11}$$

Finally we can state the refinement of Conjecture 5.1:

Conjecture 6.2. Let V be a holomorphic VOA and G a finite group of automorphisms of V. Then Conjecture 6.1 holds, and we assume that $\{V_g\}$ is an ordinary orbifold model.

1. The association $M(K,\chi) \mapsto V(K,\chi)$ induces an equivalence between the semi-simple categories Mod-$D(G)$ and V^G-Mod. Thus simple modules correspond in the two categories.
2. There is a notion of tensor product in the category V^G-Mod such that the equivalence (6.11) is one of tensor categories.

7. Concluding remarks

I. One can study part 1 of Conjecture 6.2 at a number of levels. For example it implies the following (none of them obvious):

(a) Each of the simple $D(K)$-modules $M(K,\chi)$ occurs in (6.10) with non-zero multiplicity.

(b) Each of the V^G-modules $V(K,\chi)$ is a simple module (and non-zero, according to (a)).

(c) The $V(K,\chi)$ exhaust the simple V^G-modules.

II. Part 2 of Conjecture 6.2 is very important. The very definition of the tensor product of V^G-modules remains elusive (cf. [LH], [KL], [G]). This tensor product is related to *fusion* (in the physical sense!) and the conjecture says that this aspect of the conformal field theory is captured in the Hopf algebra $D(G)$.

III. We would be amiss if we said nothing about the case in which the cocycle c of Lemma 6.1 is *not* a coboundary. In many ways this is the most interesting case.

Very briefly, one can modify $D(G)$ to obtain another associative algebra, which we might (provisionally) denote by $D_c(G)$. Multiplication is given by

$$(a \otimes e(x)) \cdot (b \otimes e(y)) = c_x(a,b)ab \otimes e(x)e(aya^{-1}).$$

This indeed defines a semisimple associative algebra and the theory of Section 2 goes through with only cosmetic changes.

The comultiplication on $D_c(G)$, however, is another story. One can only expect so-called quasi-coassociativity in general. We refer the reader to [DPR] and [BA2] to see the beautiful way in which a 3-cocycle $w \in C^3(G, \mathbf{C}^*)$ emerges. One obtains a "quasi" quantum group $D^w(G)$ (whose underlying algebra is $D_c(G)$) for which an analogue of Conjecture 6.2 is expected to be true.

References

[Ba1] P. Bantay, Orbifolds and Hopf algebra, *Phys. Lett. B* **245**(1990), 477–479.

[Ba2] P. Bantay, Orbifolds, Hopf algebras and the Moonshine, *Lett. Math. Phys.* **22**(1991), 187–194.

[D] C. Dong, Representations of the Moonshine Module vertex operator algebra, in *Proc. A.M.S. Conf. on Conformal Field Theory*, to appear.

[DM] C. Dong and G. Mason, Non-abelian orbifolds and the boson-fermion correspondence, *Comm. Math. Physics*, to appear.

[DW] R. Dijkgraaf and E. Witten, Topological gauge theories and group cohomology, *Comm. Math. Phys.* **129**(1990), 393–429.

[DPR] R. Dijkgraaf, V. Pasquier and P. Roche, Quasi-quantum groups related to orbifold models, preprint.

[Dr] V. Drinfeld, Quantum groups, in *Proc. I.C.M. at Berkeley*, (A.M.S., 1986).

[DV^3] R. Dijkgraaf, C. Vafa, E. Verlinde and H. Verlinde, The operator algebra of orbifold modules, *Comm. Math. Phys.* **123**(1989), 485–526.

[FLM] I. Frenkel, J. Lepowslcy and A. Meurman, Vertex Operator Algebras and the Monster (Academic Press, London, 1988).

[FQ] D. Freed and F. Quinn, Chern-Simons Theory with Finite Gauge Group, *Comm. Math. Physics*, to appear.

[G] M. Gaberdiel, ***Fusion in conformal field theory as the tensor product of the symmetry algebra*** (preprint, D.A.M.T.P., Cambridge, 1993).

[HKR] M. Hopkins, N. Kuhn and D. Ravenel, preprint.

[HL] Y.Z. Huang and J. Lepowsky, Toward a theory of tensor products for representations of a vertex operator algebra, preprint.

[K] V. Kac, *Infinite-dimensional Lie algebras (3rd ed.)* (Cambridge University Press, Cambridge, 1990).

[KL] D. Kazhdan and G. Lusztig, Affine Lie algebras and quantum groups, Research Notices, *Duke J. Math.* (1991), 21–29.

[KR] V. Kac and A. Raina, ***Highest weight representation of infinite-dimensional Lie algebras*** (Advanced Series in Math. Phys. Vol. 2, World Scientific, Singapore, 1984).

[L] G. Lusztig, Leading coefficients of character values of Hecke operators, in *Proc. Symp. Pure Math.* 47, Part 2 (A.M.S., 1987).

[Sw] M. Sweedler, *Hopf algebras* (W. A. Benjamin, New York, 1969).

[T] C. Thomas, *O.S.U. Lecture Notes on Elliptic Cohomology and Moonshine* (Spring 1993).

[V] E. Verlinde, *Nucl. Phys.* B 300 (1988), 360.

[W] E. Witten, *On the index of the Dirac operator in Loop Space* (Lect. Notes in Math. **1326**, Springer, Berlin, 1988).

CLOSURE PROPERTIES OF SUPERSOLUBLE FITTING CLASSES

MARTIN MENTH

Universität Würzburg, Mathematisches Institut, Am Hubland, D-97074 Würzburg, Germany

1. Introduction

In a series of articles Berger, Bryce and Cossey studied metanilpotent Fitting classes with additional closure properties ([1], [3], [4], [5]) and showed that a metanilpotent Fitting class which is either subgroup closed (S-closed) or quotient group closed (Q-closed) or saturated (E_ϕ-closed) or a Fischer class, is a formation.

Fitting classes which contain only supersoluble groups are metanilpotent. In this paper such Fitting classes, briefly named supersoluble Fitting classes, are investigated, and it is shown that they cannot be closed with respect to any of the closure operations mentioned above. Yet it is not known whether they can be subdirect product closed (R_0-closed), but there exists an example of a non-R_0-closed supersoluble Lockett class (Example 4.2).

In section 2 supersoluble Fitting classes are proved to be repellant (abstoßend) in the sense of Hauck ([11], p.318), that means: If $G \in \mathfrak{F}$, p a prime such that $G \wr C_p \in \mathfrak{F}$, then G is a p-group.

So perhaps we can summarize the results of this note in saying that supersoluble Fitting classes are incompatible with many of the usual methods for constructing new groups from given groups.

The notation is mainly standard and can be found for example in [8]. C_n denotes the cyclic group of order n, $\mathfrak{F}(G)$ the smallest Fitting class, $form(G)$ the smallest formation and $gf(G)$ the smallest saturated formation containing the group G.

Partially this article arises from the author's dissertation written at the University of Würzburg in 1992 under the direction of Prof. Dr. H. Heineken.

2. Supersoluble Fitting classes are repellant

Lemma 2.1. *Let G be a finite supersoluble group. The following are equivalent:*

(i) $\mathfrak{F}(G)$ *is supersoluble.*

(ii) $\mathfrak{F}(G) \subseteq gf(G)$.

PROOF. Trivially we only have to prove the direction from (i) to (ii). Denote by $e(p)$ the exponent of the abelian quotient group $G/Fit_p(G)$. Then $gf(G)$ is locally defined by $\mathfrak{F}_p = \emptyset$ if $p \nmid |G|$, and $\mathfrak{F}_p = \mathfrak{A}_{e(p)}$ if $p \mid |G|$, where $\mathfrak{A}_m$ is the class of all finite abelian groups of exponent dividing m. Since $\mathfrak{F}_p$ is subgroup-closed for all primes p, the formation $gf(G)$ itself is subgroup-closed. So it suffices to show that $N = N_1N_2 \in gf(G)$ whenever N is a normal product of groups $N_1, N_2 \in gf(G)$ and $N \in \mathfrak{F}(G)$. For an arbitrary prime divisor p of $|G|$, the quotient $N/Fit_p(N)$ is a normal product of $N_1Fit_p(N)/Fit_p(N)$ and $N_2Fit_p(N)/Fit_p(N)$. Since $N \in \mathfrak{F}(G)$ is supersoluble, $N/Fit_p(N)$ is abelian. Moreover $N_iFit_p(N)/Fit_p(N) \cong N_i/Fit_p(N_i) \in \mathfrak{A}_{e(p)}$, and thus $N/Fit_p(N) \in \mathfrak{F}_p$. □

Theorem 2.2. *Every supersoluble Fitting class is repellant.*

PROOF. First let G be any supersoluble group and $\mathfrak{F} = \mathfrak{F}(G)$ be the Fitting class generated by G. The saturated formation $gf(G)$ generated by G has the local definition $\mathfrak{F}_p = \emptyset$ if $p \nmid |G|$ and $\mathfrak{F}_p = form(G/Fit_p(G))$ if $p \mid |G|$.

Assume now $H \in \mathfrak{F}$, p prime, and $H \wr C_p \in \mathfrak{F}$. Since $\mathfrak{F}$ is supersoluble, we know from (2.1) that $H \wr C_p \in gf(G)$. Moreover $H \wr P \in \mathfrak{F} \subseteq gf(G)$ for all p-groups P due to [11], Theorem 5.7 and 5.8.

Suppose further that H is not a p-group. Then $|H|$ has a prime divisor q different from p. Since $H \wr C_p$ is supersoluble, H is nilpotent ([9]), hence $Fit_q(H) = H$. Now $P \in \mathfrak{F}_q$ for all p-groups P ([15], Theorem 2.b.ii), and the formation $\mathfrak{F}_q$ has infinitely many subformations, a contradiction to [2]. So H has to be a p-group.

To complete the proof we imagine any supersoluble Fitting class $\mathfrak{F}$, $H \in \mathfrak{F}$ and a prime p such that $H \wr C_p \in \mathfrak{F}$. The wreath product $G = H \wr C_p$ is supersoluble, and its base group, a direct product of copies of H, is contained in $\mathfrak{F}(G)$. So $H \in \mathfrak{F}(G)$, and the first step applied to $\mathfrak{F}(G)$ proves the assertion. □

3. Metabelian groups in supersoluble Fitting classes

Let us start with a lemma that can often be used in reductions:

Lemma 3.1. *Let $\mathfrak{F}$ be a Fitting class and $G = N \rtimes \langle s \rangle \in \mathfrak{F}$, where $N = N_1 \times \ldots \times N_k$, all N_i are $\langle s \rangle$-invariant and s induces on N_i the automorphism σ_i. Then for all $i \in \{1, \ldots, k\}$, the group $H_i = (N_0 \times N_i) \rtimes \langle s \rangle$ is also contained in $\mathfrak{F}$, where $N_0 \cong N_i$ and s induces on N_0 the automorphism σ_i^{-1}.*

PROOF. We can assume $i = 1$ and set $N^* = N_0 \times N_1 \times \ldots \times N_k$ and $\langle t \rangle \cong \langle s \rangle$. Further let $G^* = N^* \rtimes (\langle s \rangle \times \langle t \rangle)$ be a semidirect product where all factors N_i are invariant under $\langle s \rangle$ and $\langle t \rangle$, and the operation of s and t on these factors is as follows: s centralizes N_0 and operates on N_i in the same way as σ_i for $1 \leq i \leq k$; t centralizes N_1 and operates on N_i in the same way as σ_i for $2 \leq i \leq k$, and operates on N_0 as σ_1.

Since N_0 is normal in G, we have $N_0 \in \mathfrak{F}$. Therefore $\langle N^*, s \rangle \cong \langle N^*, t \rangle \cong N_0 \times G \in \mathfrak{F}$ and then $G^* \in \mathfrak{F}$. Hence the normal subgroup $H_1^* = \langle N^*, st^{-1} \rangle$ of G^* is also an $\mathfrak{F}$-group, and H_1 is isomorphic to the normal subgroup $\langle N_0 \times N_1, st^{-1} \rangle$ of H_1^*, so $H_1 \in \mathfrak{F}$. □

Notation for (3.2)–(3.4). Let p and q be different primes, e and r be natural numbers, q a divisor of $p^e - 1$, and n a primitive q-th unit root in $GF(p^e)$. Let X_r be a homocyclic group of exponent p^e and rank r, and finally $G_r = X_r \rtimes \langle s \rangle$ be a semidirect product of X_r and a cyclic group $\langle s \rangle$ of order q, where s raises all elements of X_r to the n-th power.

Lemma 3.2. *The Fitting class $\mathfrak{F}(G_r)$ generated by G_r contains all extensions of any homocyclic group X of exponent p^e by an automorphism α of X with $ord(\alpha) = q$ and $det(\alpha) = 1$.*

PROOF. The prime q is a divisor of $p^e - 1$, so $GF(p^e)$ contains a primitive q-th unit root n, and every irreducible representation of C_q over $GF(p^e)$ is linear. By Maschke's theorem, X possesses a basis $\{x_1, \ldots, x_k\}$ such that α can be written as a diagonal matrix $diag(n^{\lambda_1}, n^{\lambda_2}, \ldots, n^{\lambda_{k-1}}, n^{\lambda})$ where $\lambda = -\sum_{j=1}^{k-1} \lambda_j$.

Now let X_{k+r-1} be a homocyclic group of exponent p^e and rank $k+r-1$, and fix a basis of X_{k+r-1}. Then $\mathfrak{F}(G_r)$ contains all extensions of X_{k+r-1} by the following automorphisms:

$$\alpha_1 = diag(n^{\lambda_1}, 1, \ldots\ldots\ldots, 1, \underbrace{n^{\lambda_1}, \ldots, n^{\lambda_1}}_{r-1})$$

$$\alpha_2 = diag(1, n^{\lambda_2}, 1, \ldots\ldots, 1, \underbrace{n^{\lambda_2}, \ldots, n^{\lambda_2}}_{r-1})$$

$$\vdots$$

$$\alpha_{k-1} = diag(1, \ldots, 1, n^{\lambda_{k-1}}, 1, \underbrace{n^{\lambda_{k-1}}, \ldots, n^{\lambda_{k-1}}}_{r-1})$$

$$\alpha_k = diag(1, \ldots\ldots\ldots, 1, n^{\lambda}, \underbrace{n^{\lambda}, \ldots, n^{\lambda}}_{r-1})$$

and therefore contains the extension of X_{k+r-1} by

$$\prod_{j=1}^{k} \alpha_j = diag(n^{\lambda_1}, n^{\lambda_2}, \ldots, n^{\lambda_{k-1}}, n^{\lambda}, 1, \ldots, 1)$$

which has a normal subgroup isomorphic to $\langle X, \alpha \rangle$. □

Now we prove a result in the opposite direction: Each Fitting class that contains a non-direct extension of a homocyclic p-group by a q-group, contains also the group G_q.

Lemma 3.3. *Let α be any nontrivial automorphism of X_r, let the order of α be a power of q, and set $G = X_r \rtimes \langle \alpha \rangle$. Then $G_q \in \mathfrak{F}(G)$.*

PROOF. If $ord(\alpha) = q^m \neq 1$, then $ord(\alpha^{q^{m-1}}) = q$ and $\langle X, \alpha^{q^{m-1}} \rangle \in \mathfrak{F}(\langle X, \alpha \rangle)$, so we can suppose $ord(\alpha) = q$. As in lemma 3.2 we have a basis $\{x_1, \ldots, x_r\}$ of X_r such that $\alpha = diag(n^{\lambda_1}, \ldots, n^{\lambda_r})$. Since $\alpha \neq id$, not all the λ_i are equal to 0, let us say $\lambda_1 \neq 0$, and then without loss of generality $\lambda_1 = 1$. As a consequence of (3.1), the class $\mathfrak{F}(G)$ contains the extensions of X_q by the automorphisms

$$\begin{aligned}
\beta_1 &= diag(n, n^{-1}, 1, \ldots\ldots\ldots, 1) \\
\beta_2 &= diag(1, n^2, n^{-2}, 1, \ldots\ldots, 1) \\
&\vdots \\
\beta_{q-1} &= diag(1, \ldots\ldots, 1, n^{q-1}, n^1),
\end{aligned}$$

and so contains the extension of X_q by $\beta = \prod_{i=1}^{q-1} \beta_i = diag(n, \ldots, n)$. □

Using both (3.2) and (3.3), we obtain the non-supersolubility of Fitting classes that contain a non-direct extension of a homocyclic p-group by a q-group.

Lemma 3.4. *Let X be a homocyclic group of exponent p^e and let $G = X \rtimes Q$ be a non-direct, semidirect product of X and a q-group Q.*

a) If $q \geq 3$, then $C_{p^e} \wr C_q \in \mathfrak{F}(G)$.

b) If $q = 2$, then $\mathfrak{F}(G)$ contains the extension of X_4 by

$$\langle \alpha, \beta \rangle = \left\langle \begin{pmatrix} 0 & 0 & 1 & 0 \\ 0 & 0 & 0 & 1 \\ 1 & 0 & 0 & 0 \\ 0 & 1 & 0 & 0 \end{pmatrix}, \begin{pmatrix} -1 & & & \\ & -1 & & \\ & & 1 & \\ & & & 1 \end{pmatrix} \right\rangle.$$

c) In both cases (a) and (b) the Fitting class $\mathfrak{F}(G)$ is not supersoluble.

PROOF. There is an element $s \in Q$ that induces on X an automorphism σ with $ord(\sigma) = q$. Now $X \rtimes \langle \sigma \rangle \in \mathfrak{F}(G)$ and according to (3.3) and (3.2) the class $\mathfrak{F}(G)$ contains all extensions of a homocyclic group X of exponent p^e by $\alpha \in Aut(X)$ with $ord(\alpha) = q$ and $det(\alpha) = 1$.

a) If q is odd, then the automorphism induced on the base subgroup has determinant 1. From (3.2) we get $C_{p^e} \wr C_q \in \mathfrak{F}(G)$.

b) Since α and β are both of order 2 and have determinant 1, the groups $\langle X_4, \alpha\rangle$ and $\langle X_4, \beta\rangle$ are both in $\mathfrak{F}(G)$. Moreover $|\langle\alpha, \beta\rangle| = 8$ and therefore $\langle X_4, \alpha, \beta\rangle$ is a subnormal product of $\langle X_4, \alpha\rangle$ and $\langle X_4, \beta\rangle$. Hence $\langle X_4, \alpha, \beta\rangle \in \mathfrak{F}(G)$.

c) In case (a) the assertion is clear because supersoluble Fitting classes are repellant. The group constructed in (b) is not supersoluble, since eigenvectors of α have the form $(a, b, \lambda a, \lambda b)^T$ with $\lambda^2 = 1$ and $(a, b) \neq (0,0)$, but all of these are not eigenvectors of β. □

Theorem 3.5. *Let $\mathfrak{F}$ be a supersoluble Fitting class. Every metabelian group $G \in \mathfrak{F}$ is nilpotent.*

PROOF. There is an abelian normal subgroup N of G such that G/N is abelian. Let $Q_1N/N, \dots, Q_rN/N$ be the Sylow subgroups of G/N.

Suppose G to be nonnilpotent. Then not all the Q_i centralize N, let us say $[Q_1, N] \neq 1$. Now $\langle N, Q_1\rangle$ is nonnilpotent and contained in $\mathfrak{F}$. Furthermore Q_1 is a subnormal product of cyclic subgroups and at least one of them doesn't centralize N, so by means of reduction we can assume $G = \langle N, s\rangle$ where s is a q-element for a prime q.

Now N is a direct product of its Sylow subgroups $P_1, \dots, P_k$, and at least one of these for a prime $p \neq q$ is not centralized by s, let us say P. So s induces on P a nontrivial automorphism σ. According to (3.1) also $H = (P_0 \times P_1) \rtimes \langle s\rangle$ is contained in $\mathfrak{F}$, where $P_0 \cong P_1 \cong P$ and s induces on P_0 the automorphism σ and on P_1 the automorphism σ^{-1}. Furthermore we can restrict ourselves to the case where $ord(s) = q$.

Since $p \neq q$, by Maschke's theorem $P_0 \times P_1$ is a direct product of irreducible $\langle s\rangle$-invariant subgroups, and again using (3.1) we can suppose P_0 and P_1 to be irreducible C_q-modules. According to [10], (5.2.2), these are homocyclic. Set $p^e = exp(P_0)$.

Because $H \in \mathfrak{F}$ is supersoluble, the group $K = \langle P_0, s\rangle$ is also supersoluble. If $q > p$, then $O_q(K) \neq 1$, that is $O_q(K) = \langle s\rangle$ and $[P, s] = 1$, a contradiction. Therefore $q < p$, and K has a cyclic normal p-subgroup. Now P_0 and P_1 are cyclic, and that requires $q \mid p^e - 1$. So by (3.4) $\mathfrak{F}$ cannot be supersoluble, the final contradiction. □

4. Closure operators and supersoluble Fitting classes

Theorem 4.1. *Let $\mathfrak{F}$ be a supersoluble nonnilpotent Fitting class. Then $\mathfrak{F}$ is not closed with respect to any of the operators* Q, S, E$_\Phi$. *Moreover $\mathfrak{F}$ is not a Fischer class.*

PROOF. Starting with a supersoluble, non-nilpotent group $H \in \mathfrak{F}$ and using the Sylow tower of H and Lemma 3.1, it is easy to construct a non-nilpotent p-by-q-group $G \in \mathfrak{F}$ for suitably chosen different primes p and q. Set $N = Frat(P)$. Then N is normal in G, and G/N is a non-nilpotent, metabelian group, so $G/N \notin \mathfrak{F}$ due to (3.5). Hence $\mathfrak{F}$ cannot be Q-closed.

The other assertions follow immediately, because each saturated metanilpotent Fitting class and each metanilpotent Fischer class is a formation according to [5], Theorem 4.2 and Corollary 3.2. □

The R_0-closure cannot be excluded so easily, since R_0-closed metanilpotent Fitting classes need not be Q-closed. At least there exist non-R_0-closed supersoluble Fitting classes, for example a Fitting class presented in [13]. It is denoted by $\mathfrak{U}$ and defined as follows.

Let p be a prime, $p \equiv 1 \pmod 3$, and n be a primitive cube root of unity in $GF(p)$. Set

$$T = \langle a, b \mid a^p = b^p = [a,b,a,a] = [a,b,a,b] = [a,b,b,b] = 1 \rangle \text{ and}$$
$$U = \langle T, s \mid s^3 = 1, a^s = a^n, b^s = b^n \rangle$$

Then T is free with respect to the properties {nilpotent of class three, exponent p, two generators}, $|T| = p^5$ and $Z(U) = Z(T) \cong C_p \times C_p$.

G is called a *central product* of normal subgroups $G_1, \ldots, G_m$ if $G = G_1 \cdot \ldots \cdot G_m$ and $[G_i, G_j] = 1$ for $i \neq j$.

Now let $\mathfrak{U}_0$ be the class of all finite groups $G = XY$, where $X = O_p(G)$ and Y is a Sylow-3-subgroup of G, such that

(i) X is a central (posssibly empty) product of copies T_i of T, and

(ii) $Y/C_Y(T_i) \cong C_3$ and $T_iY/C_Y(T_i) \cong U$ for all indices i.

Denote by $O^p(G)$ the largest p-perfect normal subgroup of G.

Finally define $\mathfrak{U}$ to be the class of all groups $G \in \mathfrak{S}_p\mathfrak{S}_3$, where $O^p(G) \in \mathfrak{U}_0$. Then $\mathfrak{U}$ is the Fitting class generated by U, and $\mathfrak{U}$ is supersoluble according to [13], (4.2) and (4.3).

Example 4.2. $\mathfrak{U}$ is not R_0-closed.

PROOF. We consider a direct product $H = U \times U^\phi$ of two copies of U. The diagonal $D = \{(x, x^\phi) \mid x \in U\}$ of H is isomorphic to U. The subgroups $A = \{(x,1) \mid x \in T'\}$ and $B = \{(1, x^\phi) \mid x \in T'\}$ are normal in H, and $A \cap B = 1$. Therefore $G = \langle A, D \rangle$ is a semidirect product of A and D. Analogously $G = BD$ and $G/A \cong G/B \cong U \in \mathfrak{U}$.

G is p-perfect: $(s, s^\phi) \in O^p(G)$. Hence $[(a, a^\phi), (s, s^\phi)] = (a, a^\phi)^{n-1} \in O^p(G)$ and analogously $(b, b^\phi)^{n-1} \in O^p(G)$, so $D \leq O^p(G)$. A suitably chosen element t of $T' \setminus Z(T)$ is raised by s to the n^2-th power. Hence

$(t,1)^{n^2-1} = [(t,1),(s,s^\phi)] \in O^p(G)$, and $[(t,1),(a,a^\phi)], [(t,1),(b,b^\phi)] \in O^p(G)$, so $A \le O^p(G)$ and finally $G = O^p(G)$.

If we suppose $G \in \mathfrak{U}$, so $G \in \mathfrak{U}_0$, and $O_p(G)$ is a central product of copies T_i of T. The order of $O_p(G)$ is p^8, therefore we need exactly two copies T_1, T_2 with $|T_1 \cap T_2| = p^2$, and consequently $Z(T_1) = Z(T_2) = Z(O_p(G))$. On the other hand $Z(T) \times (Z(T))^\phi$ is a subgroup of $Z(O_p(G))$ and has order p^4, a contradiction. Therefore $G \notin \mathfrak{U}$. □

Since $\mathfrak{U}$ is a Lockett class ([14], (2.8)), it turns out to be a metanilpotent, non R_0-closed Lockett class. The following example shows that $\mathfrak{U}$ doesn't even fulfill a weaker version of the E_ϕ-closure:

Example 4.3. There is a group $G \in \mathfrak{S}_p\mathfrak{S}_3$ with the following property: $G \notin \mathfrak{U}$, but G has a normal subgroup $Z \le Z(G) \cap Frat(G)$ with $G/Z \in \mathfrak{U}$.

PROOF. Let $P = \langle a_1, b_1, a_2, b_2 \rangle$ be free with respect to the properties {four generators, exponent p, nilpotency class three}. The order of P is $p^{4+6+20} = p^{30}$ (cf. [13], section 5). Further let $\langle s \rangle$ be cyclic of order 3, and denote by H the semidirect product $P \rtimes \langle s \rangle$ with the operation $a_1^s = a_1^n$, $b_1^s = b_1^n$, $a_2^s = a_2^{n^2}$, $b_2^s = b_2^{n^2}$. Set

$$N = \langle [a_1, a_2, x], [a_1, b_2, x], [b_1, a_2, x], [b_1, b_2, x] \mid x \in \{a_1, a_2, b_1, b_2\} \rangle.$$

The triple commutators appearing here are elements of $Z(P)$ and raised to some power by s. So N is normal in H. Furthermore $|N| = p^{16}$, hence $|P/N| = p^{14}$.

Set $M = \langle N, [a_1, a_2], [a_1, b_2], [b_1, a_2], [b_1, b_2] \rangle$. We have $[a_1, a_2]^s \in [a_1, a_2]N$, and the analogous results are valid for the three other commutators in the generating system of N. Therefore M is normal in H. Now we define $G = H/N$ and $Z = M/N$. Then $Z \le Z(G)$ and $M \le P' \le Frat(P) \le Frat(H)$, so $Z = M/N \le Frat(H)/N = Frat(G)$.

Define $\overline{X} = O_p(H/M) = P/M$, $\overline{X_1} = \langle a_1M, b_1M \rangle$ and $\overline{X_2} = \langle a_2M, b_2M \rangle$. By the definition of M, the subgroups $\overline{X_1}$ and $\overline{X_2}$ are s-invariant normal subgroups of $\overline{X}$. Hence they are normal in H/M. The 3-element s raises the elements of $\overline{X_1}/\overline{X_1}'$ to the power of n and the elements of $\overline{X_2}/\overline{X_2}'$ to the power of n^2.

Furthermore $\overline{X_i}$ is nilpotent of class at most three, has exponent p and is generated by two elements. So it is isomorphic to a quotient group of T. On the other hand $\overline{X} = \overline{X_1} \cdot \overline{X_2}$ and $|\overline{X}| = p^{30-4-16} = p^{10}$, and therefore $|\overline{X_i}| = p^5$. So $\overline{X}$ is a direct product of the normal subgroups $\overline{X_i} \cong T$. Together we get $G/Z \cong H/M \in \mathfrak{U}$.

G is p-perfect: First of all H is p-perfect, because $O^p(H)$ contains the 3-element s and consequently all the elements a_1, b_1, a_2, b_2, mapped by s nontrivially. Therefore the epimorphic image G of H is p-perfect.

If we suppose $G \in \mathfrak{U}$, so $P/N = O_p(G)$ has to be a central product of normal subgroups $T_1, \ldots, T_m$ of G, all isomorphic to T. The centre of P/N has the order p^8 and $|O_p(G)/Z(O_p(G))| = p^6$. On the other hand $O_p(G)/Z(O_p(G))$ is a direct product of the groups $T_i Z(O_p(G))/Z(O_p(G))$, all having order p^3. Now $m = 2$ and then $p^{10} \geq |T_1 T_2| = |O_p(G)| = p^{14}$, a contradiction. □

This example gives a negative answer to [6], Question 8. There are earlier counterexamples G of nilpotency length three, for instance Lockett [12] or Cossey [7], Example 3.7, but no other metanilpotent counterexample is known to the author.

References

[1] Berger, T.R., Bryce, R.A. and Cossey, J., Quotient closed metanilpotent Fitting classes, *J. Austral. Math. Soc. Ser. A* **38**(1985), 157–163.

[2] Bryant, R.M., Bryce, R.A. and Hartley, B., The formation generated by a finite group, *Bull. Austral. Math. Soc.* **2**(1970), 347–357.

[3] Bryce, R.A., Subdirect product closed Fitting classes, *Bull. Austral. Math. Soc.* **33**(1986), 75–80.

[4] Bryce, R.A. and Cossey, J., Subdirect product closed Fitting classes, *Proc. Second Int. Conf. Groups, Canberra 1973* (Springer Lecture Notes **372**), 158–164.

[5] Bryce, R.A. and Cossey, J., Metanilpotent Fitting classes, *J. Austral. Math. Soc.* **17**(1974), 285–304.

[6] Cossey, J., Classes of finite soluble groups in *Proc. Second Int. Conf. Groups, Canberra 1973* (Springer Lecture Notes **372**), 226–237.

[7] Cossey, J., Products of Fitting classes, *Math. Z.* **141**(1975), 289–295.

[8] Doerk, K. and Hawkes, T., *Finite soluble groups* (de Gruyter, Berlin - New York, 1992).

[9] Durbin, J.R., Finite supersoluble wreath products, *Proc. Amer. Math. Soc.* **17**(1966), 215–218.

[10] Gorenstein, D., *Finite groups* (Harper & Row, New York - Evanston - London, 1968).

[11] Hauck, P., Fittingklassen und Kranzprodukte, *J. Algebra* **59**(1979), 313–329.

[12] Lockett, F.P., An example in the theory of Fitting classes, *Bull. London Math. Soc.* **5**(1973), 271–274.

[13] Menth, M., A Family of Fitting classes of supersoluble groups, submitted.

[14] Menth, M., Examples of supersoluble Lockett sections, *Bull. Austral. Math. Soc.* **49**(1994), 327–334.

[15] Parker, D.B., Wreath products and formations of groups, *Proc. Amer. Math. Soc.* **24**(1970), 404–408.

GROUPS ACTING ON LOCALLY FINITE GRAPHS – A SURVEY OF THE INFINITELY ENDED CASE

RÖGNVALDUR G. MÖLLER

Science Institute, University of Iceland, IS-107 Reykjavik, Iceland

Introduction

The study of infinite graphs has many aspects and various connections with other fields. There are the classical graph theoretic problems in infinite settings (see the survey by Thomassen [49]); there are special graph theoretical questions which have no direct analogues for finite graphs, such as questions about ends (see [7], [44] and the monograph [6]); Ramsey graph theory with its connections to set theory; the study of spectra of infinite graphs and random walks on infinite graphs (see the surveys [32] and [58]); the study of group actions on infinite graphs.

This survey is on the last subject, or rather on a small corner of the last subject. As is usual one concentrates on the case where the automorphism group acts transitively on the graph. The study of group actions can then be spilt up into three cases according to whether the graph under investigation has one, two or infinitely many ends. A graph has one end if there is always just one infinite component when finitely many vertices are removed from the graph. ("Component" will always mean a connected component in the graph theoretical sense.) The case of graphs with only one end is the hardest one, but in the special case of graphs with polynomial growth there are some very nice results (see [23]). The two ended case is the easiest one: roughly speaking these graphs all look like fat lines and one can say that they are very well understood (see [29] and [22]). Then there is the infinitely ended case, which is the one that this paper is all about.

Some of the motivation behind the study of graphs with infinitely many ends comes from group theory, in particular the Bass–Serre theory of group actions on trees and Stallings' Ends Theorem. In general these graphs can be said to resemble trees, and it is precisely that resemblance that one tries to extract and use when studying them. The aim of this paper is twofold: firstly to give a survey of known results and secondly to give an exposition of a new and powerful technique to capture the "treeness". It is hoped that the treatment is more or less self-contained. We will be concentrating on locally finite graphs but will also mention results for graphs with infinite valencies.

In Section 1 we define the ends of a graph and go over the basic properties concerning group actions. The new technique mentioned above is then discussed in the second section and in the third section we show how it can be used in practice by giving new proofs of some results of Woess [54] relating amenability to group actions on graphs. In Section 4 we look at the graph theoretical version of the group theoretical concept of accessibility. And, finally, in Section 5 we consider the effect of placing extra conditions either on the graph or on the action of the automorphism group.

1. Graphs and ends

We think of a graph X as a pair (VX, EX) where VX is the vertex set and EX is the set of edges. The graphs considered will be without multiple edges and loops. Unless otherwise stated our graphs are undirected. So the set of edges can be viewed as a set of two element subsets of VX. Our notation is fairly standard, but note that "$\subset$" is used to denote strict inclusion.

Let G be a group acting on a set Y. If $x \in Y$ then let G_x denote the *stabilizer of* x in G; that is, G_x is the subgroup of all elements in G that fix x. For a subset A in Y we define $G_{\{A\}}$ as the subgroup of G consisting of all elements $g \in G$ such that $gA = A$. We call the group $G_{\{A\}}$ the *setwise stabilizer of* A. The automorphism group of X is denoted by $\mathrm{Aut}(X)$, and we think of $\mathrm{Aut}(X)$ primarily as a permutation group on VX. If $\mathrm{Aut}(X)$ acts transitively on VX then the graph X is said to be *transitive*.

A graph is said to be *locally finite* if all its vertices have finite valency. Note that a connected locally finite graph has a countable vertex set. A *ray* (also called a *half-line*) in a graph X is a sequence $\{v_i\}_{i\in\mathbf{N}}$ of distinct vertices such that v_i is adjacent to v_{i+1} for all $i \in \mathbf{N}$. A *line* in X is a sequence $\{v_i\}_{i\in\mathbf{Z}}$ of distinct vertices such that v_i is adjacent to v_{i+1} for all $i \in \mathbf{Z}$. We say that a path is *simple* if all its vertices are distinct. Let $d(u, v)$ denote the minimum length of a path in X between the vertices u and v. If X is connected then d is a metric on VX.

For the rest of this section X will denote a locally finite connected graph.

1.1. Ends

There are various ways of defining the ends of a graph. The graph theoretic approach is to define the ends as equivalence classes of rays.

Definition 1. [18] Two rays R_1 and R_2 are said to be in the same *end* if there is a ray R_3 in X which contains infinitely many vertices from both R_1 and R_2.

This definition becomes very simple in the special case where X is a tree: then two rays are in the same end if and only if their intersection is a ray.

There are several ways to rephrase this definition. Clearly R_1 and R_2 are in the same end if and only if there are infinitely many disjoint paths connecting vertices in R_1 to vertices in R_2. Now it is easy to check that *being in the same end* is an equivalence relation. (The idea for the proof of transitivity is indicated on Figure 1.) The equivalence classes are called the *ends* of X and the set of ends is denoted by ΩX (in [5] and [33] the set of ends is denoted by $\mathcal{E}X$).

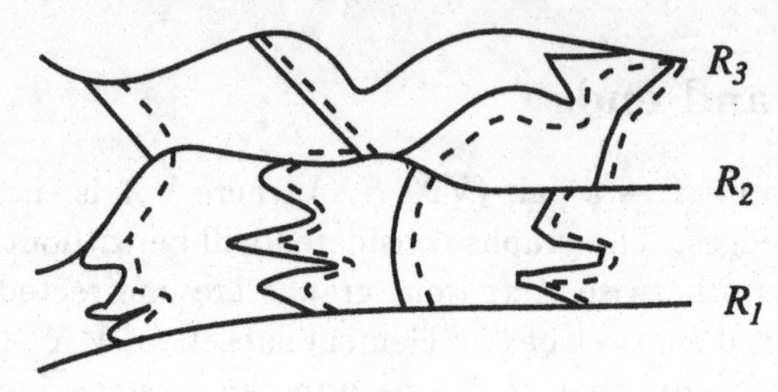

Figure 1.

Another way to rephrase the definition is to say that R_1 and R_2 are in the same end if and only if for every finite set $F \subseteq VX$ there is a path in $X \setminus F$ connecting a vertex in R_1 to a vertex in R_2. This in turn leads to yet another reformulation of the definition: two rays R_1 and R_2 are not in the same end if and only if one can find a finite set F of vertices and distinct components C_1 and C_2 of $X \setminus F$ such that C_1 contains infinitely many vertices of R_1 and C_2 contains infinitely many vertices of R_2. It is clear that a locally finite connected graph X has more than one end if and only if there is a finite set of vertices F such that $X \setminus F$ has more than one infinite component.

For a set $C \subseteq VX$ we define the *boundary* ∂C as the set of vertices in $VX \setminus C$ that are adjacent to a vertex in C. The *co-boundary* δC is defined as the set of edges that have one end vertex in C and the other one in $VX \setminus C$.

From the above definition of an end of a graph it is evident that if $C \subseteq VX$ with finite boundary and C contains infinitely many vertices from some ray R then C also contains infinitely many vertices from every ray in the same end as R. Thus it is reasonable to say that C contains the end that R is in. Let ΩC denote the set of ends that are contained in C. If $F \subseteq VX$ is finite and two ends ω and ω' are in different components of $X \setminus F$ then we say that F *separates* the ends ω and ω'.

In general one can say that this definition of the set of ends suggests that the ends describe how the graph "branches". Each end somehow represents one way of going to infinity.

Example. The infinite grid $\mathbf{Z} \times \mathbf{Z}$ does not "branch" at all and has only one end. (It is easy to find a ray in $\mathbf{Z} \times \mathbf{Z}$ that contains all the vertices. Every other ray must be in the same end as that ray.) The infinite line $\mathbf{Z}$ has two ends. On the other hand, the 3-valent regular tree T_3 clearly has a lot of "branching". It is one of the special properties of trees that given a vertex v and an end ω there is precisely one ray in ω that starts at v. Hence T_3 has $2^{\aleph_0}$ ends.

Ends come in various shapes and sizes. The main distinction is between *thick* and *thin* ends: an end is said to be thick if it contains infinitely many disjoint rays, and thin otherwise. The end of $\mathbf{Z} \times \mathbf{Z}$ is thick but the ends of $\mathbf{Z}$ and T_3 are all thin. For a thin end ω we define the *thickness* or *size* of ω, denoted by $m_1(\omega)$, as the maximum number of disjoint rays contained in ω. Halin [19] proved that if ω is thin then $m_1(\omega)$ is finite.

We say that a sequence $\{C_i\}_{i\in\mathbf{N}}$ with $\bigcap_{i\in\mathbf{N}} C_i = \emptyset$ *converges to an end* ω if all the sets in the sequence are connected and have finite boundary, and $\omega \in \Omega C_i$ for all i. The size of ω is the same as the lowest number m such that there is a sequence $\{C_i\}_{i\in\mathbf{N}}$ converging to ω such that $|\partial C_i| \leq m$ for all i.

One can also think of the ends as a boundary of the graph. This becomes clearer if we give a topological definition. This definition can be traced back to papers of Hopf and Freudenthal in the thirties and fourties (e.g. [15]).

Let $\mathcal{F}$ denote the set of all finite subsets of VX. For $F \in \mathcal{F}$ define $\mathcal{C}_F$ as the set of all infinite components of $X \setminus F$. If F_1 and F_2 are two elements of $\mathcal{F}$ such that $F_1 \subseteq F_2$ then there is a natural projection $\mathcal{C}_{F_2} \to \mathcal{C}_{F_1}$: a component of $X \setminus F_2$ being mapped to the component of $X \setminus F_1$ that contains it. So we have an inverse system in our hands. Let Ω denote its inverse limit. Now we want to identify Ω and ΩX. An element of Ω can be represented as a family $(C_F)_{F\in\mathcal{F}}$ such that if $F_1 \subseteq F_2$ then $C_{F_2} \subseteq C_{F_1}$. Given an end $\omega \in \Omega X$ it is easy to find the corresponding element in Ω: for $F \in \mathcal{F}$ we just let C_F denote the component of $X \setminus F$ that ω belongs to and then $(C_F)_{F\in\mathcal{F}}$ does the trick. Clearly the element constructed is the only element in Ω such that each of its components contains ω.

The next step is to show how we find the end corresponding to an element ω in Ω. Let $(C_F)_{F\in\mathcal{F}}$ be an element in Ω. Take a strictly increasing sequence $F_1 \subset F_2 \subset \ldots$ of finite subsets of VX such that $VX = \bigcup_{i\in\mathbf{N}} F_i$. Then $\{C_{F_i}\}_{i\in\mathbf{N}}$ is an decreasing sequence. First of all it is clear that any two ends in X are separated by some set F_i. However, one can find a ray that includes at least one vertex from ∂C_{F_i} for all $i \in \mathbf{N}$. Then there is precisely one end ω that belongs to all of the sets C_{F_i}. For any $F \in \mathcal{F}$ we can find $i \in \mathbf{N}$ such that $F \subseteq F_i$. Then $C_{F_i} \subseteq C_F$ and therefore ω is in ΩC_F.

The inverse limit construction gives a topology on ΩX. A basis of open sets for this topology is given by sets ΩC where $C \subseteq VX$ and C has finite boundary. It is easy to see that ΩX with this topology is compact. Indeed, if one puts the discrete topology on VX then one can view $VX \cup \Omega X$ as a compactification of VX. We will have occasion later on in this paper to make use of the topology introduced above, but the inverse limit construction will not be needed.

Remark. The assumption of local finiteness is not essential; ends can be defined in exactly the same manner for non-locally finite graphs. There is one important difference: the space of ends with the topology given above will not, in general, be compact if the graph is not locally finite.

1.2. Ends and automorphisms

It is clear from the definition of an end that an automorphism of X has a natural action on ΩX. As shown by Halin in his fundamental paper [20] the action on the ends gives vital clues to how the automorphism acts. The same is also evident from Tits' paper [51], where group actions on infinite trees are studied. Halin shows how automorphisms of X can be split up into three disjoint classes. Let $g \in \mathrm{Aut}(X)$. Then one of the following holds:

(i) g leaves invariant some non-empty finite subset of VX;

(ii) g fixes precisely one thick end and does not satisfy (i);

(iii) g fixes precisely two thin ends and does not satisfy (i).

Automorphisms that satisfy (ii) or (iii) are often collectively known as *translations.* For a translation g it is possible to find a line in X and some power of g that acts like a translation on the line. If X is a tree then g will act like a translation on the line. Those automorphisms that satisfy (i) are called *elliptic*, those that satisfy (ii) are called *parabolic* and those that satisfy (iii) are called *hyperbolic* (or *proper translations*). The above classification resembles the classification of automorphisms of hyperbolic space.

It is indeed quite easy to describe how one is to find an invariant line in cases (ii) and (iii). Suppose that g does not satisfy (i). Put $n = \min d(g^k v, v)$, where k ranges over $\mathbf{Z} \setminus \{0\}$ and v ranges over VX. Find $k_0 \in \mathbf{N}$ and $v_0 \in VX$ such that $n = d(g^{k_0} v_0, v_0)$, then take a path P of length n between v_0 and $g^{k_0} v_0$ and set $L = \bigcup_{i \in \mathbf{Z}} g^{ik_0} P$. It is left to the reader to show that L is a line and that g^{k_0} acts like a translation on L, (see [20, Theorem 7]).

Now one can ask about the existence of translations in $\mathrm{Aut}(X)$. The following theorem, which is a strengthened version for locally finite graphs of Theorem 1 in [25], gives the answer. First let us identify a simple fundamental property of a group acting on a graph. Let $G \leq \mathrm{Aut}(X)$. We say that G *shuffles* X if for every infinite set $C \subseteq VX$ with finite boundary and every finite set $F \subseteq VX$ there is an automorphism $g \in G$ such that $gF \subseteq C$. If X is transitive and locally finite then it is easy to see that G shuffles X. We take a vertex v in C such that the distance from v to ∂C is greater than the diameter of F as a subset of the metric space (VX, d). Then just find an automorphism $g \in G$ that maps some vertex of F to v. It is clear that $gF \subseteq C$.

Theorem 1. *Let X be a connected locally finite transitive graph. Suppose C is an infinite subset of VX with infinite complement and finite boundary. Then there is an element $g \in \mathrm{Aut}(X)$ such that $gC \subset C$ and g is of type* (iii).

PROOF. Set $G = \mathrm{Aut}(X)$. (In fact our argument works for any subgroup G of $\mathrm{Aut}(X)$ that shuffles X.) Let F be a finite connected set of vertices containing ∂C and put $C' = VX \setminus (C \cup F)$. Because G shuffles X we can find $h \in G$ such that h maps F into C. If we are lucky then we can take $g = h$, but let us suppose that this does not work. Now find h' such that $h'F \subseteq C'$. We can now ask if $h'C' \subseteq C'$ because then h'^{-1} would work, but again let us suppose not. Then $hC \not\subseteq C$ and $h'C' \not\subseteq C'$. Suppose that $F \cap hC = \emptyset$. Then, as $h(F \cup C)$ is connected, it follows that $h(F \cup C) \subseteq C$, contrary to our assumptions. (Figure 2 gives a schematic view of how things must lie.)

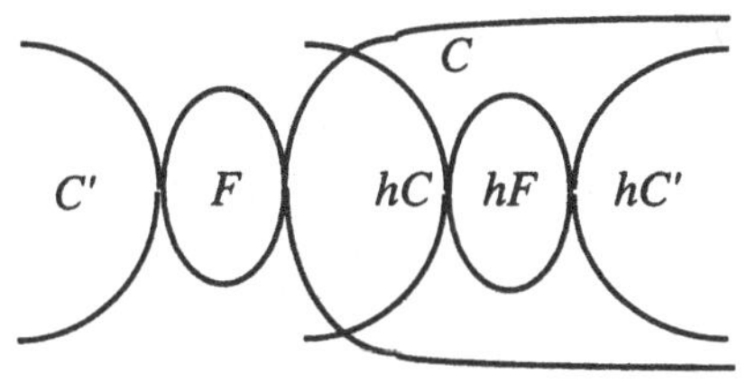

Figure 2.

We can apply the same argument with h' and C' replacing h and C. Whence we see that $F \subseteq hC \cap h'C'$. Now $F \cup C'$ is a connected subset of $X \setminus hF$ and $F \cup C'$ meets hC so $F \cup C' \subseteq hC$. Similarly $F \cup C \subseteq h'C'$. Then $h^{-1}(F \cup C) \subseteq C'$ and $h^{-1}C' \subseteq C$. Now we see that $g = h^{-1}h'^{-1}$ is as desired. It is clear that g cannot satisfy (i). For the details of how to prove the last statement in the Theorem we refer to [8, Proposition 2]. □

From this result the following corollary can be deduced.

Corollary 1. *An infinite connected locally finite transitive graph has either* $1, 2$ *or* $2^{\aleph_0}$ *ends.*

It is natural to ask if the action of $\mathrm{Aut}(X)$ on ΩX is faithful and, if not then try to identify the kernel. An automorphism g of a graph X is said to be *bounded* if there is a natural number n such that $d(v, gv) \leq n$ for all $v \in VX$. The bounded automorphisms form a subgroup of $\mathrm{Aut}(X)$, which we denote by $B(X)$.

Theorem 2. *Let X be a connected locally finite transitive graph with infinitely many ends. Then*

(i) ([16, Theorem 5]) *$B(X)$ is a locally finite group (every finitely generated subgroup is finite);*

(ii) ([33, Theorem 6]) *$B(X)$ is the kernel of the action of* $\mathrm{Aut}(X)$ *on* ΩX.

Remark. The situation as regards automorphisms of graphs that are not locally finite is similar. Often the proofs are more delicate and there are also some subtle variations from what is valid in the locally finite case. For detailed treatment see [26].

Transitive graphs of infinite valency with infinitely many ends resemble locally finite graphs strongly. This is because if X is such a graph and G acts transitively on X then G shuffles X (see [8, Theorem 3]). For example Theorem 1 and Theorem 2 remain valid for connected transitive graphs with infinitely many ends without assuming local finiteness (see [8, Theorem 4 and Theorem 6]). Corollary 1 also has an analogue: an infinite connected transitive graph of infinite valency has either 1 or at least $2^{\aleph_0}$ ends (see [8, Corollary 4 and Theorem 7]).

1.3. Ends of groups

Let G be a finitely generated group. The number of ends of G is defined as the number of ends of the Cayley graph of G with respect to some finite generating set of G (it does not depend on the choice of a generating set). Group theorists like to have an algebraic way of expressing the number of ends of a group. It was proved by Specker [47] that for a finitely generated infinite group G with finitely many ends the number of ends equals $1 + \dim H^1(G, \mathbf{Z}_2 G)$ and $\dim H^1(G, \mathbf{Z}_2 G)$ is infinite if the group has infinitely many ends (for a proof see [9]).

The structure of groups with 2 or infinitely many ends is described in the following theorems.

Theorem 3. *Let G be a finitely generated group. Then the following are equivalent:*

(i) G has precisely two ends;

(ii) G has an infinite cyclic subgroup of finite index;

(iii) G has a finite normal subgroup N such that G/N is either isomorphic to the infinite cyclic group or to the infinite dihedral group.

Next we state an extension of Stallings' Ends Theorem. This extension can be proved by combining the results in the next section with the Bass–Serre theory of group actions on trees.

Theorem 4 ([48]) *Suppose G is a finitely generated group with infinitely many ends. Then G can be written as a non-trivial free product with amalgamation $B *_C D$ where C is finite, or G can be written as a non-trivial HNN-extension $B *_C x$ where C is finite.*

2. Structure trees

As mentioned above, it is the "treeness" that one tries to extract and use when studying graphs with infinitely many ends. The "structure tree" approach has its roots in the proof of Stallings' Ends Theorem [48] (see also [4]).

The plot is to find a family of subsets of VX that is invariant under the action of $\mathrm{Aut}(X)$ and then represent them as the edge set of a tree, which $\mathrm{Aut}(X)$ acts on. First we discuss the properties of that family of sets, then how to construct the tree and the connections between the tree and the original graph. In the last part of this section we look at some simple examples of structure trees.

From now on suppose that X is a connected graph.

2.1. Tree sets

We define $\mathcal{B}X$ as the Boolean ring of all subsets of VX that have finite co-boundary. The elements of $\mathcal{B}X$ will be called *cuts*. For $n \in \mathbf{N}$ define $\mathcal{B}_n X$ as the subring of $\mathcal{B}X$ generated by those $C \subseteq VX$ with $|\delta C| \leq n$. If $C \subseteq VX$ then set $C^* = VX \setminus C$. A cut C is said to be *tight* if both C and C^* are connected.

Definition 2. We say that $E \subseteq \mathcal{B}X$ is a *tree set* if

(i) for all $e, f \in E$ we have that one of

$$e \cap f,\ e \cap f^*,\ e^* \cap f,\ e^* \cap f^*$$

is empty, (i.e. $e \subseteq f, e \subseteq f^*, e^* \subseteq f$ or $e^* \subseteq f^*$);

(ii) for all $e, f \in E$ there are only finitely many sets $g \in E$ such that $e \subset g \subset f$;

(iii) neither $\emptyset$ nor VX is in E.

If in addition the following holds then we say that E is an *undirected* tree set

(iv) if $e \in E$ then $e^* \in E$.

We will only be interested in undirected tree sets that consist of tight cuts C all of which with $|\delta C| \leq n$, for some natural number n. Let us call such a tree set *tight*.

Remark. Of course the definition of tree sets need not be restricted to subsets of $\mathcal{B}X$, but those are the only cases that we are interested in. Note also that if E is a tree set then we can always add to E the complements of the sets in E and the resulting set will also be a tree set. Thus it is not restrictive to consider only undirected tree sets.

Examples. Let X be a locally finite graph. The set of all one-vertex subsets of VX and their complements is an undirected tree set. Another obvious example of a tree set is the set of all subsets of VX that have a co-boundary consisting of only one edge. In both these examples the tree sets are invariant under $\mathrm{Aut}(X)$.

In the next section it will be convenient to think of a tree set as a partially ordered set, the ordering being given by inclusion. The *-operation is then an order reversing involution.

The difficult bit is to prove the existence of " nice" tree sets. The main result of Chapter II in [5] gives us almost all the tree sets that one could hope for and ties them nicely up with the action of the automorphisms group and the separating properties of the graph. The proof is long and technical, and will not be discussed here.

Theorem 5. ([5, Theorem II.2.20]) *Let X be a connected infinite graph and let $G \leq \mathrm{Aut}(X)$. There is a chain of G-invariant undirected tree sets $E_1 \subseteq E_2 \subseteq \ldots$ in $\mathcal{B}X$ such that all elements in E_n are tight and E_n generates $\mathcal{B}_nX$ for all n.*

Suppose that some two vertices (edges, ends) in X have the property that if an element in E_n contains one then it must contain both. Then clearly any element of the boolean ring $\mathcal{B}_nX$, which is generated by E_n, that contains one of the vertices must also contain the other. If for some two vertices (edges, ends) there is an element in $\mathcal{B}_nX$ that contains one and not the other, then there is an element in E_n that contains only one of the two vertices (edges, ends).

In applications it is often enough to find an infinite tight cut e with infinite complement such that $Ge \cup Ge^*$ is a tree set. Such a cut is called a D-*cut.* The existence of D-cuts follows from Theorem 1.1 in [11]. To end this section we state the following lemma which will come in very handy.

Lemma 1. ([50, Proposition 4.1] and [5, Lemma II.2.5]) *Let n be a natural number. For any given edge e in X there are only finitely many tight cuts C with $|\delta C| = n$ such that $e \in \delta C$. Now let E be a tight tree set, that is E consists of tight cuts, all of which have co-boundary with n or fewer edges. If $u, v \in VX$ then every descending chain in $\{e \in E \mid u \in e\}$ is finite and every chain in $\{e \in E \mid u \in e, v \notin e\}$ is finite.*

2.2. Trees, ends and automorphisms

In this section E will always denote a tight undirected tree set. For some of the things discussed here these assumptions are unnecessarily restrictive, but they are necessary to get the properties that are useful in applications to graph theory. (Note that in [33, §3] one needs to add the assumption that the tree sets discussed are tight.)

Let T be a tree with directed edges that come in pairs e, e^* with opposite directions. Thinking about T as a directed graph is purely a formal device that eases the presentation. Define a partial ordering on the edges of T such that $e \geq f$ if and only if there is a directed edge path $e = e_0, e_1, \ldots, e_n = f$ in T such that $e_i^* \neq e_{i+1}$ for all i. The *-operation acts like an order reversing involution on ET. It is easy to check that ET with this ordering and the *-operation satisfies the conditions in the definition of a tree set if "$\subseteq$" is replaced with "$\leq$".

What is wanted now is a tree T such that ET and our tree set E, ordered by inclusion, can be identified and that this identification is an order isomorphism that commutes with the *-operation.

Let Y be a directed graph such that each component of Y is isomorphic to the graph

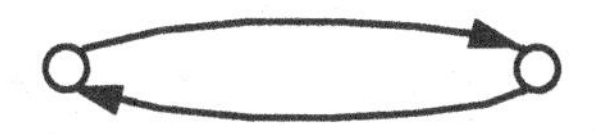

and such that there is an identification $EY \leftrightarrow E$ such that if $e \in E$ is identified with (u, v) then e^* is identified with (v, u). From now on we will not distinguish between elements of EY and E. We want to glue the components of Y together so that the resulting graph is the tree we are after and so that the ordering by inclusion of E is the same as the edge path ordering of the edge set of that tree. Let $e = (u, v)$ and $f = (x, y)$ be edges in Y (having a dual existence as elements in E). Then we identify v and x, and write $v \sim x$, if and only if $v = x$ or $f \subseteq e$ and there is no element $g \in E$ such that $f \subset g \subset e$. Let us write $f \ll e$ if $f \subset e$ and there is no $g \in E$ such that $f \subset g \subset e$. The relation $\sim$ is clearly reflexive and symmetric. We want it to be an equivalence relation so we have to prove that it is transitive. The proof is copied from [10, Theorem 2.1].

Let $e = (u, v), f = (x, y)$ and $g = (w, z)$. Suppose that $v \sim x$ and $x \sim z$. We want to show that $v \sim z$, that is $e^* \ll g$. If $e = g, e = f^*$ or $g = f^*$, then there is nothing to show, so let us assume that $f \ll e$ and $f \ll g$. We know that one of the following holds $e \subseteq g, e \subseteq g^*, e^* \subseteq g, e^* \subseteq g^*$. It is also known that $f \subset e$ and $f \subset g$, so $e \subseteq g^*$ is impossible. If $e \subseteq g$ then $f \subseteq e \subseteq g$ and because $f \ll g$ we must have either $f = e$ or $e = g$ and neither is allowed. Now suppose that $e^* \subseteq g^*$. Then $g \subseteq e$, and as in the case $e \subseteq g$ we get that $g = e$ or $g = f$, which is a contradiction. Finally we have to deal with the case $e^* \subseteq g$. We have to show that $e^* \ll g$, and then we have $v \sim z$. Suppose $e^* \subseteq h \subseteq g$. Again there are four cases. First if $f \subseteq h$ then $f \subseteq h \subseteq g$, so by assumption $f = h$ or $g = h$. If $f = h$ then we have $e^* \subseteq f \subseteq e$ which is clearly out of the question because $e \neq VX$. Then there is the possibility that $f \subseteq h^*$. Whence $f \subseteq h^* \subseteq e$ so $f = h^*$ or $e = h^*$. Now $f = h^*$ would imply that $f \subseteq g$ and $f^* \subseteq g$, which is impossible. So we must have the latter possibility, that is $e^* = h$. If $f^* \subseteq h$ then again both f and f^* are contained in g. Finally there is the possibility $f^* \subseteq h^*$. But then $h \subseteq f \subseteq e$ so $e^* \subseteq h \subseteq e$ which is impossible. Thus we must have $e^* \ll g$.

It is also clear that the ordering of E by inclusion is the same as the edge path ordering of E when E is considered as the edge set of T.

This process leaves us with a graph which we call $T = T(E)$. Now forget all about the graph Y and identify E and ET (from the construction it is obvious that $ET = EY$). Suppose that u and v are distinct vertices in T. Because one of $e \subseteq f, e \subseteq f^*, e^* \subseteq f$ or $e^* \subseteq f^*$ holds for all $e, f \in E$ we can find two edges e, f with $f \subseteq e$ such that v is an end vertex of e and u is an end vertex of f. By condition (ii) in the definition of a tree set there is a finite chain in E such that $e = e_1 \gg e_2 \gg \ldots \gg e_n = f$. This chain defines a directed path in T and both u and v lie on that path. So T is connected.

A simple cycle in T of length greater than 2 would give us a directed edge cycle $e_1, \ldots, e_n = e_1$. That would imply $e_1 \gg e_2 \gg \ldots \gg e_n = e_1$, which is impossible. So the undirected graph corresponding to T has no cycles of length greater than 2. Thus it is reasonable to call T a tree.

If E is invariant under $G \leq \mathrm{Aut}(X)$ then G has a natural action as a group of automorphisms of T.

Definition 3. If E is a tight undirected $\mathrm{Aut}(X)$-invariant tree set then we call $T = T(E)$ a *structure tree* of X.

To relate X and T more closely we have two maps $\phi : VX \to VT$ and $\Phi : \Omega X \to VT \cup \Omega T$. When T is a structure tree, the action of $\mathrm{Aut}(X)$ commutes with both ϕ and Φ.

The fundamental principle is that the edges in T should point towards what they contain, that is if $v \in e$ then e (as an edge in T) should point towards $\phi(v)$. So, if $v_0 \in VX$ and $e = (x, y)$ is an minimal element in E subject to containing v_0 then set $\phi(v_0) = y$. Of course one has to show that the choice of the minimal element e does not matter. Suppose that $f = (z, w)$ is another such element. We need to show that $e^* \ll f$, which implies $y = w$. Now recall the possibilities in condition (i) in the definition of a tree set. The condition that v_0 is in both e and f and that they are both minimal subject to this condition gives us immediately that $e^* \subseteq f$. Suppose $e^* \subset g \subset f$. By the minimality of f one sees that $v_0 \notin g$. But $g^* \subset e$, so $v_0 \notin g^*$. Here is a contradiction and therefore $e^* \ll f$. We also want to be sure that if f is some element in E containing v_0 that then f points towards $\phi(v_0) = y$. Looking over the possibilities in condition (i) in the definition of a tree set we see that either $e \subseteq f$ or $e^* \subseteq f$, and in both cases f points towards y.

It is also clear from this definition that if u and v are vertices in X and there is an element $e \in E$ that contains one and not the other then $\phi(u) \neq \phi(v)$. It should also be noted that if u and v are adjacent then $\{u, v\} \in \delta e$.

Now we define Φ in a similar manner. The situation is more involved here: it might happen that for some end ω of X that there is an infinite descending chain of elements in E all of which contain ω. But this chain will then define an end ϵ in T and we set $\Phi(\omega) = \epsilon$. If there is no such chain then we define $\Phi(\omega)$ in the same way as ϕ was defined. By the same argument as above we get that every edge in E that contains ω points towards $\Phi(\omega)$. (We say that an edge (u, v) in T points towards the end ω if v lies on the ray in ω that starts at u.) Two ends ω and ω' have different images under Φ if and only if there is an element in E that contains one but not the other.

It is often useful when studying Φ (and can be used to define Φ) to take some ray R in an end ω and consider $\phi(R)$. In general $\phi(R)$ need not be a ray but by adding in the unique simple paths in T between the images of successive vertices in R one gets a path P. Note that P need not be simple. It can be shown that if $\Phi(\omega)$ is an end then all the rays in the subgraph spanned by P are in that end. It can also be shown that if $\Phi(\omega)$ is a vertex then that vertex is the only vertex that one will go through infinitely often as one goes along the path P (see [50, §6]).

The following lemma gives further information about Φ.

Lemma 2. *The map $\Phi : \Omega X \to VT \cup \Omega T$ has the following properties.*

(i) ([33, Lemma 2]) *The restriction of Φ to $\Phi^{-1}(\Omega T)$ is bijective.*

(ii) ([33, Lemma 4]) *A vertex v in T is in the image of Φ if and only if $\Phi^{-1}(v)$ is infinite or v has infinite valency.*

(iii) ([50, Lemma 8.2]) *Let $E_1, E_2, \ldots$ be as in Theorem 5. If ω is a thin end of X then there is n such that if $T = T(E_n)$ then ω is mapped by Φ to an end of T.*

Let us just look briefly at why (i) is true. For full proofs see the references.

First injectivity. Let ω be an end of X and suppose that $\Phi(\omega)$ is an end. Take an infinite descending sequence $e_1 \gg e_2 \gg \ldots$ in E such that all the elements in the sequence contain ω. Then Lemma 1 implies that $\bigcap e_i = \emptyset$. Suppose ω' is some other end of X and F is some finite set of vertices separating ω and ω'. Then we find i such that $F \cap e_i = \emptyset$. Because e_i is connected we have that e_i is contained in some component of $X \setminus F$ and thus e_i cannot contain both ω and ω'. So $\Phi(\omega) \neq \Phi(\omega')$. To prove that Φ is surjective we take an end ϵ in T and some ray $\{v_i\}_{i \in \mathbb{N}}$ in that end. Set $e_i = (v_i, v_{i+1})$. Then the chain $e_1 \gg e_2 \gg \ldots$ is a directed edge path in T. Now we find a ray in X that includes at least one vertex from ∂e_i for all i. The end ω of X that contains this ray will clearly belong to all of the sets e_i and thus $\Phi(\omega) = \epsilon$.

The structure tree approach really comes into its own when one wants to study the action of the automorphism group on X. The following lemma relates the action of $\mathrm{Aut}(X)$ on X and ΩX to the action of $\mathrm{Aut}(X)$ on T.

Lemma 3. ([33, Corollary 1]) *Let X be a connected locally finite graph and $T = T(E)$ be some structure tree of X, where E is a tight undirected tree set.*

(i) *If $g \in \mathrm{Aut}(X)$ acts like a translation on T then g acts like a translation on X and g is hyperbolic.*

(ii) *If $g \in \mathrm{Aut}(X)$ is a translation then either g acts as a translation on T or there is an unique vertex of T fixed by g and that vertex has infinite valency.*

(iii) *If $g \in \mathrm{Aut}(X)$ is hyperbolic then there is a tight undirected tree set E_g such that g acts as a translation on $T(E_g)$.*

Let us look at the proof of (i), the others are similar. Let L be the line in T that is invariant under g. Suppose that e is an edge in L. Then δe has an infinite orbit under g, and therefore g cannot fix any nonempty finite subgraph. There are two ends of T that are fixed by g, so there are also two ends of X fixed by g, which means that g is hyperbolic.

2.3. Examples

1. First let us consider a partial portrait of some structure tree and consider what can be read directly from it.

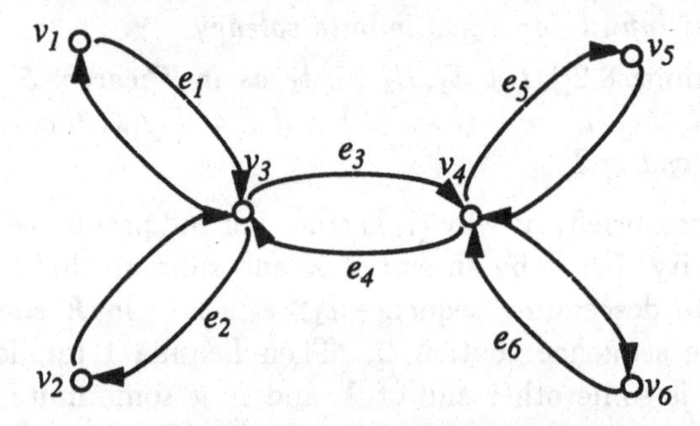

Figure 3.

Let us now look at few examples of what we can read of from Figure 3:

(i) $e_1 \gg e_3 \gg e_5$;

(ii) $e_3 = e_4^*$;

(iii) $e_3 \not\subset e_6$;

(iv) if $v \in VX$ and $v_3 = \phi(v)$ then v is contained in e_1, e_2, e_4 and e_6 but not contained in e_3 and e_5;

2. Let X be a connected regular locally finite graph. The simplest example of a tree set is the set E of all one element subsets of VX and their complements. Then $T = T(E)$ is a structure tree. In this case the structure tree looks like a star; the edges corresponding to the one element subsets pointing away from the center. It is obvious that ϕ is injective and the image of ϕ is the set of leaves of T. All the ends of X are mapped by Φ to the central vertex of T. This example shows that a structure tree of a locally finite graph need not be locally finite.

3. Now let X be a tree and let E be the set of all subsets of VX that have exactly one edge in their co-boundary. Apart from the fact that formally we consider $T(E)$ to be a directed graph the trees X and $T(E)$ are identical.

4. Let X be a graph as depicted on Figure 4; that is, X is a kind of an infinite comb with infinite teeth. Let e_1 and e_2 be cuts as shown on Figure 4.

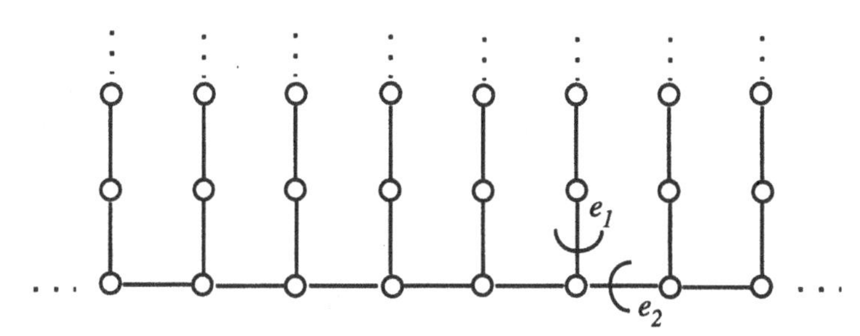

Figure 4.

Set $G = \text{Aut}(X)$. Define

$$E_1 = Ge_1 \cup Ge_1^*, \; E_2 = Ge_2 \cup Ge_2^*, \; E_3 = E_1 \cup E_2.$$

These are all tight G-invariant tree sets. The structure tree $T(E_1)$ is a star. The edges in Ge_1 point away from the center and the ends of X that correspond to the teeths of the comb are mapped to the leaves of $T(E_1)$. The two ends corresponding to the baseline are mapped by Φ to the central vertex. The tree $T(E_2)$ is a line. The maps ϕ and Φ have the same effect as contracting the teeth down to the baseline.

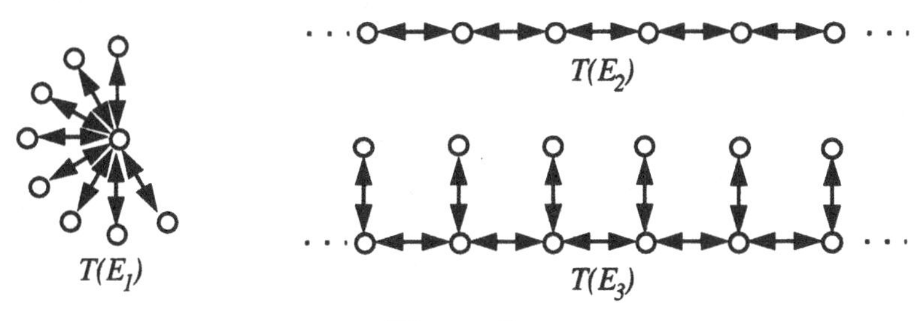

Figure 5.

On Figure 5 we see a part of $T(E_3)$, which already looks more like X than the other two. As we take bigger tree sets we get a better description of our graph.

5. Let X be the natural Cayley-graph of $\mathbf{Z}_3 * \mathbf{Z}_3$ as shown on Figure 6. Take a vertex v in X. Denote by C_v and C'_v the two components of $X \setminus \{v\}$. Set $\mathcal{C} = \{C_v, C'_v \mid v \in VX\}$ and let $\mathcal{C}^*$ denote the set of the complements of the sets in $\mathcal{C}$. Then set $E = \mathcal{C} \cup \mathcal{C}^*$, and it is easy to see that E is a tight undirected tree set. The vertices in $T(E)$ all have valencies either 2 or 3. The ones with valency 2 are in the image of ϕ and one can think of the vertices of valency 3 as corresponding to the triangles in X. The map Φ gives a bijection between ΩX and ΩT. In Section 5.1 we see that this is a part of a more general phenomenon.

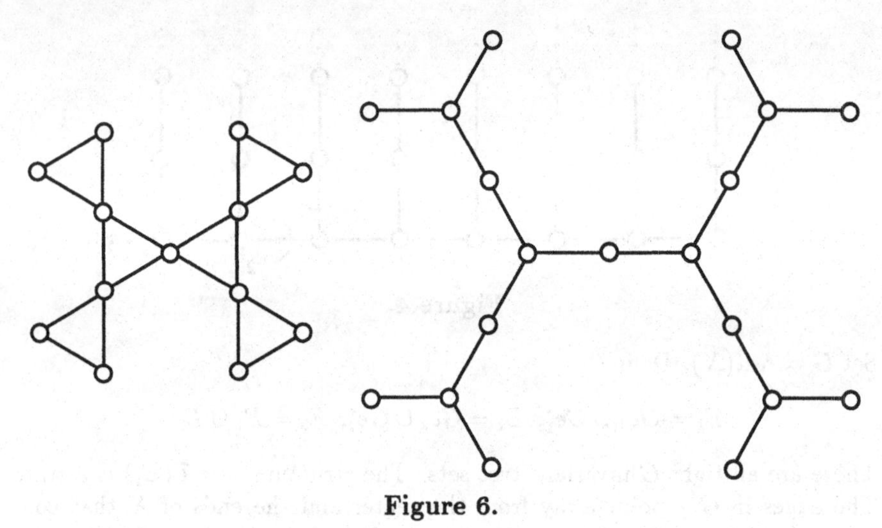

Figure 6.

3. Fixed point properties

Since the advent of the Bass–Serre theory of groups acting on trees and the paper [51] by Tits, several papers have appeared discussing generalized "fixed point properties" of group actions on trees. Recently, around the same time and independently, Nebbia [38], Pays and Valette [42], and Woess [54] arrived at what appears to be the final and fundamental truth in this matter. Of these the results of Woess are the most general because he not only solves the problem for trees but also for locally finite graphs in general. But, they are also the most difficult to prove; Woess' proofs are a real combinatorial *tour de force*. In this section we give short proofs of Woess' theorems, using the theory of structure trees.

First we do some preliminary work relating fixed point properties of group actions on trees to fixed point properties of general graphs. Then we discuss briefly the concept of amenability, which turns out to be the key to our fixed point properties. Finally we prove Woess' theorems.

3.1. Serre's property (FA)

Let X be a locally finite graph, and let $H \leq \mathrm{Aut}(X)$. The fixed point properties mentioned above are of the following types:

(a) there is a non-empty finite subgraph of X invariant under H;

(b) there is an end of X fixed by H;

(c) there is a pair of ends of X invariant under H.

First we prove a generalization of a result of Tits [51, Proposition 3.4] (see also [42, Proposition 1]). Tits proved the special case where X is a tree. We then get a sharper result: either H fixes some vertex or leaves some edge invariant or H fixes some end of T.

Theorem 6. ([54, Proposition 1]) *Let X be a connected locally finite graph, and let $H \leq \mathrm{Aut}(X)$, and suppose H contains no hyperbolic elements. Then* (a) *or* (b) *holds and, furthermore, if* (b) *holds then H fixes exactly one end.*

PROOF. Let T be some undirected structure tree of X. We know that no element of H acts like a translation on T and we can thus apply Tits' result to H acting on T. If H fixes an end of T or has a finite orbit on ET then we are clearly finished. This is because, if e is an edge in T and He is finite then $H(\delta e)$ spans a finite subgraph of X, which is invariant under H. We can thus assume, without loss of generality, that for all structure trees T of X that H fixes precisely one vertex of T and that this vertex has infinite valency in T. (If H would fix two vertices in T then the edges in the path between those two vertices would also be fixed and we would not have to do more. And, if H would fix a vertex of finite valency then H would have a finite orbit on ET.) Let $E_1 \subseteq E_2 \subseteq \ldots$ be a sequence of undirected tree sets as in Theorem 5. In each of the trees $T(E_j)$ there is a unique vertex v_j fixed by H. Since this vertex has infinite valency then, as noted before, there is a non-empty set I_j of ends of X corresponding to v_j. Clearly the sets I_j form a decreasing chain. Put $I = \bigcap I_j$. Because $\bigcup E_j$ generates the whole of $\mathcal{B}X$, we know that I is either empty or contains precisely one element. (For any two ends ω and ω' in X it is always possible to find n such that there is an element $C \in \mathcal{B}_n X$ with $\omega \in \Omega C$ and $\omega' \notin \Omega C$. Hence there is an element $e \in E$ with $\omega \in \Omega e$ and $\omega' \notin \Omega e$. Then we cannot have both ω and ω' in I_n.)

Now we use the topology on ΩX. The elements of $\mathcal{B}X$ define closed sets in the end space of X. The ends in I_j are precisely those ends that are contained in every element of E_j that points towards v_j in $T(E_j)$. Hence the set I_j is closed, since it is equal to the intersection of a family of closed sets. By assumption all the sets I_j are non-empty and ΩX is compact, so, by a standard fact about compact sets, the set I is non-empty. □

A group H acts on a tree T *without inversion* if no element of H transposes some pair of adjacent vertices. If there exists $g \in H$ and a pair of adjacent vertices $u, v \in VX$ such that $gu = v$ and $gv = u$, then we say that H acts *with inversion*.

Following Serre [45] we say that a group H has property (FA) if whenever H acts on a tree without inversion then there is some vertex of the tree fixed by H. If H acts on a tree T with inversion then we can take the barycentric subdivision T' of T and H will act on T' without inversion. Allowing actions with inversion then property (FA) is equivalent to there either being a vertex fixed by H or there being an edge invariant under H. Bass [3] studied an analogous property (FA′). If we allow actions with inversion then property (FA′) is equivalent to that every element of H either fixes a vertex or leaves some edge invariant. We can now deduce the following from the proof of Theorem 6.

Corollary 2.

(i) If a group H with property (FA) *acts on a connected locally finite graph X then either* (a) *or* (b) *holds and, if* (b) *holds then H fixes exactly one end, which is thick.*

(ii) If a group H with property (FA′) *acts on a connected locally finite graph then either* (a) *or* (b) *holds.*

There are various results on which groups do have these properties: for example, countable groups having Kazhdan's property (T) have property (FA) (see [2] and [53]) and pro-finite groups have property (FA′) (see [3]).

3.2. Amenability

Now we turn to the concept of amenability. For further information concerning amenability the reader is referred to the book by Wagon [52].

Definition 4. A group G acts *amenably* on a set Y if there is a non-negative function μ defined on the power set of Y such that

(i) $\mu(Y) = 1$;

(ii) μ is finitely additive;

(iii) μ is G-invariant.

We say that μ is a *G-invariant measure* on Y. If the left regular action of G on itself is amenable then we say that G is *amenable.*

An important variation of this definition is when Y is a topological space and instead of requiring the measure μ to be defined on all subsets we only ask for it to be defined on the Borel subsets of Y. Then we say that the action is *topologically amenable*, or just amenable if there is no danger of confusion. A topological group G is *topologically amenable* (or just, amenable) if we can find a measure μ defined on the Borel subsets of G that satisfies (i)-(iii) with respect to the left regular action of G on itself. If the topology on Y is the discrete one then topological amenability is the same as amenability. For completeness we list here the facts about amenability that we will be using.

(i) A compact topological group is amenable.

(ii) Let G be a topological group and H a closed normal subgroup of G. If both H and G/H are amenable then G is amenable.

(iii) The direct union of a directed system of amenable groups is amenable.

(iv) Abelian and soluble groups are amenable.

(v) A topological group containing a discrete non-abelian free group is not amenable. A non-abelian free group cannot act freely and amenably on a set.

To prove the first item it suffices to take μ as the Haar-measure. The second, third and fourth items are similar to [52, Theorem 10.4], and the fifth one follows also because closed subgroups of amenable groups are amenable and a non-abelian free group is not amenable.

3.3. The theorems of Woess

First we have to define a topology on $\text{Aut}(X)$. We take as a basis of neighbourhoods around the identity the pointwise stabilizers of finite subsets of VX. This is indeed just the topology of pointwise convergence on $\text{Aut}(X)$. Because we assume that X is locally finite we can easily show that the setwise stabilizers in $\text{Aut}(X)$ of non-empty finite subsets of VX are compact. Whence $\text{Aut}(X)$, endowed with this topology, is a locally compact group. For more information on this topology consult Woess' survey paper [56].

Theorem 7. ([54, Theorem 1]) *Let X be a locally finite connected graph and H a subgroup of* $\text{Aut}(X)$. *If H acts amenably on X then one of* (a), (b), (c) *holds.*

PROOF. Let μ be an H-invariant measure on VX. Let $T = T(E)$ be some structure tree of X and let ϕ be the map described in Section 2.2. Define a function ν on the power set of VT by

$$\nu(\Delta) = \mu(\phi^{-1}(\Delta)),$$

for $\Delta \subseteq VT$. Clearly ν is a H-invariant measure on VT, so H acts amenably on the tree T. By Theorem 6 we can assume that H contains a hyperbolic element h and, by Lemma 3, we may also assume that h acts like a translation on T.

Now we see that if the action of H on T satisfies none of (a), (b), (c) then the action on X will also have none of these properties. Under these conditions one can use the same method as used by Nebbia in [38], and Pays and Valette in [42] to prove that if H satisfies none of (a), (b), (c) then H would contain a free group F on two generators acting freely on T. The first step is to show that if none of (a), (b) and (c) is satisfied the we can find two elements $g, h \in H$ that act like translations on T such that the line invariant under g does not intersect the line that is invariant under h. Then one applies the "ping-pong lemma" [42, Lemma 7] to show that $\langle g, h\rangle$ is a free group freely generated by g and h and it acts freely on VT. Hence H could not act amenably, contradicting our assumption. □

From this we can deduce the following corollary which is stated in [33, Theorem 8], but had been proved earlier by H.A. Jung.

Corollary 3. *Let X be a connected locally finite graph and let $H \leq \mathrm{Aut}(X)$. If H satisfies none of* (a), (b) *or* (c) *then H contains a free group on two generators.*

Remarks. 1. For another recent proof of Nebbia's theorem and Theorem 7 see [1, Theorem 4.1 and Theorem 4.2]. That proof uses analytic properties related to amenability.

2. Theorem 6 and Corollary 3 are also valid without the assumption of local finiteness (see [27, Corollary 1.3 and Theorem 1.4]).

3. Woess [57] has proved results similar to Theorem 6 and Corollary 3 for compactifications of metric spaces that satisfy certain natural conditions concerning compatibility with the action of the isometry group. The results for ends of locally finite graphs are a special case of his results.

The converse to Theorem 7 is not true, the reason being the existence of finitely generated non-amenable groups having a Cayley-graph with only one end, see Example 2, part D in [54]. But, it is very close to being true because if we replace condition (b) with the stronger condition

(b′) there is a thin end of X fixed by H,

then we get the following theorem. Note that in the special case of trees (b′) is the same as (b).

Theorem 8. ([54, Theorem 2]) *Let X be a connected locally finite connected graph and H a subgroup of* $\mathrm{Aut}(X)$. *If one of* (a), (b′), (c) *holds then H acts amenably on X.*

PROOF. First we notice that we may assume that H is a closed subgroup of $\mathrm{Aut}(X)$. Because, if the closure of H acts amenably then surely H will act amenably and also because if H satisfies one of (a), (b′), (c) then the closure of H will also satisfy one of them. From now on we will assume that H is closed.

We then only need to show that H is (topologically) amenable because every continuous action by H on a locally compact space (in our case VX with the discrete topology) is then necessarily amenable (see [43, pp. 362-363]). It is indeed quite easy to find an H-invariant measure on VX. From the definition of the topology on H it is clear that $\{h \in H \mid hv_0 \in A\}$ is open in H. Let μ be a measure on H as described in Section 3.2, and let v_0 be a vertex in X. For $A \subseteq VX$ we set $\nu(A) = \mu(\{h \in H \mid hv_0 \in A\})$. It is easily checked that ν is an H-invariant measure on VX.

The first case arises when H satisfies (a). But then H is compact and thus amenable.

If H satisfies (b′) then, by Lemma 2, we can assume that the end ω fixed by H appears as an end in some structure tree T of X, that is $\Phi(\omega)$ is an end of T. Then H fixes the end $\Phi(\omega)$. Let $\{v_i\}_{i \in \mathbf{N}}$ be a ray in $\Phi(\omega)$ and set $e_i = (v_i, v_{i+1})$. An element of H that fixes some e_i will also fix e_j for all $j \geq i$. Then $H_{\{e_1\}} \subseteq H_{\{e_2\}} \subseteq \dots$ and all these groups are amenable because they are compact. Denote the direct union of the chain $H_{\{e_1\}} \subseteq H_{\{e_2\}} \subseteq \dots$ by H_0. Now H_0 is amenable because H_0 is the direct union of amenable groups. For every $h \in H$ there is an integer $\alpha(h)$ such that for all i large enough we have that $hv_i = v_{i+\alpha(h)}$. The map $\alpha : H \to \mathbf{Z}$ is a homomorphism with kernel H_0. So, if H does not contain any hyperbolic elements then clearly H is equal to H_0, otherwise H_0 is a closed normal subgroup of H and H/H_0 is infinite cyclic. Indeed, if $h \in H$ is a translation and h is chosen such that $\min_{v \in VT} d(v, hv)$ is as small as possible then H is a semi-direct product of H_0 and $\langle h \rangle$. In both cases we have that H is amenable.

If H satisfies (c) then find a structure tree T such that both ends ω and ω' in the pair of ends that H leaves invariant appear as ends in T. Let L be the line in T with ends ω and ω'. Then $H_{(L)}$ is a closed normal subgroup of H and it is easy to see that $H/H_{(L)}$ is either trivial, cyclic with two elements, infinite cyclic or infinite dihedral (because $H/H_{(L)}$ acts faithfully on the line L and must therefore be a subgroup of $\mathrm{Aut}(L)$, and $\mathrm{Aut}(L)$ is the infinite dihedral group). In all cases it follows that H is amenable, because $H_{(L)}$ is compact and therefore amenable, and also $H/H_{(L)}$ is amenable. □

4. Accessibility

One of the main motivations behind studying infinite graphs with more than one end is the connection with group theory that is epitomized by Stallings' Ends Theorem. In particular the so called Accessibility Conjecture has graph theoretic relevance. A counterexample to the conjecture has been found by Dunwoody [13], but the graph theoretic interpretion of accessibility is still of much interest.

Definition 5. Let X be a locally finite graph. If there is a natural number k such that any two ends in X can be separated by a set containing k or fewer vertices, then we say that X is *accessible.*

In their paper [50] Thomassen and Woess show that accessible graphs are "tree-like". This is indicated by the next result stated here, but more precise descriptions of what "tree-like" means are given in [7, §4] and [55].

Theorem 9. ([50, Theorem 7.3]) *Let X be a connected locally finite transitive graph. Then X is accessible if and only if there is a natural number n such that $\mathcal{B}_n X = \mathcal{B}X$.*

Suppose X is accessible and n is a number such that $\mathcal{B}_n X = \mathcal{B}X$. Let E_n be a tight undirected tree set generating $\mathcal{B}_n X$ and let $T = T(E)$. Then the map Φ described in Section 2.2 is injective and a thin end of X is mapped to an end of T.

So for accessible graphs it is possible to find a structure tree that represents the whole end structure of X. From this certain properties of the ends are obvious.

Corollary 4. ([50, Corollary 7.4 and Corollary 7.5]) *A connected locally finite transitive accessible graph has only countably many thick ends and there is a finite upper bound on the sizes of thin ends.*

For inaccessible graphs all this breaks down.

Theorem 10. ([50, Theorem 5.4 and Theorem 8.4]) *Let X be a connected locally finite transitive inaccessible graph.*

(i) There is a thick end ω in X such that for every natural number n there is an end ω' that cannot be separated from ω by fewer than n vertices.

(ii) For every natural number n there is a thin end in X that has size bigger than n.

We can now get a new characterization of inaccessible graphs.

Theorem 11. ([37]) *Let X be a connected locally finite transitive graph. Then X is inaccessible if and only if X has uncountably many thick ends.*

A finitely generated group is said to be accessible if its Cayley-graph, with respect to some finite generating set, is accessible. In group theory accessibility is really about decomposing groups as free products with amalgamation or as HNN-extensions, and is best understood within the framework of the Bass–Serre theory of groups acting on trees.

Definition 6. Let G be a finitely generated group. If there exists a tree T such that G acts on T with finitely many orbits and for each $e \in ET$ the group G_e is finite and for each $v \in VT$ the group G_v has at most one end, then G is said to be *accessible.*

Theorem 12. ([12]) *A finitely presented group is accessible.*

It is possible to prove accessiblity under weaker, but more technical, assumptions (see [12] and [17]).

It was generally believed for twenty years that every finitely generated group was accessible. But in 1991 Dunwoody [13] found an example of a finitely generated inaccessible group which has uncountably many thick ends. In [14] Dunwoody has also given an example of a transitive 4-valent inaccessible graph such that the automorphism group acts transitively on the edges.

Several of the results in the next section can be interpreted as saying that graphs satisfying certain extra conditions are accessible. It would be interesting to have more results along those lines. Dunwoody has in [14] extended the notion of structure trees to gain a suitable tool to study inaccessible graphs. For inaccessible graphs ordinary structure trees can only reflect a part of the end structure, but Dunwoody constructs what he calls *protrees*, which can be thought of as limits of structure trees. Thus he gets hold of the whole end structure, but protrees are not graphs and are more difficult to handle.

5. Extra conditions on the graph or the group

From Dunwoody's examples of inaccessible graphs it is evident that transitive graphs with infinitely many ends can be very badly behaved, but, as the results in this section show, it needs only relatively innocent looking extra conditions to bring them into line.

Before getting on with the job at hand we must introduce another new concept. Let G be a group acting on a set Y. For $x \in Y$ let G_x denote the stabilizer of x in G; that is, the subgroup of G consisting of all elements in G that fix x. The orbits of G_x on Y are called *suborbits.* If G acts transitively on Y then there is a 1-1 correspondence between the G_x-orbits and the G-orbits on $Y \times Y$: the suborbit $G_x y$ corresponding to the orbit $G(x, y)$. The orbits on $Y \times Y$ are called *orbitals.* When studying suborbits and orbitals it is often convenient to consider directed graphs of the form $(Y, G(x, y))$, which are called *orbital graphs.* A related idea can be used to construct new graphs on the basis of old graphs and also to clarify the structure of the graphs that we are studying. A graph $(Y, G\{x_1, y_1\} \cup \ldots \cup G\{x_n, y_n\})$ is called a *poly-orbital graph.* If X is a graph then we say that a poly orbital graph X' is *based* on X if $X' = (VX, \mathrm{Aut}(X)\{x_1, y_1\} \cup \ldots \cup \mathrm{Aut}(X)\{x_n, y_n\})$. It follows from [34,

Proposition 1] that a graph X and a connected poly-orbital graph based on it have identical end structure (the end spaces are homeomorphic). The reader that wants to know more about orbital graphs and their uses in permutation group theory is refered to [39].

5.1. Graphs with connectivity 1

The connectivity of a graph is the least number of vertices one has to remove in order for the rest of the graph to become disconnected. That a graph X has connectivity 1 means that X is connected but it is possible to find a vertex v in X such that $X \setminus \{v\}$ is not connected. Such a vertex is called a *cut-vertex.* A *block* in X is a maximal connected subgraph that has connectivity higher than 1. In most books on graph theory (e.g. [21, Chapter 3]) it is explained how a graph X with connectivity 1 is built up out of its blocks and how this is described by the block-cut-vertex tree of X. The vertex set of the block-cut-vertex tree is the union of the set of cut-vertices in X and the set of blocks of X. A block B is adjacent in the block-cut-vertex tree to a cut-vertex v if and only if v is in B. (Note that in [28] and [30] the blocks are called "lobes".) If a graph with connectivity 1 is transitive then every vertex is a cut-vertex.

Let L denote the set of blocks in X and let $\{L_i \mid i \in I\}$ be the partition of L into isomorphism types. For a block Λ in L_i denote by $\{L_i^{(j)} \mid j \in J_i\}$ the partition of the vertex set of Λ into orbits of $\mathrm{Aut}(\Lambda)$. We can of course choose this labeling of the orbits so that it is the same for all blocks in L_i. Then define $m(v, L_i^{(j)})$ as the cardinality of the set of orbits of type $L_i^{(j)}$ that contain v.

Theorem 13. ([28, Theorem 3.2]) *A graph X with connectivity 1 is transitive if and only if for all $L_i^{(j)}$ the cardinal number $m(v, L_i^{(j)})$ is the same for all $v \in VX$.*

Structure trees can be used in a very straightforward manner to describe locally finite transitive graphs X with connectivity 1. Let $\mathcal{C}_v$ denote the set of all components of $X \setminus \{v\}$, where v is a vertex in X, and let $\mathcal{C}_v^*$ denote the set of the complements of the sets in $\mathcal{C}_v$. Let

$$E = (\bigcup_{v \in VX} \mathcal{C}_v) \cup (\bigcup_{v \in VX} \mathcal{C}_v^*).$$

Then E is a tight tree set (that E is tight is ensured by the transitivity of X) and E is invariant under $\mathrm{Aut}(X)$. It is easy to convince oneself that the structure tree $T(E)$ is isomorphic to the block cut-vertex tree.

5.2. Primitivity

The group G is said to act *primitively* on a set Y if G acts transitively and the only G-invariant equivalence relations on Y are the trivial one, where all classes have size one, and the equivalence relation which has only one class, Y. The following is well known.

Problem 1. Let G be a group acting transitively on a set Y. Then the following are equivalent:

(i) G acts primitively on Y;

(ii) for all $x \in Y$ the stabilizer G_x is a maximal proper subgroup of G;

(iii) for any pair $x, y \in Y$ the graph $(Y, G\{x, y\})$ is connected.

If X is a graph then we say that X is *primitive* if $G = \mathrm{Aut}(X)$ acts primitively on VX. Condition (iii) above is very useful: it allows us to choose a different edge set without loosing connectivity. Thus we are able to bring to the surface properties that might have been hiding behind over-abundance of edges.

First we should mention a special case of Theorem 13 which gives us a complete characterisation of primitive graphs with connectivity 1.

Theorem 14. ([28, Theorem 4.2]) *For a graph X with connectivity 1 the following are equivalent:*

(i) X is primitive;

(ii) the blocks of X are primitive, pairwise isomorphic and each one has at least three vertices.

It turns out that behind every connected locally finite primitive graph with more than one end there is one with connectivity 1. The following theorem is proved with the aid of structure trees.

Theorem 15. ([36, Theorem 2 and Theorem 3]) *Suppose X is a connected locally finite primitive graph with more than one end. Then there are vertices u, v in X such that the graph $X' = (VX, G\{u, v\})$ has connectivity 1 and each block of X' has at most one end. In particular, every locally finite primitive graph is accessible.*

Thus the end structure of primitive graphs is very simple. The following result underlines further how special primitive graphs are.

Theorem 16. ([30, Proposition 1.2 and Theorem 1.3]) *For every $n = 0, 1, 3, 4, 5, \ldots$ there exists an infinite locally finite primitive graph with connectivity n, but there are no infinite locally finite primitive graphs with connectivity 2.*

Infinite graphs with no edges at all provide examples of locally finite primitive graphs with connectivity 0 and Theorem 14 tells us how to construct infinite locally finite primitive graphs with connectivity 1. Let now $n \geq 3$. Let X be the graph with connectivity 1 such that each block of X is a complete graph with n vertices and each vertex belongs to precisely 2 blocks (the graph in the fifth example in Section 2.3 is the graph you get with $n = 3$). From Theorem 14 it follows that X is primitive. Set $F = \{\{u, v\} \mid u, v \in VX$ and $d(u, v) = 2\}$. It is now left to the reader to show that the graph $X' = (VX, EX \cup F)$ has connectivity n. It should also be pointed out that X' is a poly-orbital graph.

The difficult part of the proof of the above theorem is to prove that no infinite locally finite primitive graph has connectivity 2. This is proved by elementary arguments but the proof is long and involved. Given a graph with connectivity 2 it would be nice to be able to find some kind of a description of an $\mathrm{Aut}(X)$-invariant equivalence relation on VX. It would also be nice to have some understanding of why inaccessible graphs cannot be primitive.

5.3. Transitivity conditions

First let us consider the concept of distance-transitivity.

Definition 7. A graph X is *k-distance-transitive* if for any four vertices v, v', u, u' in X with $d(v, v') = d(u, u') \leq k$ there is an automorphism $g \in \mathrm{Aut}(X)$ such that $gv = u$ and $gv' = u'$. If X is k-distance-transitive for all k then we say that X is *distance-transitive.*

Infinite locally finite connected distance-transitive graphs were classified by Macpherson [31] and, independently, by Ivanov [24]. Macpherson's proof uses D-cuts (a streamlined proof can be found in [5, Chapter II Section 3]). But Ivanov's proof uses techniques from the theory of finite distance-transitive graphs. It is difficult to see further applications of his techniques to infinite graphs.

The obvious examples of distance-transitive graphs are regular trees, but there is also a bigger closely related class of such graphs. Let k, l be natural numbers with $k \geq 1$ and $l \geq 2$. We let $X_{k,l}$ denote the infinite transitive graph with connectivity 1 where each block is a complete graph with $k + 1$ vertices and each vertex belongs to l blocks. If $k = 1$ then $X_{k,l}$ is just the l-regular tree. In the fifth example in Section 2.3 we came across $X_{2,2}$ and $X_{2,3}$ is shown on Figure 7. The graphs $X_{k,l}$ with $k \geq 1$ and $l \geq 2$ are all clearly distance-transitive.

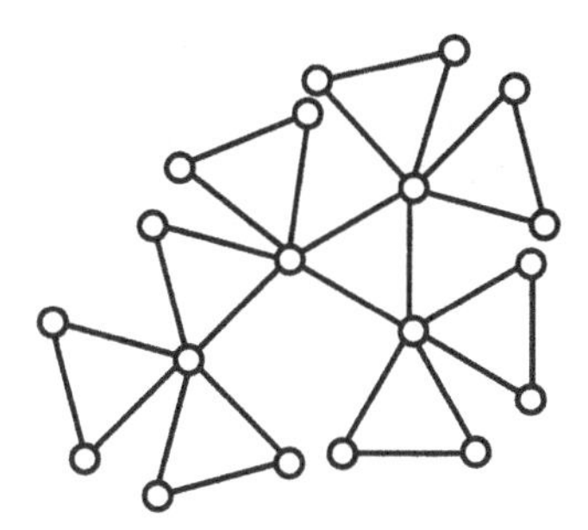

Figure 7.

Theorem 17. ([31, Theorem 1.2] and [24, Theorem 4]) *If X is an infinite connected locally finite distance-transitive graph then X is isomorphic to $X_{k,l}$ for some $k \geq 1$ and $l \geq 2$.*

Note that in particular X must have more than one end. Graphs with more than one end are very sensitive to transitivity conditions of this type.

Theorem 18. ([35]) *Let X be a locally finite connected graph with more than one end. If X is 2-distance-transitive then X is distance-transitive. In particular X must be isomorphic to $X_{k,l}$ for some $k \geq 1$ and $l \geq 2$.*

This result was conjectured by Thomassen and Woess on the basis of the following two results. The first one is an easy corollary to the above theorem. The proof given here of the latter theorem employs many of the ideas that go into the proof of Theorem 18. We say that a graph X is *k-arc-transitive* if $\mathrm{Aut}(X)$ acts transitively on the set of simple paths of length k.

Theorem 19. ([50, Theorem 3.2]) *If X is a connected locally finite 2-arc-transitive graph with more than one end, then X is a regular tree.*

Theorem 20. ([50, Theorem 3.3]) *Let X be a connected locally finite 1-arc-transitive r-regular graph, where r is a prime. If X has more than one end then X is an r-regular tree.*

PROOF OF THEOREM 19. Set $G = \mathrm{Aut}(X)$. Find a D-cut e_0 in X and set $E = Ge_0 \cup Ge_0^*$. Take a vertex v in X and let $N(v)$ denote the set of vertices that are adjacent to v. Now recall the definition of the map ϕ from Section 2.2. Of particular importance is the fact that the action of G commutes with ϕ. Set $u = \phi(v)$. Suppose that $\phi(N(v))$ is contained in some component C of $T \setminus \{u\}$. Let e be the edge in T with origin in u and pointing towards C. Then all the edges in X that have v as an end vertex are in δe but v is not in e. Now we have that $\{v\}$ would be a component of e^*, contradicting our original assumption that e is a D-cut. So $\phi(N(v))$ is dispersed over more than one component of $T \setminus \{u\}$.

From 1-arc-transitivity it follows that G_v acts transitively on $N(v)$. Then G_v, which is a subgroup of G_u, acts transitively on the components of $T \setminus \{u\}$ that contain elements from $\phi(N(v))$. Hence for every two such components C and C' we have that the number of vertices of $N(v)$ mapped by ϕ to C and C' is the same. Because the valency of v is a prime this number must be 1.

Now we have to prove that X does not have any cycles. Let us look what happens to a simple path $v_0, v_1, \ldots, v_n$ when it is mapped to T by ϕ. Because X is 1-arc-transitive there is a number c such that for every pair of adjacent vertices v and v' in T we have that $d_T(\phi(v), \phi(v')) = c$. First of all it is clear that $\phi(v_2)$ cannot be in the same component of $T \setminus \{\phi(v_1)\}$ as $\phi(v_0)$. So

$$d_T(\phi(v_0), \phi(v_2)) = d_T(\phi(v_0), \phi(v_1)) + d_T(\phi(v_1), \phi(v_2)) = 2 \cdot c.$$

Note that $\phi(v_0), \ldots, \phi(v_{k-1})$ all belong to the same component of $T \setminus \{\phi(v_k)\}$. It is now easily proved by induction that $d_T(\phi(v_0), \phi(v_k)) = k \cdot c$. From this it follows that X has no cycles because if $k > 1$ then v_0 and v_k cannot be adjacent in X. □

It is tempting to think that some kind of a classification or description can be found for 1-arc-transitive graphs, but Dunwoody has given an example of an inaccessible 4-regular 1-arc-transitive graph so there is probably not much hope for any such results.

It is also natural to ask about what happens if we put extra conditions on the action of the automorphism group on the ends, and in [56] Woess asks for a classification of connected locally finite graphs with more than one end where the automorphism group acts transitively on the ends. His question was answered independently by Möller [33] and Nevo [40]. In both cases the proofs use structure trees. Examples of such graphs are distance-transitive graphs and, indeed, all other examples are very similar to them. Before we state the result in full details we must first define some notation. A tree T is a bipartite graph, that is the vertex set can be partitioned into two disjoint sets so that no two vertices in the same set are adjacent. The automorphism group respects this partition. Let $\mathrm{Aut}^+(T)$ denote the subgroup of $\mathrm{Aut}(T)$ that stabilizes this partition. A tree is *semi-regular* if the vertices in each part of the partition all have the same valency.

Theorem 21. *Let X be a connected locally finite graph with infinitely many ends. Suppose X is end-transitive. Then there is an $\mathrm{Aut}(X)$-congruence π with finite classes such that X/π is isomorphic to a connected component of a poly-orbital graph arising from the action of $\mathrm{Aut}^+(T)$ on VT, where T is a semi-regular locally finite tree, or there is a non-empty finite subset of VX*

invariant under Aut(X)*. In particular, if there is no non-empty* Aut(X)*-invariant finite set of vertices then* Aut(X) *acts with finitely many orbits on* VX.

This result has been applied by Nevo in [41] to study random walks and harmonic analysis on such graphs.

Woess [56] also asks for information about graphs X where there is an end ω such that the stabilizer in Aut(X) of ω acts transitively on VX. Let us now look at an example of such a graph [46, Example 2 and Lemma 9]. Let T be a regular tree, and let ω be an end of T. Now add edges to T such that $\{u, v\}$ becomes an edge if the distance between u and v in T is precisely 2, and v lies on the ray starting at u and belonging to ω or *vice versa*. The resulting graph, X, will have the desired properties and the end ω is fixed by Aut(X). This example is fairly descriptive of the general situation.

Theorem 22. *Let X be a locally finite connected graph with infinitely many ends. Put $G =$ Aut(X). Assume there is an end ω of X such that G_ω acts transitively on VX.*

- *(i)* ([34, Theorem 2]) *Then G acts transitively on $\Omega X \setminus \{\omega\}$. In particular, if ω is not fixed by G then G acts transitively on ΩX.*
- *(ii)* ([34, Theorem 4]) *Then there is a G-congruence π with finite classes on VX such that X/π is a poly-orbital graph arrising from the action of a transitive group H on a locally finite distance transitive graph X_0. There is also an end ω_0 in X_0 such that H_{ω_0} is transitive on VX_0.*

References

[1] S. Adams and R. Lyons, Amenability, Kazhdan's property and percolation for trees and equivalence relations, *Israel J. Math.* **75**(1991), 341–370.

[2] R. Alperin, Locally compact groups acting on trees and property (T), *Monatsh. Math.* **93**(1982), 261–265.

[3] H. Bass, Some remarks on group actions on trees, *Comm. Algebra* **4**(1976), 1091–1126.

[4] G.M. Bergman, On groups acting on locally finite graphs, *Ann. Math.* **88**(1968), 335–340.

[5] W. Dicks and M.J. Dunwoody, *Groups acting on graphs* (Cambridge University Press, Cambridge 1989).

[6] R. Diestel, *Graph decompositions– a study in infinite graph theory* (Oxford University Press, Oxford 1990).

[7] R. Diestel, The end structure of a graph: recent results and open problems, *Discrete Math.* **100**(1992), 313–327.

[8] R. Diestel, H.A. Jung and R.G. Möller, On vertex transitive graphs of infinite degree, *Arch. Math.* **60**(1993), 591–600.

[9] M.J. Dunwoody, The ends of finitely generated groups, *J. Algebra* **12**(1969), 339–344.

[10] M.J. Dunwoody, Accessibility and groups of cohomological dimension one, *Proc. London Math. Soc. (3)* **38**(1979), 193–215.

[11] M.J. Dunwoody, Cutting up graphs, *Combinatorica* **2**(1982), 15–23.

[12] M.J. Dunwoody, The accessibility of finitely presented groups, *Invent. Math.* **81**(1985), 449–457.

[13] M.J. Dunwoody, An inaccessible group, in *The Proceedings of Geometric Group Theory 1991* (L.M.S. Lecture Notes Series, Cambridge University Press).

[14] M.J. Dunwoody, Inaccessible groups and protrees, preprint, The University of Southampton, 1992.

[15] H. Freudenthal, Über die Enden diskreter Räume und Gruppen, *Comm. Math. Helv.* **17**(1944), 1–38.

[16] C.D. Godsil, W. Imrich, N. Seifter, M.E. Watkins and W. Woess, A Note on Bounded Automorphisms of Infinite Graphs, *Graphs Combin.* **5**(1989), 333–338.

[17] J.R.J. Groves and G.A. Swarup, Remarks on a technique of Dunwoody, *J. Pure Appl. Algebra* **75**(1991), 259–269.

[18] R. Halin, Über unendliche Wege in Graphen, *Math. Ann.* **157**(1964), 125–137.

[19] R. Halin, Über die Maximalzahl fremder unendlicher Wege in Graphen, *Math. Nachr.* **30**(1965), 119–127.

[20] R. Halin, Automorphisms and Endomorphisms of Infinite Locally Finite Graphs, *Abh. Math. Sem. Univ. Hamburg* **39**(1973), 251–283.

[21] F. Harary, *Graph Theory* (Addison–Wesley, 1969).

[22] W. Imrich and N. Seifter, A note on the growth of transitive graphs, *Discrete Math.* **73**(1989), 111–117.

[23] W. Imrich and N. Seifter, A survey on graphs with polynomial growth, *Discrete Math.* **95**(1991), 101–117.

[24] A.A. Ivanov, Bounding the diameter of a distance-regular graph, *Soviet Math. Doklady* **28**(1983), 149–152.

[25] H.A. Jung, A note on fragments of infinite graphs, *Combinatorica* **1**(1981), 285–288.

[26] H.A. Jung, Some results on ends and automorphisms of graphs, *Discrete Math.* **95**(1991), 119–133.

[27] H.A. Jung, On finite fixed sets in infinite graphs, preprint, The Technical University Berlin 1991.

[28] H.A. Jung and M.E. Watkins, On the structure of infinite vertex-transitive graphs, *Discrete Math.* **18**(1977), 45–53.

[29] H.A. Jung and M.E. Watkins, Fragments and automorphisms of infinite graphs, *Europ. J. Combin.* **5**(1984), 149–162.

[30] H.A. Jung and M.E. Watkins, The connectivities of locally finite primitive graphs, *Combinatorica* **9**(1989), 261–267.

[31] H.D. Macpherson, Infinite distance transitive graphs of finite valency, *Combinatorica* **2**(1982), 63–69.

[32] B. Mohar and W. Woess, A survey on spectra of infinite graphs, *Bull. London Math. Soc.* **21**(1989), 209–234.

[33] R.G. Möller, Ends of graphs, *Math. Proc. Cambridge Philos. Soc* **111**(1992), 255–266.

[34] R.G. Möller, Ends of graphs II, *Math. Proc. Cambridge Philos. Soc.* **111**(1992), 455–460.

[35] R.G. Möller, Distance-transitivity in infinite graphs, *J. Combin. Theory Ser. B*, to appear.

[36] R.G. Möller, Primitivity and ends of graphs, *Combinatorica*, to appear.

[37] R.G. Möller, Accessibility and ends of graphs, (in preparation).

[38] C. Nebbia, Amenability and Kunze-Stein property for groups acting on a tree, *Pacific J. Math.* **135**(1988), 371–380.

[39] P.M. Neumann, Finite permutation groups, edge-coloured graphs and matrices in *Topics in group theory and computation*, (Academic Press, 1977).

[40] A. Nevo, A structure theorem for boundary-transitive graphs with infinitely many ends, *Israel J. Math.* **75**(1991), 1–20.

[41] A. Nevo, Boundary theory and harmonic analysis on boundary transitive graphs, preprint, The Hebrew University, 1991.

[42] I. Pays and A. Valette, Sous-groupes libres dans les groupes d'automorphismes d'arbres, *Enseign. Math.* **37**(1991), 151–174.

[43] J.P. Pier, *Amenable locally compact groups* (Wiley, New York 1984).

[44] N. Polat, Topological aspects of infinite graphs, in *Cycles and rays* (NATO ASI Ser. C, Kluwer Academic Publishers, Dordrecht 1990), 197–220.

[45] J.-P. Serre, *Trees* (Springer 1980).

[46] P.M. Soardi and W. Woess, Amenability, unimodularity, and the spectral radius of random walks on infinite graphs, *Math. Z.* **205**(1990), 471–486.

[47] E. Specker, Die erste Cohomologiegruppe von Überlagerungen und Homotopieeigenschaften dreidimensionaler Mannigfaltigkeiten, *Comm. Math. Helv.* **23**(1949), 303–333.

[48] J.R. Stallings, On torsion free groups with infinitely many ends, *Ann. Math.* **88**(1968), 312–334.

[49] C. Thomassen, Infinite graphs, in *Selected topics in graph theory 2* (Academic Press, London 1983), 129–160.

[50] C. Thomassen and W. Woess, Vertex-transitive graphs and accessibility, *J. Combin. Theory Ser. B*, to appear.

[51] J. Tits, Sur le groupe des automorphismes d'un arbre, in *Essays on topology and related topics (Mémoires dédiés à G. de Rham)* (Springer, 1970), 188–211.

[52] S. Wagon, *The Banach-Tarski Paradox* (Cambridge University Press, Cambridge, 1985).

[53] Y. Watatani, Property T of Kazhdan implies property FA of Serre, *Math. Japon.* **27**(1982), 97–103.

[54] W. Woess, Amenable group actions on infinite graphs, *Math. Ann* **284**(1989), 251–265.

[55] W. Woess, Graphs and groups with tree-like properties, *J. Combin. Theory Ser. B* **47**(1989), 361–371.

[56] W. Woess, Topological groups and infinite graphs, *Discrete Math.* **95**(1991), 373–384.

[57] W. Woess, Fixed sets and free subgroups of groups acting on metric spaces, preprint, University of Milano.

[58] W. Woess, Random walks on infinite graphs and groups– a survey on selected topics, preprint, University of Milano.

AN INVITATION TO COMPUTATIONAL GROUP THEORY

J. NEUBÜSER

Lehrstuhl D für Mathematik, RWTH, Templergraben 64, 52062 Aachen, Germany

Introduction

Throughout the second week of the Groups '93 meeting a workshop on Computational Group Theory (CGT for short) took place. The underlying mathematical methods were described in four series of lectures ('Finitely Presented Groups', 'Collection Methods', 'Permutation Groups', 'Representations') and three single lectures ('Cohomology Groups', 'Matrix Groups', 'Groups, Graphs and Designs'). The practical aspect was present in a lecture 'Introduction to GAP', a course 'Programming in GAP' and exercises using GAP during the afternoons. It is a pleasure to thank the large number of colleagues who helped running this workshop.

During the first week, I gave a plenary talk which had the same title as this paper. It had the dual function to give information on the plan and intention of the workshop, but also in a wider sense to discuss the present state of CGT and its role in group theory at large. In parts the talk reflected rather personal viewpoints. Since in that it differs from papers normally read in a conference, I had doubts, if it should at all go into the proceedings. The editors have encouraged me to write it up, so here it is, much in the form of the talk delivered.

CGT may be described as comprising the development, analysis, implementation and use of group theoretical algorithms. The first two depend on and draw from the corresponding part of group theory and hence teaching the methods of CGT is best organized following the lines of the theory - it definitely was organized that way in our workshop. However, a good deal of the strength of CGT for application to problems in present day group theoretical research stems from the fact that in the presently available CGT program systems such as Cayley [BC91] or GAP [S+93] a broad variety of methods is at hand - such as it is in a researcher's mind. I want to start by illustrating this point by a case study taken from the Aachen Diplom thesis of Alexander Hulpke [Hul93] which obtains a rather detailed analysis of a fairly big group using just standard methods, in this case in GAP - and, of course, the mathematical insight into the relevant theory and the available tools. In a second section I want to sketch some history and some highlights of CGT which will naturally lead to the last section with a - in parts critical - discussion of the present state and problems of the field.

1. A case study

The main task of this Diplom thesis was the detailed analysis and re-implementation in GAP of an algorithm for the determination of the character table of a finite group from its class multiplication coefficients, a method that had been proposed by J. D. Dixon [Dix67] and improved and implemented in Cayley by G. Schneider [Sch90]. Originally just to test this implementation, some hitherto unknown character tables of maximal subgroups of sporadic simple groups were determined in the thesis. One of these is a soluble maximal subgroup of order 3 265 173 504 in Fi_{23}. This table has in fact been used meanwhile for the determination of the 5-modular character table of Fi_{23} by G. Hiss.

I shall describe an analysis of this subgroup, listing and commenting the GAP commands used, in order to demonstrate the ease of using such a program system. At certain stages printing of a result of a computation is suppressed by a ';;'. Also at some stages the GAP function `time` is called which prints the CPU time used by the last command (in ms) in order to give an idea how long such an computation needs (on a DEC station).

The Cambridge ATLAS [CCN+85] provides a list of maximal subgroups of Fi_{23}, which for the first class of maximal subgroups contains:

Order	Index	Structure	Character
129 123 503 308 800	31 671	$2 \cdot Fi_{22}$	1a+782a+30 888a

This tells us that Fi_{23} has a (faithful) transitive permutation representation of degree 31 671, the character of which is the sum of three irreducibles 1a, 782a, and 30 888a. The corresponding line for our subgroup, called S in the sequel, contains

Order	Index	Structure	
3 265 173 504	1 252 451 200	$3_+^{1+8} \cdot 2_-^{1+6} \cdot 3_+^{1+2} \cdot 2S_4$	N($3B$)

giving us the information that it can be obtained as normalizer of an element of class $3B$. In addition to this information we use:

- a generating set (of two elements) for Fi_{23} in a permutation representation of degree 31 671, which we got from R.A. Wilson,
- the character table of Fi_{23}, available in the GAP library of character tables, which among many others contains all character tables of the ATLAS.

We start by reading in this character table and determining the permutation character of degree 31 671 as the sum of those three irreducibles, which (as in the ATLAS) have numbers 1, 2, and 6 in the list of characters:

```
gap> c:=CharTable("Fi23");;
```

```
gap> permchar:=Sum(c.irreducibles{[1,2,6]});;
gap> permchar[1];
31671
```

Fi_{23} has four classes of elements of order 3. The following commands tell us that these four classes have numbers 5, 6, 7, and 8 in the character table and that they can be discriminated by the value of the permutation character:

```
gap> Filtered([1..98],i->c.orders[i]=3);
[ 5, 6, 7, 8 ]
gap> permchar{last};
[ 351, 324, 135, 27 ]
```

So elements in class $3B$ (the second class of elements of order 3) have 324 fixed points.

The table head of the character table of Fi_{23}, either in the ATLAS or from the GAP library of character tables, gives us information about our chances of finding an element from class $3B$ in a random search. The centralizer of an element of class $3B$ is of order 1 632 586 752, that is, we have a probability of 1 : 1 632 586 752 that a random element of Fi_{23} is in class $3B$. However from the power map we see that elements of class $3B$ can be obtained i.a. as powers of elements of classes $36A$, $36B$, and $27A$, which are selfcentralizing, so that we have an almost 10 % chance to hit an element of one of these 3 classes. The following GAP routine (which uses that classes in the character table of Fi_{23} are ordered by increasing order of elements) tells us that altogether we have more than a 16 % chance to find an element with a power in class $3B$ by a random search.

```
gap> roots:=[6];;
>    for i in [1..Length(c.classes)] do
>      if ForAny(c.powermap,j->j[i] in roots) then
>        AddSet(roots,i);
>      fi;
>    od;
gap> roots;
[ 6, 15, 17, 19, 21, 33, 34, 35, 36, 43, 45, 46, 47, 52, 66,
  67, 68, 69, 70, 71, 72, 83, 87, 93, 94 ]
gap> Sum(roots,i->1/c.centralizers[i]);
257647/1594323
```

We prepare the search for such elements in $G := Fi_{23}$ by calculating a stabilizer chain

$$G = G_0 > G_1 > \ldots > G_n = \langle 1 \rangle,$$

where $G_i = Stab_{G_{i-1}}(i)$, and coset representatives of G_i in G_{i-1} for each $i = 1, \ldots, n$. Since we know the order of Fi_{23} we can supply it to the function

StabChain, which can use it as a stopping criterion for random methods which is much faster for such big groups than the deterministic Schreier-Sims Algorithm.

```
gap> Read("fischer23.grp");                  # taken from a local file
gap> StabChain(Fi23,rec(size:=4089470473293004800,
> random:=100));;time;
154834
```

The time function tells us that this took about two and a half minutes.

We can now use a GAP function Random which chooses for each i a random coset representative of G_i in G_{i-1} and multiplies these. We search for elements that have powers in class $3B$, using the test that the number of fixed points is indeed 324, by the following self-explanatory GAP routine.

```
gap> opdom:=PermGroupOps.MovedPoints(Fi23);;
gap> found:=false;; g:=();; h:=();;
gap> repeat
>    g:=Random(Fi23);
>    if Order(Fi23,g) mod 3 = 0 then
>      h:=g^(Order(Fi23,g)/3);
>    if Number(opdom, i->i^h=i) = 324 then
>       found:=true;
>      fi;
>    fi;
>  until found;time;
41931
```

When this stops, we check that the element h is indeed of order 3 and has 324 fixed points. We also note, that we found a root g of h and not an element of order 3, as we had already suspected.

```
gap> Order(Fi23,h);Number(opdom,i->i^h=i);
3
324
gap> Order(Fi23,g);
27
```

We now set up a subgroup H generated by h and call a GAP function to compute its normalizer in Fi_{23}. We see that finding the normalizer does take about 21 minutes.

```
gap> H:=Subgroup(Fi23, [h]);;
gap> N:=Normalizer(Fi23, H);;time;Size(N);
1263599
3265173504
```

So we now have our maximal subgroup of order 3 265 173 504 as a permutation group N of degree 31 671. It would be advantageous to have a faithful permutation representation of this group of smaller degree. As a simple attempt to go in this direction, we determine the orbits of N:

```
gap> orb:=Orbits(N,opdom);;List(orb,Length);
[ 11664, 19683, 324 ]
```

The representation on the orbit of length 324 cannot be faithful since it consists of fixed points of h, however the representation on the second largest orbit is faithful:

```
gap> P:=Operation(N,orb[1]);;
gap> MakeStabChainRandom(P);;time;Size(P);
73432
3265173504
```

However since we know S to be soluble, we prefer to work with a polycyclic presentation, called an AG-group by tradition [LNS84]. We obtain this by

```
gap> A:=AgGroup(P);;time;Size(A);
1101031
3265173504
gap> A.name:="Fi23M7";;
```

in about 18 minutes. Again for safety (see Section 3 of this paper) we have checked the order. We may now get information such as the derived series in rather short time (5 seconds):

```
gap> ds:=DerivedSeries(A);;time;Length(ds);
4953
11
gap> List(ds,i->Collected(Factors(Size(i))));
[ [ [ 2, 11 ], [ 3, 13 ] ], [ [ 2, 10 ], [ 3, 13 ] ],
  [ [ 2, 10 ], [ 3, 12 ] ], [ [ 2, 8 ], [ 3, 12 ] ],
  [ [ 2, 7 ], [ 3, 12 ] ],
  [ [ 2, 7 ], [ 3, 10 ] ], [ [ 2, 7 ], [ 3, 9 ] ],
  [ [ 2, 1 ], [ 3, 9 ] ],
  [ [ 3, 9 ] ], [ [ 3, 1 ] ], [ [ 1, 1 ] ] ]
```

This confirms some of the information about the structure of the group quoted above from the ATLAS. However using methods described e.g. in [CNW90] we can obtain further information about the splitting of extensions, e.g. we can determine representatives of the conjugacy classes of complements of the second last group in the derived series which is an extra special group of order 3^9.

```
gap> compcl:=List([1..8],i->Complementclasses(ds[i],ds[9]));;
gap> List(compcl,Length);
[ 3, 3, 1, 1, 9, 3, 1, 1 ]
```

We can also determine (in less than a minute) the conjugacy classes of elements using methods from [MN89]:

```
gap> conjcl:=ConjugacyClasses(A);;time;Length(conjcl);
49309
181
```

We see that there are 181 classes. Finally we can determine the character table of this group, using Alexander's implementation of the Dixon/Schneider method.

```
gap> C:=CharTable(A);;time;
18365368
```

This is indeed the longest computation of the whole run, taking a little over 5 hours. Printing a 181 × 181 table is not what we want, but we can use GAP to extract information from the character table. We may for instance ask for the maximal degree of an irreducible character.

```
gap> Maximum(List(C.irreducibles,i->i[1]));
18432
```

Or we may ask for the extension of the rationals generated by the character values:

```
gap> irrat:=Filtered(Union(C.irreducibles),i->not IsRat(i));
[ -4*E(8)-4*E(8)^3, -2*E(8)-2*E(8)^3, -E(8)-E(8)^3,
  E(8)+E(8)^3, 2*E(8)+2*E(8)^3, 4*E(8)+4*E(8)^3 ]
gap> alpha:=irrat[4];
E(8)+E(8)^3
gap> X(Rationals).name:="x";;
gap> Polynomial(Rationals,MinPol(alpha));
x^2 + 2
gap> Polynomial(Rationals,MinPol(E(8)));
x^4 + 1
```

We may also ask for the kernel of a character and in this particular case identify it with the third last group in the derived series.

```
gap> char144:=First(C.irreducibles,i->i[1]=144);;
gap> KernelChar(char144);
[ 1, 2, 3, 4, 5, 6, 7, 8 ]
```

```
gap> Sum(last,i->c.classes[i]);
39366
gap> List(ConjugacyClasses(A){last2},
>         x->x.representative in ds[8]);
[ true, true, true, true, true, true, true, true ]
```

2. Some history and highlights

As we have seen, the case study described in the last section can be performed

- in a day or two (with only a few hours in attention; the character table can be done overnight)
- on standard equipment (a UNIX workstation in this case, a PC under DOS would do as well)
- using standard methods, available in CGT program systems (GAP in this case).

In this section I want to outline some steps in the development of CGT that led to this situation, not going into much detail.

Prehistory (before 1953). Not only did group theory start with the very computational problem of the solvability of quintics by radicals but also (hand-) computation of groups from Mathieu's discovery of the first sporadics to Hölder's determination of the groups of order pqr made up a good deal of last century's group theory. However Dehn's formulation of the word, conjugacy, and isomorphism problems for finitely presented groups in 1911 [Deh11] (even though it had precursors) may be thought of as the beginning of the prehistory proper of CGT, focussing attention on the request for group theoretical algorithms. Two points are worth noting:

- The challenge came from topology, not from inside group theory, and indeed even now people using groups outside group theory are strong "customers" of CGT.
- Novikov's proof of the algorithmic unsolvability of the word problem put an end to the hope that group theory could fulfil Dehn's request. Soon after came proofs of the non-existence of algorithms that could **decide** if a finitely presented group is trivial, finite, abelian, etc. (see G. Baumslag's book [Bau93] for a vivid description).

Nevertheless in 1936 J. A. Todd and H.S.M. Coxeter [TC36] provided at least a systematic method to attempt to show finiteness by "coset enumeration" and today CGT provides a whole bunch of methods for the investigation of finitely presented groups that in frequent use have proved quite powerful, even though they are not decision algorithms.

Of the period predating the use of real computers I want to mention three more events:

- In 1948 H. Zassenhaus described an algorithm for the classification of space groups [Zas48] that much later has been put to very practical use.
- In 1951 M. H. A. Newman in a talk at the 'Manchester University Computer Inaugural Conference' [New51] proposed to use probabilistic methods for getting some insight into the vast number of groups of order 256. The title of the talk: "The influence of automatic computers on mathematical methods" is remarkable for that time, and the fact that the (56 092) groups of that order were only determined in 1989 by E.A. O'Brien [O'B91] should be noted.
- Perhaps most notable because of its foresight however is a quotation from a proposal of A. Turing in 1945 to build an electronic computer: "There will positively be no internal alterations to be made even if we wish suddenly to switch from calculating the energy levels of the neon atom to the enumeration of groups of order 720" (see [Hod85], page 293).

Early history (1953–1967) of CGT may be thought of as starting with the first implementations of computing methods for groups. The first, about 1953, that I am aware of, are a partial implementation of the Todd-Coxeter method by B. Haselgrove on the EDSAC II in Cambridge (see [Lee63]) and of methods for the calculation of characters of symmetric groups by S. Comet on the BARK computer in Stockholm [Com54]. Other areas of group theory were tried soon after. E.T. Parker and P.J. Nicolai made an - unsuccessful - search for analogues of the Mathieu groups [PN58], 1959 programs for calculating the subgroup lattice of permutation groups [Neu60] and a little later for polycyclicly presented 2-groups were written. Other methods and special investigations followed, programs of this time were written mostly in machine code, use of all kinds of trickery to save storage space and (although not quite as urgently) computing time was crucial. The end of this period, in which CGT started to unfold but had hardly contributed results to group theory that would greatly impress group theorists, is roughly marked by the Oxford conference "Computational Problems in Abstract Algebra" in 1967 [Lee70]. Its proceedings contain a survey of what had been tried until then [Neu70] but also some papers that lead into the

Decade of discoveries (1967–1977). At the Oxford conference some of those computational methods were presented for the first time that are now, in some cases varied and improved, work horses of CGT systems: Sims' methods for handling big permutation groups [Sim70], the Knuth-Bendix method for attempting to construct a rewrite system from a presentation [KB70], variations of the Todd-Coxeter method for the determination of presentations of subgroups [Men70]. Others, like J. D. Dixon's method for the determination of the character table [Dix67], the p-Nilpotent-Quotient method of I. D. Macdonald [Mac74] and the Reidemeister-Schreier method of G. Havas [Hav74] for subgroup presentations were published within a few years from that conference.

However at least equally important for making group theorists aware of CGT were a number of applications of computational methods. I mention three of them: The proof of the existence of Lyons' sporadic simple group by C.C. Sims in 1973, using his permutation group methods [Sim73], the determination of the Burnside group $B(4,4)$ of order 2^{422} by M. F. Newman and G. Havas using an extension of the original p-Nilpotent-Quotient method [New76], and the determination of the (4783 isomorphism classes of) space groups of 4-dimensional space [BBN+78], using not only Zassenhaus' algorithm but also the possibility to find all subgroups of the maximal finite subgroups of $GL(4,\mathbf{Z})$ using the programs for the determination of subgroup lattices.

This progress encouraged attempts to design CGT systems in which various methods could be used without having to translate data from one program to the other. By 1974 a first "Aachen – Sydney Group System" was operational [Can74], and in 1976 John Cannon published "A Draft Description of the Group Theory Language Cayley" [Can76], which may be thought of as a turn to

Modern Times. The claim that since about 1977 the development of CGT is speeding up rapidly can be justified by looking at four aspects: results, methods, systems and publicity.

Results. Concrete computational results, of which again I list only some, are as follows: Sims proved the existence of the Babymonster of order 4 154 781 481 226 426 191 177 580 544 000 000 as a permutation group of degree 13 571 955 000 using very special implementations based on his permutation group techniques [Sim80], the existence of Janko's J_4 was proved using computational techniques for modular representations [Nor80]. p-Nilpotent Quotient techniques are now strong enough to show, for example, that the class 18 factor of the restricted Burnside group $R(2,7)$ has order 7^{6366} [HNV-L90]. Following a proposal of M. F. Newman [New77], E.A. O'Brien implemented a p-group generation program sufficient to classify the (58760 isomorphism classes of) groups of order 2^n, $n \leq 8$ [O'B91]. Making and checking the Cambridge "Atlas of Finite Groups" [CCN+85], probably the most widely used group theoretical table, involved a great deal of implementation and use of group theoretical programs (in particular for working with characters), and the same holds for books listing Brauer trees of sporadics [HL89] or perfect groups [HP89] as well as for other listings, e. g. of primitive or transitive permutation groups.

Methods. These and many more concrete computations were made possible by a large number of new methods and their integration into general and specialized systems.

I have mentioned already the p-group generation method, which builds on the p-Nilpotent Quotient Algorithm (pNQ) and links to the recent exploration of the possibility to classify families of p-groups of constant coclass with the help of space groups [L-GN80]. Another very recent offspring of the pNQ is a (proper) Nilpotent Quotient Algorithm stepping down the lower central series of a finitely presented group. A number of proposals have been made and a couple of them implemented recently for finding soluble factor groups of finitely presented groups, using concepts such as cohomology groups, modular representations, or Gröbner bases [L-G84],[Ple87],[Sim90].

Working via homomorphic images has become the method of choice for handling polycyclicly presented finite soluble groups [LNS84] but also has become indispensable for handling permutation groups [KL90].

The methods for the investigation of permutation groups have almost undergone a revolution, bringing in structure theory such as the O'Nan-Scott classification of primitive groups, or even the classification of finite simple groups, now, for example, allowing the determination of a composition series of groups of degrees in some cases up into the hundred thousands. It is particularly interesting to note that some of these new methods for permutation groups, which have now become very practical, too, first were brought in through rather theoretical discussions of the complexity of permutation group algorithms. ([Neu86], [Kan85], [KT88], [BLS88], [BCF$^+$ar], [BCFS91], [BS92], to mention just a small selection of many papers on this subject.)

A broad variety of methods, many interactive, are available for working with representations and characters [NPP84],[LP91].

A long neglected, but now very rapidly growing branch of CGT are methods for the study of matrix groups over finite fields [NP92], making strong use of the Aschbacher classification [Asc84].

Systems. A program system comprises various components: storage management, a problem-oriented language for both interactive use and for writing programs, that can call system functions (and possibly directly access data), a library of functions that can be applied to the objects studied, and libraries of such objects. The first general system for CGT, developed in the mid-70-ties, was the Aachen-Sydney Group System, for which J. Cannon provided storage management (stackhandler) and language (Cayley). Functions were written in Fortran. This system developed into the Cayley System [BC91], the functions of which were semiautomatically translated to C about 1987. Cayley has been remodelled recently under the name MAGMA [CP93] and extended to a general Computer Algebra system reaching beyond CGT.

Another general system for CGT called GAP (**G**roups, **A**lgorithms, and **P**rogramming) was started in Aachen in 1986. Its design is influenced by the computer algebra system Maple: GAP has a kernel, written in C, containing storage management, language interpreter (in the future also a compiler) for the GAP language and basic time-critical functions. The large majority of the GAP functions are written in the GAP language and so can be understood, checked and altered, a point on which I want to comment in the last section.

Both Cayley and GAP have found wide use. With Cayley more exact figures can be given, since it must be licensed for a certain fee, J. Cannon reports over 200 licenses by 1990 [Can91]. GAP can be obtained free of charge via ftp and handed on so that no full control of its spread exists. An indication are the about 240 members of the GAP-forum and more than 350 reports of installation that we obtained.

During the late 70-ties and the 80-ties a number of very specialized systems have been written which became absorbed or outdated by Cayley and GAP, some others complement the scopes of the two general systems, among them Quotpic [HR93], giving a very good visualisation for the steps in the calculation of factor groups of finitely presented groups, MOLGEN [GKL92] for the application of group theoretical methods to the construction of graphs, in particular representing the structure of organic compounds, and MOC [LP91] for the construction of modular character tables.

Publicity. Already the demand for CGT systems indicates the widespread use of algorithmic methods in research (and increasingly also teaching) of group theory. The presence of CGT in general meetings on group theory, such as all four Groups St Andrews meetings 1981, 1985, 1989, and now 1993, as well as the growing frequency of specialized meetings on CGT are further indications. While the first meeting fully devoted to CGT was held in Durham 1982 [Atk84], further ones were Oberwolfach 1988 and 1992, DIMACS 1991 [FK93] and 1994. In addition to a number of surveys (see in particular [CH92] for a recent one with many references) two monographs on parts of the field have recently appeared. While Greg Butler [But91] tries to introduce Computer Science students with no preknowledge of group theory to computational methods for permutation groups, Charles Sims [Sim94] gives an authoritative account of methods for the investigation of finitely presented groups, emphasizing common features of various approaches.

3. Some concerns

Summing up what has been reported in the preceding sections, one may come to very optimistic conclusions. Computational Group Theory has provided evidence of its power, many of its methods are generally and comfortably available and widely used. And for the future: Computers are getting faster, storage space bigger, both cheaper, methods better, program systems more comprehensive and easier to use. So the future is all gleaming with promise! Is it? I have some concerns.

(i) I mentioned 1967, the year of the Oxford Conference, as the time of breakthrough for CGT. That same year Huppert's book "Endliche Gruppen I" appeared. It is still on my shelf and as useful as 27 years ago, while none of the programs that we proudly talked about in Oxford is running any more. Of course Huppert built on more than a hundred years of research in Group Theory and almost a hundred years experience of writing books about it, while Computational Group Theory and the art of implementing its methods were still in their infancy. However, I have to admit: While Huppert's book will most likely still be on your shelves in 2021, after the next 27 years, one must have serious doubts if you will still be able to use GAP (or MAGMA) then. Computers and computer languages are still changing rapidly. It is, for instance, very much in vogue to bet on parallel computers as the tool of the future, but it is by no means clear to me which of the many models of parallel computers and operating systems and languages for them will make it. And we are still lacking safe methods to preserve the huge amount of work that has gone, and is going, into the development of systems such as GAP or MAGMA for even a foreseeable time span. We do not even have defined standards to save the mathematical facts – character tables, group classifications etc. – that have already been created by use of CGT in a way that guarantees for many years the possibility to use and to check them.

(ii) I had the privilege of getting involved with the proofreading of Huppert's book. Almost every page has been read, rewritten and reread several times over with immense care. Nevertheless the last edition of the book has almost three pages of errata. I have to admit for GAP that we did not have the manpower and time to do checking of any comparable intensity and I doubt that the situation is much better with other systems. But worse: On page 128 Huppert's book refers to the Schreier conjecture as to the "Schreibweise Vermutung". A typo like that in a textbook causes at best a smile by the reader, in a program it may cause some unpredictable action of the computer. To make one point clear: This is not reviving the old prejudices that one cannot trust results obtained by computer calculations. Modern computers are by orders of magnitude more reliable in doing computations by rules than human brains. When we talk about bugs we talk about human mistakes made in setting up these rules (i. e. programs) which are of exactly the same nature as mistakes that occur in proofs. In program systems such as GAP with presently about 50 000 lines of C-code in the kernel, 120 000 lines of GAP-code in the program library, and a manual of about 1000 pages, bugs are practically unavoidable. Among these, those that cause the system to crash or produce obvious nonsense are annoying, the real dangerous ones are those that produce wrong output that still looks possible or even plausible at first sight. What can be done about this by system developers?

Programs have always been tested by running examples: it is a problem of mathematical insight and imagination to choose a sufficiently representative set of such test examples and a problem of manpower to choose it big enough. We will come back to this with point (iv).

Use of modern programming languages that allow or even enforce more transparent implementation of algorithms has certainly helped very much to avoid and to find bugs – in the same way as modern standards of formalizing and formulating proofs have done this with writing mathematics. However it does not totally eliminate the problem – again the same has to be said of standards of writing proofs.

Finally on this topic: Providing corrections to a book is an easy matter: I use my 1967 copy of Huppert's book with a copy of those 3 pages of errata; if I want, I just need a pencil to mark them on the margin. What about removing bugs from a program? If source code is interpreted, it is still reasonably easy to apply a patch. If a (compiled) executable is used, either, if available, the source must be corrected and then recompiled or, if not, a new executable must be got and installed. Each is a much more cumbersome procedure that needs much more assistance from the side of the system developer.

(iii) You can read Sylow's Theorem and its proof in Huppert's book in the library without even buying the book and then you can use Sylow's Theorem for the rest of your life free of charge, but – and for understandable reasons of getting funds for the maintenance, the necessity of which I have pointed out in (i) and (ii) – for many computer algebra systems license fees have to be paid regularly for the total time of their use. In order to protect what you pay for, you do not get the source, but only an executable, i. e. a black box. You can press buttons and you get answers in the same way as you get the bright pictures from your television set but you cannot control how they were made in either case.

With this situation two of the most basic rules of conduct in mathematics are violated: In mathematics information is passed on free of charge and everything is laid open for checking. Not applying these rules to computer algebra systems that are made for mathematical research (as is the case practically exclusively with systems for CGT) means moving in a most undesirable direction. Most important: can we expect somebody to believe a result of a program that he is not allowed to see? Moreover: do we really want to charge colleagues in Moldava several years of their salary for a computer algebra system? And even: if O'Nan and Scott would have to pay a license fee for using an implementation of their ideas about primitive groups, should not they in turn be entitled to charge a license fee for using their ideas in the implementation?

(iv) The preceding discussion describes a dilemma. On one hand CGT systems are as useful, in fact even more indispensible, for giving algorithmic ideas an impact on the progress of our knowledge on concrete groups as are books for the dissemination of theoretical insight. On the other hand it should have become clear that development and maintenance of such systems pose more problems and involve more continuously needed manpower than writing a book. I hope also to have given good reason, why I do not think that license fees are a desirable solution. Most systems have obtained some support through research grants, – in the case of GAP we gratefully acknowledge such support by the Deutsche Forschungsgemeinschaft – but typically such support is given during a restricted time for original development rather than continuously for maintenance.

I do not have a patent remedy for the problem but let me close by describing first what we try with GAP and then what in my view is needed on a broader scale if CGT is to continue to develop as well as it has since Oxford 1967.

(v) Our policy: Both the (C-) kernel and GAP library of functions are distributed with full source, free of charge, through anonymous ftp. We provide patches at intervals of a few months and release new versions of the system about every 9–12 months. Some C-programs written by other teams, devoted to special problems and tuned for these to high efficiency, are linked to GAP as 'share libraries' that can be called from GAP but remain under the responsibility of their authors.

We maintain an electronic 'GAP-forum' not only for discussion of the use of GAP in research and teaching but also (and at least as important) for bug reports of users that are thus made generally known immediately, and we encourage users to join the forum. In addition, being helped by some experienced users, we try to provide advice with technical problems, e. g. installation problems, that are sent to an address 'GAP-trouble'.

The kernel of GAP contains time-critical parts of the system, such as storage management, language interpreter and basic functions. It is clear that most users of GAP neither know nor want to know much about the methods used in the kernel, they rightly just want to rely on them. Therefore the kernel is kept as small as possible so that its development and maintenance can be managed by very few people who are highly experienced with system building (at present in the very first place Martin Schönert). On the other hand the fact that the much larger GAP library is written in the more transparent GAP language allows users to play an active part in the development of GAP. They can and should – as always when using a program – exercise all their expertise to check critically if results "look right", if they have doubts they can look at the code, trying to locate or even to correct mistakes, but in any case they can and should relate their doubts through the GAP-forum to the whole community of users and to the developers of the system. In fact this scheme has worked quite well during the last few years and has helped considerably improving the reliability of GAP.

It goes almost without saying that the relative ease of reading and writing GAP language also gives users a much better chance to adapt an existing function or to write a new one for their particular problem and in fact this is also done frequently now. It would be most desirable, however, if in many more cases some extra effort would be spent in preparing such "private" programs for general use and making them available to the public. Again with GAP we try to provide some organisational help for such publicizing of programs. Of course the ability to control, adapt, or amend functions in GAP presupposes at least some basic knowledge about group theoretical algorithms, which should therefore start to enter group theory courses, not at all to replace but to complement some parts of the theory.

(vi) I have emphasized the central role that we attribute to cooperation in our 'GAP policy'. There still remains a large amount of work to be done by rather few who develop and maintain GAP. It is important to realize that such work, if it is to live up to the state of the art, must be performed in close contact with the progress of group theory and with group theoretical problems that are presently studied. Progress in CGT in scope but also in efficiency of implementations has come far more from a better understanding of the underlying mathematics than from better implementation techniques. (These rather have their importance for reliability, flexibility, and portability of the code.) That is, the development of a group theory system is a job for group theorists (some of whom are also good at system programming) rather than for professional system programmers. But then it must also be realized that the work of such people must be recognized as contributing to mathematics. I still often enough see papers by authors who for this paper have successfully used GAP (or MAGMA or some other systems) and refer to this use by saying: "...and then by computer calculation we obtained...". Would you quote a theorem from a paper that has an author and a title by saying: "... and then by browsing through our library we became aware of the following fact ..."? Of course if we wish to change this habit, on the part of the system developers we have to be more careful in attributing contributions to individuals (with GAP we try at least) and also we have to find ways of certifying contributions to systems similar to established methods with publishing papers.

Summing up, in my view it is necessary to avoid everything that would separate work on the design **and** implementation of algorithms from other mathematical work. Rather we have to make every possible effort to adjust to habits and to adopt rules of conduct that are common practice in other parts of mathematics. For this Computational Group Theory needs the closest cooperation with all other parts of Group Theory.

It is this that the title of the paper wants to express: I want to invite you to Computational Group Theory, which should not be considered as a group theoretical fool's paradise where shiny black boxes spit out character tables and cohomology groups, rather it should be considered as a field that needs a lot of tending, but is also worth your help in tending it.

References

[Asc84] M. Aschbacher, On the maximal subgroups of the finite classical groups, *Invent. Math.* **76**(1984), 469–514.

[Atk84] M.D. Atkinson (ed.), *Computational Group Theory, Durham, 1982* (Academic Press, 1984).

[Bau93] G. Baumslag, *Topics in Combinatorial Group Theory* (Lectures in Mathematics, ETH Zürich. Birkhäuser, 1993).

[BBN+78] H. Brown, R. Bülow, J. Neubüser, H. Wondratschek and H. Zassenhaus, *Crystallographic groups of four-dimensional space* (Wiley-Interscience, New York, 1978).

[BC91] W. Bosma and J. Cannon, *A handbook of Cayley functions* (Computer Algebra Group, Sydney, Australia, 1991).

[BCF+ar] L. Babai, G. Cooperman, L. Finkelstein, E. Luks and A. Seress, Fast monte carlo algorithms for permutation groups, *J. Comput. System Sci.*, to appear.

[BCFS91] L. Babai, G. Cooperman, L. Finkelstein and A. Seress, Nearly linear time algorithms for permutation groups with a small base, in *Proc. International Symposium on Symbolic and Algebraic Computation, Bonn* (ISSAC '91, 1991), 200–209.

[BLS88] L. Babai, E. Luks and A. Seress, Fast management of permutation groups, in *Proc. 29th IEEE Symp. on Foundations of Computer Science* (1988), 272–282.

[BS92] R. Beals and A. Seress, Computing composition factors of small base groups in almost linear time, in *Proc. 24th ACM Symp. on the Theory of Computing (1992)* (1992), 116–125.

[But91] G. Butler, *Fundamental algorithms for permutation groups* (Lecture Notes in Computer Science **559**, Springer, Berlin, 1991).

[Can74] J. Cannon, A general purpose group theory program, in [New74], 204–217.

[Can76] J. Cannon, A draft description of the group theory language Cayley in [Jen76], 66–84.

[Can91] J. Cannon, A bibliography of Cayley citations, *SIGSAM Bull.* **25**(1991), 75–81.

[CCN+85] J.H. Conway, R.T. Curtis, S.P. Norton, R.A. Parker and R.A. Wilson, *ATLAS of finite groups* (Oxford University Press, 1985).

[CH92] J. Cannon and G. Havas, Algorithms for groups, *Austral. Comput. J.*

27(1992), 51–60.

[CNW90] F. Celler, J. Neubüser and C.R.B. Wright, Some remarks on the computation of complements and normalizers in soluble groups, *Acta Appl. Math.* **21**(1990), 57–76.

[Com54] S. Comet, On the machine calculation of characters of the symmetric group, in *Tolfte Skandinaviska Matematikerkongressen* (M. Riesz (ed.), Lund, 1954. Hakan Ohlssons Boktryckeri, Lund), 18–23.

[CP93] J. Cannon and C. Playoust, *An Introduction to MAGMA* (School of Mathematics and Statistics, University of Sydney, 1993).

[Deh11] M. Dehn, Über unendliche diskontinuierliche Gruppen, *Math. Ann.* **71**(1911), 116–144.

[Dix67] J.D. Dixon, High speed computation of group characters, *Numer. Math.* **10**(1967), 446–450.

[FK93] L. Finkelstein and W.M. Kantor (eds.), *Groups and computation, Proc. DIMACS Workshop, October 1991* (AMS-ACM, 1993).

[GKL92] R. Grund, A. Kerber, and R. Laue, MOLGEN, ein Computeralgebrasystem für die Konstruktion molekularer Graphen, *Com. Math. Chem.* **27**(1992).

[Hav74] G. Havas, A Reidemeister-Schreier program, in [New74], 347–356.

[HL89] G. Hiss and K. Lux, *Brauer trees of sporadic groups* (Oxford University Press, 1989).

[HNV-L90] G. Havas, M.F. Newman, and M.R. Vaughan-Lee, A nilpotent quotient algorithm for graded Lie rings, *J. Symbolic Comput.* **9**(1990), 653–664.

[Hod85] A. Hodges, *Alan Turing, The Enigma of Intelligence* (Unwin Paperbacks, 1985).

[HP89] D.F. Holt and W. Plesken, *Perfect groups* (Oxford University Press, 1989).

[HR93] D.F. Holt and S. Rees, A graphics system for displaying finite quotients of finitely presented groups, in [FK93], 113–126.

[Hul93] A. Hulpke, *Zur Berechnung von Charaktertafeln* (Diplomarbeit, Lehrstuhl D für Mathematik, Rheinisch Westfälische Technische Hochschule, 1993).

[Jen76] R.D. Jenks (ed.), *SYMSAC 1976* (Yorktown Heights, NY, 1976, Association Computing Machinery, New York).

[Kan85] W. Kantor, Sylow's theorem in polynomial time, *J. Comput. System Sci.* **30**(1985), 359–394.

[KB70] D.E. Knuth and P.B. Bendix, Simple word problems in universal algebras, in [Lee70], 263–297.

[KL90] W. Kantor and E.M. Luks, Computing in quotient groups, in *Proceedings of the 22nd ACM Symposium on Theory of Computing, 1990* (1990), 524–534.

[KT88] W. Kantor and D. Taylor, Polynomial-time versions of Sylow's theorem,

J. Algorithms **9**(1988), 1–17.

[Lee63] J. Leech, Coset enumeration on digital computers, *Proc. Cambridge Philos. Soc.* **59**(1963), 257–267.

[Lee70] J. Leech (ed.), *Computational Problems in Abstract Algebra, Oxford, 1967* (Pergamon Press, Oxford, 1970).

[L-G84] C.R. Leedham-Green, A soluble group algorithm, in [Atk84], 85–101.

[L-GN80] C.R. Leedham-Green and M.F. Newman, Space groups and groups of prime-power order I, *Arch. Math.* **35**(1980), 193–202.

[LNS84] R. Laue, J. Neubüser, and U. Schoenwaelder, Algorithms for finite soluble groups and the SOGOS system, in [Atk84], 105–135.

[LP91] K. Lux and H. Pahlings, Computational aspects of representation theory of finite groups, in *Representation Theory of Finite Groups and Finite-Dimensional Algebras* (G.O. Michler and C.M. Ringel (eds.), Progress in Mathematics 95, Birkhäuser, 1991), 37–64.

[Mac74] I.D. Macdonald, A computer application to finite *p*-groups, *J. Austral. Math. Soc.* **17**(1974), 102–112.

[Men70] N.S. Mendelsohn, Defining relations for subgroups of finite index of groups with a finite presentation, in [Lee70], 43–44.

[MN89] M. Mecky and J. Neubüser, Some remarks on the computation of conjugacy classes of soluble groups, *Bull. Austral. Math. Soc.* **40**(1989), 281–292.

[Neu60] J. Neubüser, Untersuchungen des Untergruppenverbandes endlicher Gruppen auf einer programmgesteuerten elektronischen Dualmaschine, *Numer. Math.* **2**(1960), 280–292.

[Neu70] J. Neubüser, Investigations of groups on computers. (in [Lee70]), 1–19.

[Neu86] P.M. Neumann, Some algorithms for computing with finite permutation groups, in *Proceedings of Groups – St. Andrews 1985* (E.F. Robertson and C.M. Campbell (eds.), London Math. Soc. Lecture Note Ser. **121**, Cambridge University Press, 1986), 59–92.

[New51] M.H.A. Newman, The influence of automatic computers on mathematical methods, in *Manchester University Computer Inaugural Conference* (F.C. Williams (ed.), Manchester, 1951, Tillotsons (Bolton) Ltd., Bolton, England), 13–15.

[New74] M.F. Newman (ed.), *Second International Conference on the Theory of Groups, Canberra, 1973* (Lecture Notes in Math. **372**, Springer, Berlin, 1974).

[New76] M.F. Newman, Calculating presentations for certain kinds of quotient groups in [Jen76], 2–8.

[New77] M.F. Newman, Determination of groups of prime-power order, in *Group theory, Proc. Miniconf., Austral. Nat. Univ., Canberra, 1975* (R.A. Bryce, J. Cossey, and M.F. Newman (eds.), Lecture Notes in Math. **573**, Springer, Berlin, 1977), 73–84.

[Nor80] S. Norton, The construction of J_4, *Proc. Sympos. Pure Math.* **37**(1980), 271–277.

[NP92] P.M. Neumann and C.E. Praeger, A recognition algorithm for special linear groups, *Proc. London Math. Soc.* **65**(1992), 555–603.

[NPP84] J. Neubüser, H. Pahlings and W. Plesken, CAS; design and use of a system for the handling of characters of finite groups, in [Atk84], 195–247.

[O'B91] E.A. O'Brien, The groups of order 256, *J. Algebra* **143**(1991), 219–235.

[Ple87] W. Plesken, Towards a soluble quotient algorithm, *J. Symbolic Comput.* **4**(1987), 123–127.

[PN58] E.T. Parker and P.J. Nicolai, A search for analogues of Mathieu groups, *Math. Tables Aids Comput.* **12**(1958), 38–43.

[S+93] M. Schönert et al., *GAP 3.3* (Lehrstuhl D für Mathematik, Rheinisch Westfälische Technische Hochschule Aachen, 1993).

[Sch90] G.J.A. Schneider, Dixon's character table algorithm revisited, *J. Symbolic Comput.* **9**(1990), 601–606.

[Sim70] C.C. Sims, Computational methods in the study of permutation groups, in [Lee70], 169–183.

[Sim73] C.C. Sims, The existence and uniqueness of Lyons' group, in *Finite groups '72, Proceedings of the Gainesville conf., Univ. of Florida, Gainesville, Florida, 1972* (T. Gagen, M.P. Hale and E.E. Shult (eds.), North-Holland, Amsterdam, 1973), 138–141.

[Sim80] C.C. Sims, How to construct a baby monster, in *Finite simple groups II* (Proc. Symp., Durham, 1978, M.J. Collins (ed.), Academic Press, 1980), 339–345.

[Sim90] C.C. Sims, Implementing the Baumslag-Cannonito-Miller polycyclic quotient algorithm, *J. Symbolic Comput.* **9**(1990), 707–723.

[Sim94] C.C. Sims, *Computation with finitely presented groups* (Cambridge University Press, 1994).

[TC36] J.A. Todd and H.S.M. Coxeter, A practical method for enumerating cosets of a finite abstract group, *Proc. Edinburgh Math. Soc. (2)* **2**(1936), 26–34.

[Zas48] H. Zassenhaus, Über einen Algorithmus zur Bestimmung der Raumgruppen, *Comment. Math. Helv.* **21**(1948), 117–141.

ON SUBGROUPS, TRANSVERSALS AND COMMUTATORS

MARKKU NIEMENMAA and ARI VESANEN

Department of Mathematics, University of Oulu, 90570 Oulu, Finland

1. Introduction

A great deal of work has been done to investigate the situation that a group G can be written as a product of two subgroups K and H. N. Ito [9] has shown that whenever K and H are abelian then G is soluble. Also, if G is finite and H and K are nilpotent then G is soluble (Wielandt [24], Kegel [10]). In the 1970's and 1980's several special cases of this type were considered; for a good selection of the kind of results that were obtained one can look in Amberg [1] and the references given there.

Now it is interesting to notice that if H is a subgroup of G then the natural way to combine H and G is to write $G = AH$ where A is a left transversal to H in G. It is surprising that the influence of the properties of H and its transversals on the structure of G has been studied so little (of course, transversals have been used very efficiently in the construction of the transfer homomorphism). In this survey we shall consider the situation that $G = AH = BH$ and the left transversals A and B are connected by the commutator condition $[A, B] \leq H$. We investigate the solubility of G as well as the situation in some finite simple groups. The situation and the conditions that we study arise in a natural way from some problems in loop theory and quasigroup theory. Thus we also give some applications of our results in the final section of our survey. These problems were already considered by the first author in the 1989 St Andrews Conference (see the survey article in the Proceedings of the conference [17]) and now we want to show the development that has taken place during the last four years. Our notation is standard [6], [3].

2. Connected transversals

Let H be a subgroup of G and let A and B be two left transversals to H in G. We say that A and B are H-connected if $[a, b] \in H$ whenever $a \in A$ and $b \in B$. If A is a left transversal to H in G and $[A, A] \leq H$ then we say that A is H-selfconnected. Now H-connected transversals are in fact both left and right transversals ([12], Lemmas 2.1 and 2.2) so we may simply speak of transversals. If $G' \leq H$ then H clearly has a pair of connected transversals. On the other hand, if G is simple then it is not so easy to find a proper subgroup H with connected transversals. To illustrate the situation we give two examples:

1) The alternating group A_5 has a subgroup H of order 12. If we choose A to be a cyclic group of order five then A is an H-selfconnected transversal.
2) If we consider the group A_4 and $H = \langle(1\ 2)(3\ 4)\rangle$ then there exist no connected transversals to H in A_4.

If $H = 1$ then everything is trivial: $G = A = B$ and $G' = 1$, hence G is abelian. Now what happens if H is cyclic? The situation that G is finite was first considered by Kepka and Niemenmaa [12] and later on Niemenmaa and Rosenberger [16] studied the case that G is locally finite. Finally, Kepka and Niemenmaa [14] managed to prove the following theorem which settles the problem for any G. In the theorem $L_G(H)$ means the core of H in G (the largest normal subgroup of G contained in H).

Theorem 1. *Let H be a cyclic subgroup of G and let $L_G(H) = 1$. If there exist H-connected transversals A and B then $A = B$ is an abelian group and $G'' = 1$.*

This result naturally means that if H is cyclic with $L_G(H) \neq 1$ then $G''' = 1$. The next step is to study the situation where H is abelian. For finite G Kepka and Niemenmaa [13] showed that if H is abelian then G is soluble. Quite recently, the same authors [15] managed to prove (by using Zorn's Lemma)

Theorem 2. *Let H be a finite abelian subgroup of a group G and let A and B be H-connected transversals in G. Then G is soluble.*

In 1992 Niemenmaa [18] showed that a finite group is soluble provided that H is nilpotent of odd order or $|H| = 8$. The general case remains open: If H is a nilpotent subgroup of a finite group G does it then follow that G is soluble?

Let us now step outside the class of nilpotent groups and assume that H is a nonabelian subgroup of a finite group G and $|H| = 6$. We shall prove that G is soluble; the standard technique here is to assume that G is a minimal counterexample. It finally turns out that G is a finite simple group which has a self-centralizing subgroup of order three. By using a result of Feit and Thompson [5] we conclude that either $G \cong PSL(2,7)$ or $G \cong A_5$. Now H has to be maximal in G and thus $PSL(2,7)$ is not possible. In the case of A_5 we have maximal subgroups of order six but these subgroups do not possess connected transversals.

Now we can state our second problem (for finite G): Assume that $|H| = pq$, where p and q are different primes, H is not abelian and there exist H-connected transversals A and B. Is it then true that G is soluble?

After this we turn our attention to finite simple groups. If H is a proper subgroup of a simple group G and H possesses connected transversals then $L_G(H) = 1$ and H is maximal in G ([12], Lemma 2.6). In [20] and [21] the groups $PSL(2,q)$ were investigated by Vesanen from this point of view. These groups are suitable for the study of connected transversals, because their maximal subgroups are well-known. By using the results of Ihringer [8] we conclude that the centralizer ring corresponding to G is commutative and from the well-known properties of the centralizer ring (see [23], Chapter V) it follows that we can reduce our investigation to maximal subgroups of small index. If we denote the number of conjugacy classes of G by C, then we have the following inequality ([20], Lemma 6.6)

$$|G:H| \leq (C-1)|H|+1.$$

If $G = PSL(2,q)$ we know that $C \leq (q+5)/2$ and we get a lower bound on the order of the subgroup with connected transversals. Combinatorial considerations and direct calculations can be applied to the subgroups having a small index to get ([20], Theorem 6.11): If $q > 59$ is a power of an odd prime, then there exist no connected transversals to any proper subgroup of the group $PSL(2,q)$. Similar considerations can be applied to the case $q = 2^n$ (cf. [21]) and collecting the information in the papers mentioned we get

Theorem 3. *Let H be a proper subgroup of the group $PSL(2,q)$, where $q = 2^n \geq 4$ or $q > 59$ is an odd prime power. If there exist H-connected transversals A and B, then $q = 2^n$, $|H| = q(q-1)$ and $A = B$ is a cyclic subgroup of order $q+1$.*

It can also be shown [19] that in $PSL(2,7) \cong GL(3,2)$ there exist connected transversals A and B only to those maximal subgroups that are isomorphic to S_4 and then $A = B$ is a cyclic subgroup of order seven.

3. Applications

One of the main problems in loop theory is to decide which groups are isomorphic to multiplication groups of loops (the reader is advised to consult [3], [12], [13] and [17] about the relation between groups and loops and the basic concepts of loop theory). The starting point of applications is the following theorem ([12], Theorem 4.1).

Theorem 4. *A group G is isomorphic to the multiplication group of a loop if and only if there exist a subgroup H and H-connected transversals A and B satisfying $L_G(H) = 1$ and $G = \langle A, B\rangle$.*

Now the theory of connected transversals and the knowledge of maximal subgroups of finite simple groups allows us to investigate simple groups as possible multiplication groups of loops. Drapal and Kepka [4] proved that the alternating group A_n is (in its natural permutation representation) the multiplication group of a loop if $n \geq 6$. It is also known, as Liebeck [11] proved as a by-product of the classification of finite simple Moufang loops, that the triality group $D_4(q)$ is isomorphic to the multiplication group of a Paige loop. A direct application of Theorem 3 gives

Theorem 5. *The group $PSL(2,q)$ is not isomorphic to the multiplication group of a loop if $q = 2^n$ or if $q > 59$ is an odd prime power.*

From the previous section it also follows that $PSL(2,7) \cong GL(3,2)$ is not isomorphic to the multiplication group of a loop. However, $PSL(2,9) \cong A_6$ which implies that there exists a loop whose multiplication group is isomorphic to $PSL(2,9)$.

Remark. We can also show that $PSL(2,q)$ is not isomorphic to the multiplication group of a quasigroup if $q > 59$ is a power of an odd prime ([20], Corollary 7.3). This result is somewhat surprising, because Ihringer [7] has shown that $PSL(2,2^n)$ is isomorphic to the multiplication group of a quasigroup.

If Q is a loop, then we write $M(Q)$ for the multiplication group of Q. The stabilizer of the neutral element of Q is denoted by $I(Q)$ and we say that $I(Q)$ is the inner mapping group of Q (of course $I(Q) \leq M(Q)$). Now the properties of $M(Q)$ and $I(Q)$ reflect the properties of Q. In 1946 Bruck [2] showed that: 1) if Q is centrally nilpotent then $M(Q)$ is soluble; 2) if $M(Q)$ is nilpotent, then Q is centrally nilpotent. Case 2) was studied more closely by Vesanen [22] and he was able to prove

Theorem 6. *If $M(Q)$ is a finite nilpotent group, then Q is a direct product of centrally nilpotent p-loops.*

Remark. It follows from Theorem 6 that a finite nilpotent group G is isomorphic to the multiplication group of a loop if and only if its Sylow subgroups are isomorphic to multiplication groups of loops. We wish to point out that there exist centrally nilpotent loops which are not direct products of p-loops and, consequently, have non-nilpotent (but soluble) multiplication groups (for details, see [2], Chapter III).

The influence of $I(Q)$ on the structure of Q was investigated by Kepka and Niemenmaa [13], [15]. They first showed that if H is an abelian subgroup of a finite group G such that there exist H-connected transversals A and B in G and $G = \langle A, B \rangle$, then H is subnormal in G. From this we easily get the following

Corollary. *If Q is a finite loop such that $I(Q)$ is abelian, then Q is centrally nilpotent.*

References

[1] B. Amberg, Infinite factorized groups, in *Groups -Korea 1988* (Lecture Notes in Mathematics **1398**, Springer Verlag, 1989), 1–24.

[2] R. H. Bruck: Contributions to the theory of loops, *Trans. Amer. Math. Soc.* **60**(1946), 245–354.

[3] R. H. Bruck, *A Survey of Binary Systems* (Springer -Verlag, Berlin–Heidelberg–New York, 1971, third ed.).

[4] A. Drapal and T. Kepka, Alternating groups and latin squares, *Europ. J. Combin.* **10**(1989), 175–180.

[5] W. Feit and J. G. Thompson, Finite groups which contain a self-centralizing subgroup of order 3, *Nagoya Math. J.* **21**(1962), 185–197.

[6] B. Huppert, *Endliche Gruppen I* (Springer -Verlag, Berlin–Heidelberg–New York, 1967).

[7] T. Ihringer, On multiplication groups of quasigroups, *Europ. J. Combinatorics* **5**(1984), 137–141.

[8] T. Ihringer, Quasigroups, loops and centralizer rings -Contributions to general algebra, 3, in *Proceedings of the Vienna Conference* (1984).

[9] N. Ito, Über das Produkt von zwei Abelschen Gruppen, *Math. Z.* **62**(1955), 400–401.

[10] O. H. Kegel, Produkte nilpotenter Gruppen, *Arch. Math.* **12**(1961), 90–93.

[11] M. W. Liebeck, The classification of finite simple Moufang loops, *Math. Proc. Cambridge Philos. Soc.* **102**(1987), 33–47.

[12] M. Niemenmaa and T. Kepka, On multiplication groups of loops, *J. Algebra* **135**(1990), 112–122.

[13] M. Niemenmaa and T. Kepka, On connected transversals to abelian subgroups in finite groups, *Bull. London Math. Soc.* **24**(1992), 343–346.

[14] M. Niemenmaa and T. Kepka, On loops with cyclic inner mapping groups, *Arch. Math.* **60**(1993), 233–236.

[15] M. Niemenmaa and T. Kepka, On connected transversals to abelian subgroups, *Bull. Australian Math. Soc.*, to appear.

[16] M. Niemenmaa and G. Rosenberger, On connected transversals in infinite groups, *Math. Scand.* **70**(1992), 172 - 176.

[17] M. Niemenmaa, Problems in loop theory for group theorists, in *Groups St. Andrews 1989* Vol. 2 (London Mathematical Society Lecture Note Series, **60**, 1991), 396–399.

[18] M. Niemenmaa, Transversals, commutators and solvability in finite groups, unpublished manuscript.

[19] M. Niemenmaa and A.Vesanen, On connected transversals in the group $PSL(2,7)$, *J. Algebra*, to appear.

[20] A. Vesanen, On Connected Transversals in $PSL(2,q)$, *Ann. Acad. Sci. Fenn. Ser. A I Math. Dissertationes* **84**(1992).

[21] A. Vesanen, The group $PSL(2,q)$ is not the multiplication group of a loop, *Comm. Algebra*, to appear.

[22] A. Vesanen, On p-groups as loop groups, *Arch. Math.*, to appear.

[23] H. Wielandt, *Finite permutation groups* (Academic Press, New York–London, 1968).

[24] H. Wielandt, Über Produkte von nilpotenten Gruppen, *Illlinois J. Math.* **2**(1958), 611–618.

INTERVALS IN SUBGROUP LATTICES OF FINITE GROUPS[1]

P.P. PÁLFY

Mathematical Institute of the Hungarian Academy of Sciences, H-1364 Budapest, Pf. 127, Hungary
E-mail: h1134pal@huella.bitnet

1. Introduction

In this survey paper we shall consider two closely related lattice representation problems, one from group theory, another from universal algebra. We shall put more emphasis on the group theoretical approach; an excellent survey with more universal algebraic flavour has been written by Thomas Ihringer [7].

Various problems concerning subgroup lattices have been studied already since the thirties, see the monograph of Michio Suzuki [25]. Our question deals with a relatively new aspect of this area; it concerns intervals in subgroup lattices. By an *interval* $[H; G]$ in the subgroup lattice of a group G we mean the lattice of all subgroups of G containing the given subgroup H. The motivation for the following main problem will be given below.

Problem 1.1. Which finite lattices can be represented as intervals in subgroup lattices of finite groups?

Historically the problem originates from universal algebra, from the description of congruence lattices of algebraic structures. (For concepts from universal algebra see [12]; we shall also give a somewhat detailed discussion of the universal algebraic background in Section 7.) One of the first highly sophisticated construction techniques in this field was developed by G. Grätzer and E. T. Schmidt in their fundamental paper [4].

Theorem 1.2. (Grätzer and Schmidt [4]) *Every algebraic lattice is isomorphic to the congruence lattice of some algebra.*

Since every finite lattice is algebraic, every finite lattice can be represented as a congruence lattice of an algebra. However, the original construction in [4], as well as the more recent ones by Pudlák [18] and Tůma [27] all yield infinite algebras even if the lattice is finite. Hence it is natural to ask:

Problem 1.3. Which finite lattices can be represented as congruence lattices of finite algebras? (See [3], Problem 13.)

[1]Research supported by Hungarian National Foundation for Scientific Research grant no. 1903.

It is easy to see that the congruences remain the same if we replace the set of operations of the algebra by the set of its unary polynomial functions. So, if we wish, we may restrict our attention to unary algebras $(A; F)$, where F is a set, or even a semigroup under composition, of unary operations over A. In order to introduce the group theoretic problem, we make a seemingly very restrictive assumption on the unary algebra. Namely, we consider unary algebras $(A; G)$, where the operations G form a transitive permutation group acting on A. Now let us fix an element $a \in A$ arbitrarily. It is easy to see (cf., e.g., [16], Lemma 3):

Lemma 1.4. *The congruence lattice of the unary algebra $(A; G)$ is isomorphic to the interval $[G_a; G]$, where G_a denotes the stabilizer of a.*

It is usual to identify the set of permuted elements A with the coset space of the stabilizer G_a. Having made this identification, the blocks of a congruence relation (i.e. a G-invariant partition) corresponding to a subgroup $G_a \leq K \leq G$ become exactly the cosets of K in G. Hence the congruence lattice of a transitive permutation group considered as a unary algebra is just an interval in the subgroup lattice of the group. So our main problem (1.1) is just another way of formulating the following special case of (1.3):

Problem 1.5. Which finite lattices can be represented as congruence lattices of finite unary algebras $(A; G)$, where G is a transitive permutation group acting on A?

Let us mention that if we drop the assumption of finiteness then the additional requirement that the unary operations are all permutations does not yield any further restriction on the lattice. An ingenious construction by Jiří Tůma shows that the Grätzer–Schmidt theorem still holds under these circumstances.

Theorem 1.6. (Tůma [27]) *Every algebraic lattice is isomorphic to an interval in the subgroup lattice of an infinite group.*

Returning to the finite problems we should mention that although only a few lattices are known to be representable as congruence lattices of finite algebras or as intervals in subgroup lattices of finite groups, it is still possible that every finite lattice has such a representation. However, my personal feeling is that finiteness does impose some restriction on the structure of the congruence lattice.

In (1.5) we have restricted the general problem to a very special case. However, there are certain lattices (details will be given in Section 7) such that, if they can be represented as congruence lattices of finite algebras with arbitrary operations, then they also have representations as congruence lattices of unary algebras where the operations form a transitive permutation group, i.e. these lattices occur as intervals in subgroup lattices of finite groups if and only if they are representable as congruence lattices of finite algebras. Based on this Pavel Pudlák and I [16] were able to show the equivalence of the following two statements. (But as I have already mentioned, I believe that both of them are false.)

Theorem 1.7. (Pálfy and Pudlák [16]) *The following are equivalent:*

(1) Every finite lattice can be represented as an interval in the subgroup lattice of a finite group.

(2) Every finite lattice can be represented as the congruence lattice of a finite algebra.

So it would be enough to find a finite lattice which cannot be isomorphic to an interval in the subgroup lattice of any finite group, then there would exist a — maybe different — finite lattice which would not have any representation as congruence lattice of a finite algebra. In the following survey, mostly we omit the proofs, we just outline some of the constructions and give detailed proofs only for some results not published elsewhere.

In Section 2 we deal with representations of distributive lattices. Section 3 summarizes the results concerning lattices of height two. This is the most thoroughly studied class of lattices with respect to the representation problems (1.1) and (1.3). In Section 4 we mention representations for two infinite sequences of lattices: the partition and the quasiordering lattices. Section 5 deals with the closure properties of the class $\mathcal{L}$ of lattices isomorphic to intervals in subgroup lattices of finite groups: $\mathcal{L}$ is closed with respect to taking duals and finite direct products. In Section 6 we restrict our attention to soluble groups. It turns out that this is a significant restriction. Finally, in Section 7 we give a more detailed universal algebraic background together with some interesting results related to (1.3).

2. Distributive lattices

It is always the easiest to find representations for distributive lattices. This is also the case concerning our problem.

Theorem 2.1. *Every finite distributive lattice is representable as an interval in the subgroup lattice of some finite group.*

For a representation in the subgroup lattice of $GL(n,q)$ with n and q sufficiently large see Tůma [26], cf. Section 4 below. Another way of proving this result is provided by the following obvious lemma.

Lemma 2.2. *Let D denote the diagonal subgroup $\{(g,g) \in G \times G | g \in G\}$. Then the interval $[D; G \times G]$ is isomorphic to the lattice of normal subgroups of G.*

Now (2.1) follows from the fact that every finite distributive lattice can be represented as the lattice of normal subgroups of a finite group, see Silcock [23] or Pálfy [14].

3. Lattices of height two

Having settled the case of distributive lattices it is natural to turn to modular lattices. However, here we must face great difficulties just considering the simplest lattices. Following Goralčík [2] we concentrate our attention on the lattices of height two. Let M_n denote the lattice consisting of a greatest, a smallest and n pairwise incomparable elements (see Figure 1).

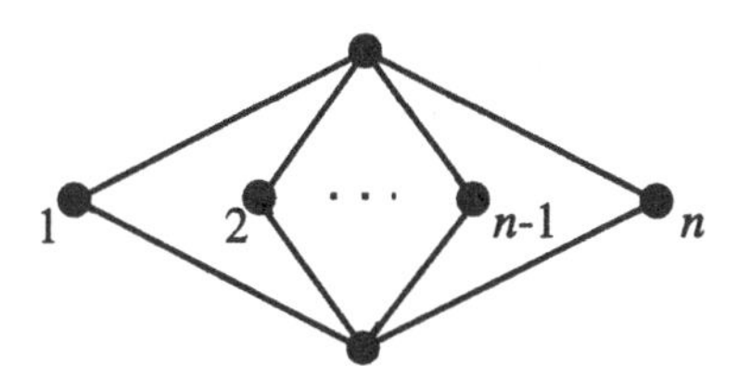

Figure 1.

As M_1 and M_2 are distributive, we will assume that $n \geq 3$. First we are going to construct some examples.

Example 3.1. For any prime-power q the lattice M_{q+1} can be represented as an interval in the subgroup lattice of a finite group.

Let V be a 2-dimensional vector space over the q-element field F. Let $G = \{x \mapsto \lambda x + v | \lambda \in F^*, v \in V\}$ and $H = \{x \mapsto \lambda x | \lambda \in F^*\}$. Then it is easy to check that each subgroup K in the interval $[H; G]$ has the form $K = \{x \mapsto \lambda x + w | \lambda \in F^*, w \in W\}$ for some subspace W of V. Hence $[H; G] \cong M_{q+1}$.

There are several results showing that under certain additional assumptions $[H; G] \cong M_n$ forces $n - 1$ to be a prime-power. Here is a sample of these:

Theorem 3.2. *Let G be a finite group, H a subgroup of G not containing any nontrivial normal subgroup of G. Assume that $[H; G] \cong M_n$ for some $n \geq 3$. Then $n - 1$ is a prime-power provided one of the following holds:*

(1) G has a nontrivial Abelian normal subgroup. (Pálfy and Pudlák [16])

(2) *G has two distinct minimal normal subgroups. (Köhler [8])*

(3) *For each $g \in G$ $\langle H, g\rangle$ is a proper subgroup of G. (Ihringer [6])*

(4) *There exist three pairwise permuting intermediate subgroups, i.e. $H < K_i < G$, $i = 1,2,3$, such that $K_iK_j = K_jK_i, 1 \leq i < j \leq 3$. (Pálfy and Saxl [17])*

For quite some time it had been expected that no other M_n's can be represented by finite groups. So it came as a great surprise when Walter Feit found an occurence of M_7 in the subgroup lattice of the alternating group $Alt(31)$.

Example 3.3. (Feit [1]) There are subgroups H_7 and H_{11} in $Alt(31)$ such that $[H_n; Alt(31)] \cong M_n, n = 7, 11$.

For H_7 one chooses a solvable subgroup of order $31 \cdot 5$ (this is unique up to conjugacy). Then H_7 is contained in its normalizer of order $31 \cdot 15$ and in six subgroups isomorphic to $PSL(5,2)$. Similarly, H_{11} of order $31 \cdot 3$ is contained in its normalizer and in ten subgroups isomorphic to $PSL(3,5)$.

However, these examples are quite exceptional as the following result shows.

Proposition 3.4. (Pálfy [15]) *Let p be a prime, H a subgroup of $Alt(p)$ such that $[H; Alt(p)] \cong M_n$. Then $n = 1,2,3,5,7$ or 11. In the latter two cases $p = 31$.*

The most important development concerning our problem has been achieved by Andrea Lucchini, who has found representations for a new infinite sequence of M_n's.

Example 3.5. (Lucchini [10]) For any prime-power q the lattice M_{q+2} can be represented as an interval in the subgroup lattice of a finite group. Furthermore, M_n with $n = (q^t + 1)/(q + 1) + 1$, where t is an odd prime number, can be represented as well.

Let us sketch the construction yielding M_{q+2}. Take $S = PSL(q^2 - 1, q^2)$ and the following automorphisms of $S \times S$: $\gamma(X, Y) = (X^{-t}, Y^{-t})$, where X^{-t} denotes the inverse-transpose matrix of X; $\phi(X, Y) = (\bar{X}, \bar{Y})$, where $\bar{X}$ denotes the image of X under the involutory automorphism of the underlying q^2-element field ($x \mapsto x^q$); and $\tau(X, Y) = (Y, X)$. (We are a bit sloppy here, more precisely we should consider matrices up to a scalar multiple. It is easy to see that this does not cause any trouble.) The three automorphisms γ, ϕ and τ commute with each other and all have order two. Take the group of automorphisms $A = \{1, \gamma\tau, \phi\tau, \gamma\phi\}$. Now let $G = (S \times S)A$. Furthermore, let K be the subgroup of S stabilizing a decomposition of the underlying vector space into a 1-dimensional and a (q^2-2)-dimensional subspace. Finally, let $H = \{(k,k)|k \in K\}A$. The claim is that $[H;G] \cong M_{q+2}$. In fact, the intermediate subgroups are the following: $(K \times K)A$, and for each $\lambda \in GF(q^2)^*$ with $\lambda^{q-1}=1$ the subgroup $\{(X, T_\lambda^{-1}XT_\lambda)|X \in PSL(q^2-1,q^2)\}A$, where T_λ denotes the diagonal matrix $diag(\lambda,1,\ldots,1)$.

So presently the small values of n for which no finite representation of M_n has yet been found are $n = 16, 23, 35, 36, 37, 40, \ldots$

4. Partition and quasiordering lattices

In this short section we quote just two positive results from a paper of Jiří Tůma [26].

All partitions of an n-element set form a lattice which will be denoted by Π_n. Let $k_1,\ldots,k_n$ be natural numbers such that no two partial sums of them are equal (for example, let $k_i = 2^{i-1}$). Let $K = \sum k_i$, $G = Sym(K)$ and $H = Sym(k_1) \times \ldots \times Sym(k_n)$. One can check directly:

Proposition 4.1. *With the above defined groups we have* $[H;G] \cong \Pi_n$.

Another important sequence of lattices is formed by the quasiordering lattices Ω_n, $n = 1,2,\ldots$. A quasiordering is a reflexive, transitive relation. Ω_n is the lattice of all quasiorderings (or, equivalently, all topologies) on an n-element set.

Theorem 4.2. (Tůma [26]) *Every quasiordering lattice Ω_n can be represented as an interval in the subgroup lattice of a finite group.*

Tůma gives the following construction. As above let $k_1,\ldots,k_n$, $k_i \geq 2$ be natural numbers such that no two partial sums of them are equal. Let $K = \sum k_i$ and $q \geq 3n-2$ be a prime-power. Then let $G = GL(K,q)$ and $H = GL(k_1,q) \times \ldots \times GL(k_n,q)$. It is proved in [26] that $[H;G] \cong \Omega_n$. Since every finite distributive lattice can be embedded as an interval into a suitable quasiordering lattice Ω_n, (2.1) can be obtained as a corollary of (4.2).

5. Closure properties

Consider the class $\mathcal{L}$ of all finite lattices that are representable as intervals in subgroup lattices of finite groups. A very useful construction of Hans Kurzweil shows that $\mathcal{L}$ is closed under dualization (turning the lattices upside-down).

Theorem 5.1. (Kurzweil [9]) *For any interval $[H;G]$ in the subgroup lattice of a finite group G there is a finite group $\hat{G}$ and a subgroup $\hat{H}$ such that $[\hat{H};\hat{G}]$ is isomorphic to the dual of $[H;G]$.*

The construction is the following. Let $|G : H| = n$ and represent G on the right cosets of H, so assume $G \leq Sym(n)$ with stabilizer H. Take an arbitrary nonabelian simple group S and form the wreath product $\hat{G} = S \wr G$ and let $\hat{H} = diag(S) \times G$, where $diag(S) = \{(s, \ldots, s) | s \in S\}$. The construction is based on the fact that $[diag(S); S^n]$ is isomorphic to the dual of the partition lattice Π_n.

Based on (5.1) we can show that $\mathcal{L}$ is also closed under forming (finite) direct products.

Proposition 5.2. *If each lattice L_i, $i = 1, \ldots, n$ is representable as an interval in the subgroup lattice of a finite group then their direct product $\prod L_i$ is also isomorphic to an interval in the subgroup lattice of some finite group.*

Applying the construction in the proof of (5.1) we can find suitable wreath products $\hat{G}_i$ and subgroups $\hat{H}_i$ such that $L_i \cong [\hat{H}_i; \hat{G}_i]$. It can be checked that for these groups we have $[\prod \hat{H}_i; \prod \hat{G}_i] = \prod[H_i; G_i]$. (Indeed, it is almost trivial if we use distinct simple groups S_i in each wreath product, but it is still true if the same simple group S serves as the base group in all wreath products $\hat{G}_i = S \wr G_i$.)

Trivially, $\mathcal{L}$ is closed under taking intervals. However, it cannot be closed under taking sublattices unless it contains all finite lattices. Indeed, we have seen in (4.1) that all partition lattices Π_n belong to this class, and by a celebrated result of Pudlák and Tůma [20], solving an old problem of Whitman [28], every finite lattice can be embedded as a sublattice into a suitable finite partition lattice. This result can be formulated in the following way as well:

Theorem 5.3. (Pudlák and Tůma [20]) *Every finite lattice can be embedded (as a sublattice) into the subgroup lattice of a finite group.*

6. Soluble groups

It is not possible to represent every finite lattice as an interval in the subgroup lattice of a finite soluble group. Though distributive lattices do have such representations, already not every M_n can be represented. Further important necessary conditions can be found in a work of Hermann Heineken [5]. We shall also show by an example, that the class of finite lattices representable as intervals in subgroup lattices of finite soluble groups, is not closed under dualization.

Theorem 6.1. *Every finite distributive lattice can be represented as an interval in the subgroup lattice of some finite soluble group.*

Pálfy [14] constructs finite soluble groups with prescribed distributive lattice of normal subgroups. In virtue of (2.2) it supplies an isomorphic interval in the subgroup lattice of the direct square of the group.

The following is a special case of (3.2.1).

Theorem 6.2. *If M_n, $n \geq 3$ occurs as an interval in the subgroup lattice of a finite soluble group, then $n-1$ is a prime-power.*

Now we are going to prove simple necessary conditions for intervals in subgroup lattices of finite soluble groups. Let G be soluble and let A be a minimal normal subgroup of G, then A is Abelian. Let M be an arbitrary subgroup of G not containing A. Then it is easy to see that M is maximal in G if and only if $AM = G$. In this case $A \cap M = 1$. We shall consider intervals $[H; G]$, where H is not maximal in G and does not contain any nontrivial normal subgroup of G. Then the above observation yields $H < AH < G$.

Lemma 6.3. *Let G be a finite soluble group, A a minimal normal subgroup of G, and $H < K < G$ subgroups such that $\langle AH, K \rangle = G$. Then*

(1) K is a maximal subgoup of G;

(2) $AH \cap K = H$;

(3) $[AH; G] \cong [H; K]$.

PROOF. AK is a subgroup containing both AH and K, hence $AK = G$. Since K is a proper subgroup, $A \not\leq K$, hence by the previous considerations K is maximal. We know also that $A \cap K = 1$ in this case, so we obtain $AH \cap K = (A \cap K)H = H$. Finally, the inclusion preserving mappings $X \mapsto X \cap K$ and $Y \mapsto AY$ between the intervals $[AH; G]$ and $[H; K]$ are inverses to each other as follows. Let $AH \leq X \leq G$, then $X = X \cap AK = A(X \cap K)$. Conversely, if $H \leq Y \leq K$, then $AY \cap K = (A \cap K)Y = Y$. □

First we shall use (6.3) to derive one of the necessary conditions obtained by Heineken generalizing (6.2). Then (6.3) will be used again in order to construct an interval whose dual cannot be isomorphic to an interval in the subgroup lattice of any finite soluble group.

Theorem 6.4. (Heineken [5]) *Let G be a finite soluble group, $H < G$ and assume that the intersection of any two maximal subgroups in the interval $[H; G]$ is H. Then the number of maximal subgroups in $[H; G]$ is 1, 2 or $q+1$, where q is a prime-power. Furthermore, with at most one exception H is maximal in each maximal subgroup in the interval $[H; G]$.*

PROOF. We prove the second statement of the theorem. We may assume wlog that H is core-free and the interval contains at least two maximal subgroups. Let A be a minimal normal subgroup of G. Then we have $H < AH < G$. By our assumption there is a unique maximal subgroup $M \geq AH$. Let $K > H$ be any maximal subgroup different from M. Now $A \leq M$, so $A \not\leq K$ and $K < AK = G$ by the maximality of K. Furthermore, $H = M \cap K \geq AH \cap K \geq H$, so equality holds. Moreover, we also have $MK \geq (AH)K = AK = G$, hence $AH \doteq M$ follows. Now (6.3.3) can be applied and it yields $[H;K] \cong [AH;G] = [M;G]$, i.e. H is maximal in K. The statement about the number of maximal subgroups could be verified the same way as in the published proof of (3.2.1). □

Example 6.5. The lattice on Figure 2 cannot be represented as an interval in the subgroup lattice of a finite soluble group, though its dual lattice has such a representation.

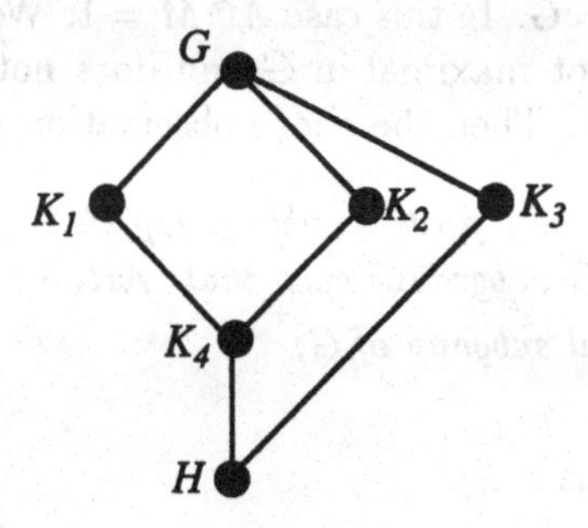

Figure 2.

PROOF. Assume the contrary, then for a minimal normal subgroup A of G we have $AH = K_i$ for some $i = 1, 2, 3, 4$. However, (6.3) will fail for each choice. If $AH = K_1$ then $\langle K_1, K_2 \rangle = G$ but $K_1 \cap K_2 \neq H$ contrary to (2). The case $AH = K_2$ is symmetric. If $AH = K_3$ then $\langle K_3, K_4 \rangle = G$, contrary to (1). Finally, if $AH = K_4$ then $\langle K_4, K_3 \rangle = G$ but $[K_4; G] \not\cong [H; K_3]$, contrary to (3).

The dual lattice occurs as an interval, for example, in the subgroup lattice of the wreath product $G = D_p \wr C_2$, where p is a prime, D_p denotes the dihedral group of order $2p$ and C_2 the cyclic group of order two. Namely, let H be the square of a 2-element subgroup of D_p. It is easy to check that it is contained in exactly the following subgroups: $H \cong C_2 \times C_2$, $D_p \times C_2$, $C_2 \times D_p$, $D_p \times D_p$, $\mathbf{N}_G(H) = C_2 \wr C_2$ and G. □

Notice that in virtue of (5.1) the lattice on Figure 2 does occur as an interval in the subgroup lattice of a (nonsoluble) finite group.

7. Universal algebraic background

By an algebraic structure, or simply an *algebra* we mean a pair $(A; F)$, where A is a nonempty set and F is a system of finitary operations on A, i.e.

mappings $A^n \to A$, $n \in \{0, 1, 2, \ldots\}$. An equivalence relation θ over A is called a *congruence* of the algebra $(A; F)$ if for any $f \in F$, $f : A^n \to A$ and elements $a_1, \ldots, a_n; b_1, \ldots, b_n \in A$ such that $a_i \,\theta\, b_i$, $i = 1, \ldots, n$ we have $f(a_1, \ldots, a_n) \,\theta\, f(b_1, \ldots, b_n)$. The intersection of any number of congruences of an algebra is again a congruence, hence all congruences of an algebra form a complete lattice, $Con(A; F)$, the *congruence lattice* of the algebra. There is a not so obvious property of congruence lattices caused by the fact that the basic operations of algebras are finitary by definition. An element x of a complete lattice is called *compact* if $x \leq \bigvee_{i \in I} y_i$ for an arbitrary join of elements y_i implies the existence of a finite subset $J \subset I$ such that already $x \leq \bigvee_{j \in J} y_j$ holds. For arbitrary elements $a, b \in A$ the smallest congruence having a and b in the same congruence class is called the *principal congruence* collapsing a and b. The principal congruences are compact elements in the congruence lattice and every congruence is a join of principal congruences. Hence in every congruence lattice each element is the join of compact elements. Such lattices are called *algebraic lattices.* (1.2) says that there are no further restrictive properties of congruence lattices, any algebraic lattice can be represented as a congruence lattice.

A simplification of Problem (1.3) can be achieved by considering only unary algebras, i.e. algebras where each operation has one variable. This reduction is made possible by the observation that the congruences will not change if we replace the original operations by the *unary polynomial functions* of the algebra, i.e. those functions which can be expressed using one variable symbol, constant symbols for each element of A and the basic operations of the algebra. If all these functions — apart from the constants — happen to be permutations, then the congruences are simply the systems of blocks of imprimitivity for the group formed by these permutations.

For certain lattices we have shown in [16] that if they have a representation as the congruence lattice of a finite algebra then they must have a representation such that the algebra is unary and its operations form a transitive permutation group. This result was later generalized by Ralph McKenzie [11]. We state here a simplified version.

Theorem 7.1. *Let L be a finite simple lattice such that the intersection of the maximal elements of L (the so-called coatoms) is the least element of L (often denoted by 0). If L is isomorphic to the congruence lattice of some finite algebra, then in the minimal such representation of L the nonconstant unary polynomial functions of the algebra are all permutations.*

Some further conditions, easily satisfied, imply that these permutations form a transitive group (see [16]). Hence for a certain class of lattices the universal algebraic representation problem (1.3) is actually equivalent to the group theoretic one (1.1). Fortunately, this class of lattices is wide enough to contain all finite lattices as filters (top intervals) in them. This serves as the basis for (1.7).

Since the lattices M_n, $n \geq 4$ share the above properties, the problems (1.1) and (1.3) are equivalent for them. Parallel to (3.2) there are universal algebraic assumptions which also imply that $n - 1$ is a prime-power, if a certain type of representation of M_n exists. Here we need the concept of permuting congruences. The product of two congruences is defined in the usual way: $\alpha \circ \beta = \{(a, b) \in A \times A | \exists c \in A : a\,\alpha\,c, c\,\beta\,b\}$. This is usually *not* an equivalence relation. We say that the congruences $\alpha, \beta \in Con(A; F)$ *permute* if $\alpha \circ \beta = \beta \circ \alpha$. If any two conguences of every algebra in a variety of algebras permute, then we say that the variety is *congruence permutable.* (For example the variety of groups is congruence permutable.)

Theorem 7.2. *Let $(A; F)$ be a finite algebra. Assume that $Con(A; F) \cong M_n$ for some $n \geq 3$. Then $n-1$ is a prime-power provided one of the following holds:*

(1) The algebra $(A; F)$ generates a congruence permutable variety. (Quackenbush [21])

(2) There exist three pairwise permuting nontrivial congruences of $(A; F)$. (Pálfy and Saxl [17])

(3) Each principal congruence is a proper congruence of $(A; F)$. (Ihringer [6], Sauer, Stone and Weedmark [22] and [24])

(4) The algebra has only one operation (of arbitrary arity). (McKenzie [11])

Presently, however, constructions for representing M_n's with $n - 1$ not a prime-power are available only through the group theoretic approach, see (3.3) and (3.5).

For the closure properties of the class of lattices isomorphic to congruence lattices of finite algebras, we have results similar to those in Section 5. This class is also closed under dualization — as was shown, generalizing the idea of Hans Kurzweil, by his student Raimund Netter [13]; and for direct products as well — as was announced in the paper of Jiří Tůma [26]. The congruence lattice of a quotient algebra is a filter in the congruence lattice of the original algebra. Moreover, double dualization yields that we can take ideals, hence intervals as well.

We finish the paper by mentioning a deep and interesting result of Pavel Pudlák and Jiří Tůma. Notice that finite distributive lattices satisfy the assumption of the theorem.

Theorem 7.3. (Pudlák and Tůma [19]) *Let L be a finite lattice such that both L and its congruence lattice $Con(L)$ have the same number of join-irreducible elements. Then L can be represented as a congruence lattice of a finite algebra.*

I do not know, whether these lattices have representations as intervals in subgroup lattices of finite groups as well.

References

[1] Feit, W., An interval in the subgroup lattice of a finite group which is isomorphic to M_7, *Algebra Universalis* **17**(1983), 220–221.

[2] Goralčík, P., Problem, *Colloq. Math. Soc. J. Bolyai* **17**(1975), 604.

[3] Grätzer, G., *Universal Algebra*, 2nd ed. (Springer Verlag, Berlin–Heidelberg–New York, 1979).

[4] Grätzer, G. and E.T. Schmidt, Characterizations of congruence lattices of abstract algebras, *Acta Sci. Math. (Szeged)* **24**(1963), 34–59.

[5] Heineken, H., A remark on subgroup lattices of finite soluble groups, *Rend. Sem. Mat. Univ. Padova* **77**(1987), 135–147.

[6] Ihringer, T., A property of finite algebras having M_n's as congruence lattices, *Algebra Universalis* **19**(1984), 269–271.

[7] Ihringer, T., Congruence lattices of finite algebras: The characterization problem and the role of binary operations, *Algebra Berichte* **53** (Fischer Verlag, München, 1986).

[8] Köhler, P., M_7 as an interval in a subgroup lattice, *Algebra Universalis* **17**(1983), 263–266.

[9] Kurzweil, H., Endliche Gruppen mit vielen Untergruppen, *J. reine angew. Math.* **356**(1985), 140–160.

[10] Lucchini, A., M_{q+2} as an interval in the subgroup lattice, preprint, 1992.

[11] McKenzie, R., Finite forbidden lattices (Lecture Notes Math., **1004**, 1983),176–205.

[12] McKenzie, R.N., G.F. McNulty and W.F. Taylor, *Algebras, Lattices, Varieties, I* (Wadsworth & Brooks/Cole, Pacific Grove, CA, 1987).

[13] Netter, R., Eine Bemerkung zu Kongruenzverbänden, preprint, 1986.

[14] Pálfy, P.P., Distributive congruence lattices of finite algebras, *Acta Sci. Math. (Szeged)* **51**(1987), 153–162.

[15] Pálfy, P.P., On Feit's examples of intervals in subgroup lattices, *J. Algebra* **116**(1988), 471–479.

[16] Pálfy, P.P. and P. Pudlák, Congruence lattices of finite algebras and intervals in subgroup lattices of finite groups, *Algebra Universalis* **11**(1980), 22–27.

[17] Pálfy, P.P. and J. Saxl, Congruence lattices of finite algebras and factorizations of groups, *Comm. Algebra* **18**(1990), 2783–2790.

[18] Pudlák, P., A new proof of the congruence lattice representation theorem, *Algebra Universalis* **6**(1976), 269–275.

[19] Pudlák, P. and J. Tůma, Yeast graphs and fermentation of algebraic lattices, *Colloq. Math. Soc. J. Bolyai* **14**(1974), 301–341.

[20] Pudlák, P. and J. Tůma, Every finite lattice can be embedded in a finite

partition lattice, *Algebra Universalis* **10**(1980), 74–95.

[21] Quackenbush, R.W., A note on a problem of Goralčík, *Colloq. Math. Soc. J. Bolyai* **17**(1975) 363–364.

[22] Sauer, N., M.G. Stone and R.H. Weedmark, Every finite algebra with congruence lattice M_7 has principal congruences, in Lecture Notes Math. **1004** (1983), 273–292.

[23] Silcock, H.L., Generalized wreath products and the lattice of normal subgroups of a group, *Algebra Universalis* **7**(1977), 361–372.

[24] Stone, M.G. and R.H. Weedmark, On representing M_n's by congruence lattices of finite algebras, *Discrete Math.* **44**(1983), 299–308.

[25] Suzuki, M., *Structure of a group and the structure of its lattice of subgroups* (Springer, Berlin, 1956).

[26] Tůma, J., Some finite congruence lattices I, *Czech. Math. J.* **36**(1986), 298–330.

[27] Tůma, J., Intervals in subgroup lattices of infinite groups, *J. Algebra* **125**(1989), 367–399.

[28] Whitman, P.M., Lattices, equivalence relations and subgroups, *Bull. Amer. Math. Soc.* **52**(1946), 507–522.

AMALGAMS OF MINIMAL LOCAL SUBGROUPS AND SPORADIC SIMPLE GROUPS

CHRISTOPHER PARKER* and PETER ROWLEY†

* Sonderforschungsbereich 343, Universität Bielefeld, 33615 Bielefeld, Germany
†Department of Mathematics, UMIST, P.O. Box 88, Manchester M60 1QD, England

1. Introduction

Let H be a finite group and let p be a prime number dividing $|H|$. For $S \in Syl_p H$, put $\mathcal{L}oc(S) = \{N_H(R) \mid 1 \neq R \leq S\}$. A ***p*-local subgroup** of H is a subgroup of H which is in $\mathcal{L}oc(S)$ for some $S \in Syl_p H$.

The modern era of p-local analysis (meaning the study of various collections of p-local subgroups) may be considered to have its origin in Thompson's thesis [T1] (see also [T2] and [T3]). There he introduced what is now known as the Thompson order and subsequent refinements of this idea were important in the proof of the Odd Order Theorem [FT]. Questions concerning p-local subgroups frequently play a role in important problems about finite groups. This has been particularly true of the work which resulted in the classification of the finite simple groups.

In 1980 Goldschmidt inaugurated the amalgam method when in [Gol] he used amalgams to study a configuration arising in the N-group paper. More recently the amalgam method has been deployed in the revision of the simple group classification (see, for example, [S1]).

The basic situation in which the amalgam method applies is given in

Hypothesis 1.1. G is a group containing distinct finite subgroups P_1 and P_2 such that

(i) $G = \langle P_1, P_2 \rangle$; and

(ii) $P_1 \cap P_2$ contains no non-trivial normal subgroups of G.

Often it will also be the case that P_1 and P_2 are p-local subgroups of G (with G a finite group) and in such cases the amalgam method is a body of techniques for investigating the chief factors of P_i within $O_p(P_i)$ $(i = 1, 2)$, given certain information about $P_i/O_p(P_i)$. The ultimate aim of such an analysis is usually to obtain specific information about the structure of $P_1 \cap P_2$.

Let $S \in Syl_p H$. We define two sets of subgroups of H as follows.

$$\mathcal{L}(S) = \{P \leq H \mid S \leq P \text{ and } S \neq O_p(P) \neq 1\}.$$
$$\mathcal{ML}(S) = \{P \in \mathcal{L}(S) \mid P = \langle S^P \rangle \text{ and } S \text{ is contained in a unique maximal subgroup of } P\}.$$

The groups in $\mathcal{ML}(S)$ are referred to as minimal local subgroups (of H). If, for example, $P/O_p(P)$ is a group of Lie type of (Lie) rank 1 or $P/O_2(P) \cong S_5$, (with $p = 2$), then P is a minimal local subgroup.

Next we introduce a subset of $\mathcal{L}(S)$.

$$\mathcal{P}(S) = \{P \in \mathcal{L}(S) \mid \text{there exists a } p\text{-component } K \text{ of } P \text{ such that } P = \langle K, S\rangle = \langle S^P\rangle\}.$$

(K is a p-component of P if and only if $K \trianglelefteq \trianglelefteq P$ and either

(i) $K = K'$ and $K/O_p(K)$ is quasisimple; or

(ii) $K/O_p(K)$ is a q-group, q a prime, $K = O^p(K)$ and $KS \in \mathcal{ML}(S)$.)

Motivation for the work presented here is to be found in an idea of Gomi [Gom], which was reformulated in [S2;(2.9)] as follows:

Lemma 1.2. *Suppose*

(i) for every $P \in \mathcal{L}(S)$, $C_P(O_p(P)) \leq O_p(P)$; and

(ii) S is contained in at least two maximal p-local subgroups of H.

Then there exists $P_1 \in \mathcal{ML}(S)$ and $P_2 \in \mathcal{P}(S)$ such that $O_p(\langle P_1, P_2\rangle) = 1$.

If, furthermore, we have $\mathrm{core}_{\langle P_1,P_2\rangle}(P_1 \cap P_2) \leq S$ (a case of frequent interest), then, by Lemma 1.2, $\mathrm{core}_{\langle P_1,P_2\rangle}(P_1 \cap P_2) = 1$. Hence Hypothesis 1.1 holds for $G = \langle P_1, P_2\rangle$ and so, if suitable amalgam results are available, we obtain structural information about $P_1 \cap P_2$ and S. This is typical of the type of situation in which amalgam results are brought to bear in the current revision of the simple group classification.

We assume, from now on, that Hypothesis 1.1 holds. Because our aim is to study P_i, ($i = 1, 2$) there is no loss in generality in supposing $G = P_1 *_{P_1 \cap P_2} P_2$ (the free amalgamated product of P_1 and P_2 over $P_1 \cap P_2$). Also, we identify P_1, P_2 and $P_1 \cap P_2$ with their images in G.

In order to state one of the main results in [PR] we define certain classes of groups.

$$\mathcal{L}\mathrm{ie(even)} := \{L_2(2^n), U_3(2^n), SU_3(2^n), Sz(2^{2n+1}) \mid n \in \mathbb{N}\}.$$

$$\mathcal{D} := \{\mathrm{Dih}(2p) \mid p \text{ is an odd prime}\}.$$

($\mathrm{Dih}(2n)$ being the dihedral group of order $2n$).

$$\mathcal{L} := \mathcal{L}\mathrm{ie(even)} \cup \mathcal{D} \cup \{S_5\}.$$

$$\mathcal{S} := \{L \mid F^*(L)/Z(F^*(L)) \text{ is a sporadic simple group and } Z(F^*(L)) \text{ has odd order}\}.$$

Theorem 1.3. ([PR]) *Suppose P_1 and P_2 are finite subgroups of G which satisfy the following conditions:*

(i) $O^{2'}(P_1)/O_2(O^{2'}(P_1)) \in \mathcal{L}$ *and* $O^{2'}(P_2)/O_2(O^{2'}(P_2)) \in \mathcal{S}$;
(ii) $Syl_2(B) \subseteq Syl_2(P_1) \cap Syl_2(P_2)$ *where* $B := P_1 \cap P_2$;
(iii) $P_i = O^{2'}(P_i)B$, *for* $i = 1, 2$; *and*
(iv) $C_{P_i}(O_2(P_i)) \leq O_2(P_i)$, *for* $i = 1, 2$.

Then $b \in \{1, 2\}$.

We define the parameter b below; for the moment we note that a small value of b implies restrictions on the structure of $O_2(P_i)$, $(i = 1, 2)$. Observe that Theorem 1.3 is applicable to the situation depicted in Lemma 1.2 when $P_1/O_2(P_1) \in \mathcal{L}$ and $P_2/O_2(P_2) \in \mathcal{S}$. Further, we remark that eight of the sporadic simple groups possess configurations which satisfy the conditions of Theorem 1.3: $\cdot 2$, $\cdot 1$, Fi_{22}, Fi_{23}, Fi'_{24} $(b = 1)$, J_4, BM, M $(b = 2)$ and J_4 (this time with $b = 1$, $P_1/O_2(P_1) \cong S_5$).

Much of the amalgam method is concerned with using the action of G upon the *coset graph* Γ to deduce information about P_1 and P_2. The coset graph Γ is defined as follows: the vertices $V(\Gamma)$ are $\{P_i g \mid g \in G, i = 1, 2\}$ with the edge set, $E(\Gamma)$, being $\{(P_i g, P_j h) \mid i \neq j \text{ and } P_i g \cap P_j h \neq \emptyset\}$. Evidently G acts upon the graph Γ by right multiplication. By Serre [Se] Γ is a tree; the following basic result may be found in [Gol].

Lemma 1.4.

(i) *G acts faithfully on Γ and has two orbits on* $V(\Gamma)$.
(ii) *For* $\alpha \in V(\Gamma)$, $G_\alpha := stab_G(\alpha)$ *is G-conjugate to one of* P_1 *and* P_2.
(iii) *For* $(\alpha, \beta) \in E(\Gamma)$, $G_{\alpha\beta} := stab_G((\alpha, \beta))$ *is G-conjugate to* $P_1 \cap P_2$.

We note that the fact that G acts faithfully on Γ is a consequence of Hypothesis 1.1(ii) and we also see that we may determine the structure of the P_i $(i = 1, 2)$ by studying G_α for $\alpha \in V(\Gamma)$.

Before coming to the ideas around which much of the amalgam arguments revolve, we list some notation. Suppose $\delta \in V(\Gamma)$ and $(\delta, \lambda) \in E(\Gamma)$. Then,

$$\Delta(\delta) := \{\gamma \in V(\Gamma) \mid \gamma \text{ is adjacent to } \delta\},$$
$$O(\mathcal{L}) := \{\gamma \in V(\Gamma) \mid G_\gamma \text{ is } G\text{-conjugate to } P_1\},$$
$$O(\mathcal{S}) := \{\gamma \in V(\Gamma) \mid G_\gamma \text{ is G-conjugate to } P_2\},$$
$$L_\delta := O^{2'}(G_\delta),$$
$$Q_\delta := O_2(L_\delta),$$
$S_{\delta\lambda}$ is a fixed Sylow 2-subgroup of $G_{\delta\lambda}$,
$$Z_\delta := \langle \Omega_1(Z(S_{\delta\lambda}))^{L_\delta} \rangle,$$
$$V_\delta := \langle Z_\lambda{}^{L_\delta} \rangle \text{ and}$$
$$U_\delta := \langle V_\lambda^{L_\delta} \rangle.$$

Observe that $O(\mathcal{L})$ and $O(\mathcal{S})$ are the two orbits of G on $V(\Gamma)$. Moreover, for $\delta \in V(\Gamma)$, L_δ is transitive on $\Delta(\delta)$. (Because we may suppose $\delta \in \{P_1, P_2\}$, and then $\Delta(\delta)$ consists of the cosets of B in P_i, $i \in \{1,2\}$. So G_δ is transitive on $\Delta(\delta)$ whence, by hypothesis (iii) of Theorem 1.3, L_δ is transitive on $\Delta(\delta)$.) Therefore, we also have $Z_\delta = \langle \Omega_1(Z(S_{\delta\lambda})) | \lambda \in \Delta(\delta)\rangle$ and $V_\delta = \langle Z_\lambda | \lambda \in \Delta(\delta)\rangle$.

Now we define b, the *critical distance* of Γ. Let $d(\ ,\ )$ be the standard distance function on Γ. Then

$$b = \min_{\tau,\rho \in V(\Gamma)} \{d(\tau,\rho) \mid Z_\tau \not\leq Q_\rho\}.$$

If $(\tau,\rho) \in V(\Gamma) \times V(\Gamma)$ is such that $d(\tau,\rho) = b$ and $Z_\tau \not\leq Q_\rho$, then (τ,ρ) is called a *critical pair*. We denote the set of all critical pairs of Γ by $\mathcal{C}$. Whenever we use the notation $(\alpha,\alpha') \in \mathcal{C}$ we label the unique path in Γ joining α and α' as follows:

$$\alpha, \beta, \alpha+2, \ldots, \alpha'-1, \alpha'.$$

Lemma 1.5. *Let $\delta, \lambda, \mu \in V(\Gamma)$. Then*

(i) $1 \neq Z_\delta \leq \Omega_1(Z(Q_\delta))$;

(ii) $b \geq 1$; and

(iii) If $d(\lambda,\mu) < b$, then $[Z_\lambda, Z_\mu] = 1$ and, if $d(\lambda,\mu) \leq b$, then $Z_\lambda \leq L_\mu$.

PROOF. Let $\tau \in \Delta(\delta)$. Since $S_{\delta\tau} \neq 1$, clearly $Z_\delta \neq 1$, and hypothesis (iv) of Theorem 1.3 implies that $\Omega_1(Z(S_{\delta\tau})) \leq \Omega_1(Z(Q_\delta))$. Hence $Z_\delta \leq \Omega_1(Z(Q_\delta))$ and we also have $b \geq 1$. The first part of (iii) follows using (i) and the definition of b. In the case when $d(\lambda,\mu) = b$ we have

$$Z_\lambda \leq Q_{\mu-1} \leq O^{2'}(G_\mu) = L_\mu$$

(where $\mu - 1 \in \Delta(\mu)$ and $d(\lambda, \mu-1) = b-1$), from which we get the second part of (iii). □

The groups Z_δ, V_δ and (to a lesser extent) U_δ are the focus of many of our arguments, which make extensive use of the set $\mathcal{C}$. Let $(\alpha,\alpha') \in \mathcal{C}$. Then by Lemma 1.5(iii) $Z_\alpha \leq G_{\alpha'}$ and $Z_{\alpha'} \leq G_\alpha$. Likewise, we also get $V_\beta \leq G_{\alpha'}$ and $V_{\alpha'} \leq G_\beta$. Thus Z_α and $Z_{\alpha'}$ normalize each other and V_β and $V_{\alpha'}$ normalize each other. The direction our arguments take depend crucially on particular properties of $GF(2)$-representations of groups in $\mathcal{L}$ and $\mathcal{S}$. For example, suppose $b > 2$. Then V_β and $V_{\alpha'}$ are elementary abelian (by Lemma 1.5 (iii)) and therefore $[[V_{\alpha'}, V_\beta], V_\beta] = 1$. That is to say, V_β acts quadratically upon $V_{\alpha'}$ and accordingly great use is made of the results in [MS] concerning quadratic modules for groups in $\mathcal{S}$. Another feature of $GF(2)$-modules which arises in the proof of Theorem 1.3 is the codimension of fixed point spaces for certain 2-groups. In this connection the next result, which may be proved by induction on n, is often used. Assume that X is an H-operator group with $X = X_n \geq X_{n-1} \geq ... \geq X_0 = 1$ a chain of H-invariant subgroups such that $X_{i-1} \trianglelefteq X_i$ for $i = 1, \ldots, n$.

Lemma 1.6. *For $i = 1, \ldots, n$ put $\overline{X}_i = X_i/X_{i-1}$. Then,*

$$\prod_{i=1}^{n}[\overline{X}_i : C_{\overline{X}_i}(H)] \leq [X : C_X(H)].$$

Typically we use Lemma 1.6 with $X = Z_\delta$ or V_δ and H some 2-subgroup of G_δ. Then bounds for $[X : C_X(H)]$ and module data allow us to probe the G_δ-chief factors of Z_δ or V_δ.

When X is an H-operator group we use $\eta(H, X)$ to denote the number of non-central H-chief factors in X.

For $(\alpha, \alpha') \in \mathcal{C}$ we have the following fundamental subdivision:

$$[Z_\alpha, Z_{\alpha'}] \neq 1 \text{ (the non-commuting case)}$$
$$[Z_\alpha, Z_{\alpha'}] = 1 \text{ (the commuting case)}$$

For the moment we just remark that the commuting case is invariably problematical. The points mentioned above will be further illustrated later in this paper. In the next section we discuss the non-commuting case and give in detail a portion of the proof which highlights a recurring theme in the proof of Theorem 1.3–namely that the case $P_1/O_2(P_1) \cong S_5$ is often anamolous. In Section three we outline the strategy for dealing with the commuting case and present a part of the proof in which we must examine a $b = 3$ situation. These arguments give an indication of why b being small restricts the structure of $O_2(P_i)$ $i = 1, 2$.

2. The non-commuting case

In this section we give an overview of the proof of the following result:

Theorem 2.1. *If $(\alpha, \alpha') \in \mathcal{C}$ is such that $[Z_\alpha, Z_{\alpha'}] \neq 1$, then $b = 2$.*

We begin with some general remarks about the non-commuting case. Let $(\alpha, \alpha') \in \mathcal{C}$ be such that $[Z_\alpha, Z_{\alpha'}] \neq 1$. Then, by Lemma 1.5(i), $Z_{\alpha'} \not\leq Q_\alpha$ and therefore, $(\alpha', \alpha) \in \mathcal{C}$ as well. Further, we have $C_{S_{\alpha\beta}}(Z_\alpha) = Q_\alpha$ (for $C_{S_{\alpha\beta}}(Z_\alpha) > Q_\alpha$ together with the assumed structure of L_α/Q_α would force $C_{L_\alpha}(Z_\alpha) = L_\alpha$, contrary to $[Z_\alpha, Z_{\alpha'}] \neq 1$). Likewise $C_{S_{\alpha'\alpha'-1}}(Z_{\alpha'}) = Q_{\alpha'}$. Thus $\eta(L_\alpha, Z_\alpha) \neq 0 \neq \eta(L_{\alpha'}, Z_{\alpha'})$ and this observation is one of the main reasons why the non-commuting case is more tractable than the commuting case. By Lemma 1.5(iii) and hypothesis (ii) of Theorem 1.3, $Z_{\alpha'} \leq S_{\alpha\beta}$, hence

$$C_{Z_{\alpha'}}(Z_\alpha) = Z_{\alpha'} \cap C_{S_{\alpha\beta}}(Z_\alpha) = Z_{\alpha'} \cap Q_\alpha$$

and similarly,

$$C_{Z_\alpha}(Z_{\alpha'}) = Z_\alpha \cap Q_{\alpha'}.$$

Since $(\alpha, \alpha'), (\alpha', \alpha) \in \mathcal{C}$, without loss of generality we may suppose that $|Z_\alpha Q_{\alpha'}/Q_{\alpha'}| \leq |Z_{\alpha'} Q_\alpha / Q_\alpha|$. So we have $[Z_\alpha : Z_\alpha \cap Q_{\alpha'}] \leq |Z_{\alpha'} Q_\alpha/Q_\alpha|$ and thus

$$[Z_\alpha : C_{Z_\alpha}(Z_{\alpha'})] \leq |Z_{\alpha'} Q_\alpha / Q_\alpha|. \qquad (*)$$

Therefore, Z_α is what is termed an FF-module for L_α/Q_α. None of the groups in $\mathcal{S}$ possess FF-modules and therefore $\alpha \in O(\mathcal{L})$. The FF-modules (over $GF(2)$) for groups in $\mathcal{L}$ imply that the inequality (*) becomes an equality which enables us, by symmetry, to deduce that $Z_{\alpha'}$ is also an FF-module for $L_{\alpha'}/Q_{\alpha'}$. Hence $\alpha' \in O(\mathcal{L})$ as well. Now, as $U_3(2^n), SU_3(2^n), Sz(2^{2n+1})$ do not have any FF-modules (over $GF(2)$), we have established

Lemma 2.2.

(i) $|Z_{\alpha'} Q_\alpha/Q_\alpha| = [Z_\alpha : C_{Z_\alpha}(Z_{\alpha'})] = [Z_{\alpha'} : C_{Z_{\alpha'}}(Z_\alpha)] = |Z_\alpha Q_{\alpha'}/Q_{\alpha'}|$;

(ii) $L_\alpha/Q_\alpha \cong S_5$ *or* $L_2(2^n)$, $n \in \mathbb{N}$; *and*

(iii) b *is even.*

We suppose Theorem 2.1 is false, and argue for a contradiction. In view of Lemma 2.2(iii) we must have $b \geq 4$. A relatively short argument (making essential use of $b \geq 4$) disposes of the case when $L_\alpha/Q_\alpha \cong L_2(2^n)$ $(n \in \mathbb{N})$ and when $L_\alpha/Q_\alpha \cong S_5$ with Z_α a natural S_5-module. (The natural modules for $L_2(2^n)$ and the natural module for S_5 behave in the same way from the point of view of amalgam arguments.) We move to the closing stages of the proof of Theorem 2.1 and see how the case $L_\alpha/Q_\alpha \cong S_5, Z_\alpha$ an orthogonal S_5-module and $|Z_\alpha Q_{\alpha'}/Q_{\alpha'}| = 2$ is dealt with. (An orthogonal S_5-module is, by definition, a $GF(2)S_5$-module isomorphic to the 4-dimensional irreducible summand of the permutation module on five letters.) Note that here we must have $Z_\alpha Q_{\alpha'}/Q_{\alpha'} = \langle t_{\alpha'} \rangle$ where $t_{\alpha'}$ is a transposition of $S_5 \cong L_{\alpha'}/Q_{\alpha'}$ (with an analogous statement for $Z_{\alpha'} Q_\alpha / Q_\alpha$).

For $\tau \in O(\mathcal{L})$ and $(\tau, \xi) \in E(\Gamma)$, the unique maximal subgroup of L_τ containing $S_{\tau\xi}$ will be denoted by $M_{\tau\xi}$. So $M_{\tau\xi}/Q_\tau \cong S_4$ and therefore $M_{\tau\xi}/O_2(M_{\tau\xi}) \cong L_2(2)$. Now put $H = \langle M_{\alpha\beta}, L_\beta \rangle$ and $Q_H = \mathrm{core}_H(L_\beta \cap M_{\alpha\beta})$. It can be shown that H/Q_H satisfies the hypotheses of Theorem 1.3 with $M_{\alpha\beta}/O_2(M_{\alpha\beta}) \cong L_2(2)$ and $L_\beta/Q_\beta \in \mathcal{S}$; in particular, $M_{\alpha\beta} \not\leq L_\beta$. Taking advantage of work already completed on this special case we deduce that

2.1.1. the following hold:

(i) $[\mathrm{core}_{L_\beta}(O_2(M_{\alpha\beta})) : Q_H] \leq 2$; and

(ii) $\mathrm{core}_{L_\gamma}(O_2(M_{\alpha+2\beta})) \cap \mathrm{core}_{L_\beta}(O_2(M_{\alpha\beta})) = Q_H$ where γ is any vertex in $\Delta(\alpha+2)$ with $M_{\alpha+2\gamma} = M_{\alpha+2\beta}$.

We next put $Y_\beta := \bigcap_{\lambda\in\Delta(\beta)} Z_\lambda$ $(= \mathrm{core}_{L_\beta} Z_\alpha)$ under the spotlight. Now $Z_\alpha \neq Y_\beta$ (as $\mathrm{core}_G G_{\alpha\beta} = 1$ by Hypothesis 1.1(ii)) and we know that $Z_\beta \leq Y_\beta$. Thus, as $|Z_\alpha| = 2^4$, $|Y_\beta| = 2, 2^2$ or 2^3. We can quickly dispose of the possibility $|Y_\beta| = 2^3$. Suppose $|Y_\beta| = 2^3$ holds. Then, clearly, as $Y_\beta \leq Z_{\alpha+2}, Q_{\alpha+2} \cap Q_\beta$ centralizes Y_β. From $Y_\beta \trianglelefteq G_\beta$ we have $Y_\beta \trianglelefteq S_{\alpha\beta}$. A simple calculation in the orthogonal S_5-module shows that no non-trivial subgroup of $S_{\alpha\beta}/Q_\alpha$ can centralize a subgroup of Z_α which has order 2^3 and is normal in $S_{\alpha\beta}$. This therefore forces $Q_{\alpha+2} \cap Q_\beta \leq Q_\alpha$. But then $Z_{\alpha'} \leq Q_{\alpha+2} \cap Q_\beta \leq Q_\alpha$, contrary to $(\alpha', \alpha) \in \mathcal{C}$. So $|Y_\beta| \neq 2^3$.

Next we examine the case $|Y_\beta| = 2^2$, using the following subgraph of Γ:

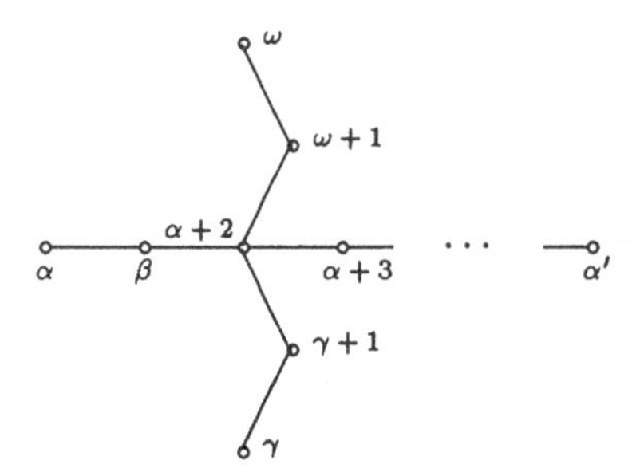

where $\beta, \omega+1, \gamma+1$ are three distinct vertices of $\Delta(\alpha+2)$ for which $M_{\alpha+2\beta} = M_{\alpha+2\omega+1} = M_{\alpha+2\gamma+1}$.

Using (2.1.1) we obtain the following result:

2.1.2. If $|Y_\beta| = 2^2$ then, up to a relabelling of vertices, we have

(i) $Z_{\alpha'} \leq \mathrm{core}_{L_{\gamma+1}}(O_2(M_{\alpha+2\gamma+1}))$; and

(ii) $(\alpha, \alpha'), (\omega, \alpha') \in \mathcal{C}$, $|Z_\omega Q_{\alpha'}/Q_{\alpha'}| = |Z_\alpha Q_{\alpha'}/Q_{\alpha'}| = 2$ with $Z_\omega Q_{\alpha'} \neq Z_\alpha Q_{\alpha'}$.

With (2.1.2) to hand we can show that $|Y_\beta| = 2^2$ leads to a contradiction. Since $Z_{\alpha'} Q_\alpha / Q_\alpha = \langle t_\alpha \rangle$, where t_α is a transposition of L_α/Q_α, $Z_{\alpha'} \nleq O_2(M_{\alpha\beta})$. Hence, as $Q_H \leq O_2(M_{\alpha\beta})$, $Z_{\alpha'} \nleq Q_H$. Therefore, combining (2.1.1)(i) and (2.1.2)(i) yields $[Z_{\alpha'} Q_H : Q_H] = 2$, whence $[Z_{\alpha'} : Z_{\alpha'} \cap Q_H] = 2$. Since $[Z_{\alpha'} : Z_{\alpha'} \cap O_2(M_{\alpha\beta})] = 2$ and $Z_{\alpha'} \cap O_2(M_{\alpha\beta}) \geq Z_{\alpha'} \cap Q_H$, this yields $Z_{\alpha'} \cap O_2(M_{\alpha\beta}) = Z_{\alpha'} \cap Q_H$. Now, as $Q_\alpha \leq O_2(M_{\alpha\beta})$, $Z_{\alpha'} \cap Q_\alpha \leq Z_{\alpha'} \cap O_2(M_{\alpha\beta})$ and therefore $[Z_{\alpha'} : Z_{\alpha'} \cap Q_\alpha] = 2$ yields $Z_{\alpha'} \cap Q_\alpha = Z_{\alpha'} \cap O_2(M_{\alpha\beta})$. Since, by (2.1.2)(ii), $(\omega, \alpha') \in \mathcal{C}$ with $|Z_\omega Q_{\alpha'}/Q_{\alpha'}| = 2$ $(= |Z_{\alpha'} Q_\omega / Q_\omega|)$, the same arguments apply, giving

$$\begin{aligned} Z_{\alpha'} \cap Q_\alpha &= Z_{\alpha'} \cap O_2(M_{\alpha\beta}) = Z_{\alpha'} \cap Q_H \\ &= Z_\omega \cap O_2(M_{\omega\omega+1}) = Z_{\alpha'} \cap Q_\omega. \end{aligned}$$

Then $Z_\alpha Q_{\alpha'}/Q_{\alpha'}$ and $Z_\omega Q_{\alpha'}/Q_{\alpha'}$ centralize the same hyperplane of $Z_{\alpha'}$ and so (by properties of orthogonal S_5-modules) $Z_\alpha Q_{\alpha'} = Z_\omega Q_{\alpha'}$, which is against (2.1.2)(ii). So we have shown that $|Y_\beta| \neq 2^2$ and hence

2.1.3. $|Y_\beta| = 2$.

We now bring two further subgroups into the picture

$$H_\beta := \langle [Z_\alpha, S_{\alpha\beta}, S_{\alpha\beta}]^{L_\beta} \rangle; \text{ and } X_\alpha := \langle H_\beta^{L_\alpha} \rangle.$$

It follows readily from (2.1.3) that

2.1.4. we have

(i) $\eta(L_\beta, H_\beta) \neq 0$;

(ii) $\eta(L_\beta, V_\beta) \geq 2$; and

(iii) $\eta(L_\alpha, X_\alpha) \geq 3$.

As we shall see in a moment, the sizes of centralizers of 2-groups on modules which have quadratically acting 2-groups are incompatible with (2.1.4).

2.1.5. $X_\alpha \leq Q_{\alpha'-2}$ ($\leq G_{\alpha'-1}$).

If $X_\alpha \not\leq Q_{\alpha'-2}$, then, by the construction of X_α, there must exist $\alpha - 1 \in \Delta(\alpha)$ and $\alpha - 2 \in \Delta(\alpha - 1)$ such that

$$J_{\alpha-2\alpha-1} := [Z_{\alpha-2}, S_{\alpha-2\alpha-1}, S_{\alpha-2\alpha-1}] \not\leq Q_{\alpha'-2}.$$

Clearly, then, $Z_{\alpha-2} \not\leq Q_{\alpha'-2}$ and therefore, $(\alpha-2, \alpha'-2) \in \mathcal{C}$. Considering the action of $Z_{\alpha'-2}$ upon $Z_{\alpha-2}$, and using properties of the orthogonal S_5-module gives us

$$[J_{\alpha-2\alpha-1}, Z_{\alpha'-2}] = 1,$$

whence $J_{\alpha-2\alpha-1} \leq C_{S_{\alpha'-3\alpha'-2}}(Z_{\alpha'-2}) = Q_{\alpha'-2}$, a contradiction. This verifies (2.1.5).

From (2.1.5) we have

$$H_{\alpha-1} \leq X_\alpha \leq G_{\alpha'-1},$$

and so $[H_{\alpha-1}, V_{\alpha'-1}, V_{\alpha'-1}] = 1$. In particular, $V_{\alpha'-1} \cap Q_\alpha$ is a subgroup of $L_{\alpha-1}$ which acts quadratically upon $H_{\alpha-1}$. Now (2.1.4)(i) and results on modules for groups in $\mathcal{S}$ yield that

2.1.6. $|(V_{\alpha'-1} \cap Q_\alpha)Q_{\alpha-1}/Q_{\alpha-1}| \leq 2^4$.

If $X_\alpha \leq L_{\alpha'}$, then $[X_\alpha : X_\alpha \cap Q_{\alpha'}] \leq 2^2$ and so $[X_\alpha : C_{X_\alpha}(Z_{\alpha'})] \leq 2^2$. But then (using Lemma 1.6) $\eta(L_\alpha, X_\alpha) \leq 2$, against (2.1.4)(iii). Thus $X_\alpha \not\leq L_{\alpha'}$ and so we may find an $\alpha - 1 \in \Delta(\alpha)$ and $\alpha - 2 \in \Delta(\alpha - 1)$ such that

$$J_{\alpha-2\alpha-1} := [Z_{\alpha-2}, S_{\alpha-2\alpha-1}, S_{\alpha-2\alpha-1}] \not\leq Q_{\alpha'-1}.$$

Since $V_{\alpha'-1} \cap Q_{\alpha-1} \leq S_{\alpha-2\alpha-1}$,

$$[J_{\alpha-2\alpha-1}, V_{\alpha'-1} \cap Q_{\alpha-1}] \leq [Z_{\alpha-2}, S_{\alpha-2\alpha-1}; 3],$$

and so $|[J_{\alpha-2\alpha-1}, V_{\alpha'-1} \cap Q_{\alpha-1}]| \leq 2$. Let $t \in J_{\alpha-2\alpha-1} \backslash Q_{\alpha'-1}$. Then

$$[V_{\alpha'-1} \cap Q_{\alpha-1} : C_{V_{\alpha'-1} \cap Q_{\alpha-1}}(t)] \leq 2$$

and hence, using (2.1.6), we obtain

$$[V_{\alpha'-1} : C_{V_{\alpha'-1}}(t)] \leq 2.2^4.2^2 = 2^7.$$

Because $\eta(L_\beta, V_\beta) \geq 2$ this, together with module data, forces

$$L_\beta / C_{L_\beta}(V_\beta / [V_\beta, Q_\beta]) \cong \mathit{Aut}\ M_{22}.$$

As a result we can now use more specific module data to lower the bound in (2.1.6) to 2^2. Consequently,

$$[V_{\alpha'-1} : C_{V_{\alpha'-1}}(t)] \leq 2.2^2.2^2 = 2^5.$$

However, module data and Lemma 1.6 then yield $\eta(L_{\alpha'-1}, V_{\alpha'-1}) \leq 1$, contradicting (2.1.4)(ii). With this contradiction our analysis of the case $L_\alpha/Q_\alpha \cong S_5$, Z_α an orthogonal S_5-module and $|Z_\alpha Q_{\alpha'}/Q_{\alpha'}| = 2$ is complete.

3. The commuting case

The main result here is

Theorem 3.1. *If $(\alpha, \alpha') \in \mathcal{C}$ and $[Z_\alpha, Z_{\alpha'}] = 1$, then $b = 1$.*

The first step in proving this theorem is to show that $\alpha \in O(\mathcal{S})$ for $(\alpha, \alpha') \in \mathcal{C}$. Thus to complete the proof of Theorem 3.1 we must then show that $b = 1$. So we assume $b > 1$ and argue for a contradiction. Let $(\alpha, \alpha') \in \mathcal{C}$ be such that $[Z_\alpha, Z_{\alpha'}] = 1$. A major part of the analysis of this situation is contained in

Lemma 3.2. $L_\beta/Q_\beta \not\cong S_5$.

We join the proof of Lemma 3.2 at the point where we are moving in for the kill. So we are assuming that $L_\beta/Q_\beta \cong S_5$ holds (as well as $b > 1$) with the aim of uncovering a contradiction. So far we have deduced that $b = 3$ (so $\alpha + 2 = \alpha' - 1$) and further pertinent information available to us at this point is gathered in (3.2.1). For $\lambda \in O(\mathcal{S})$ and $\delta \in O(\mathcal{L})$ we define two subgroups F_λ and Y_δ as follows: F_λ is a subgroup of L_λ such that $F_\lambda \geq Q_\lambda$ and $F_\lambda/Q_\lambda = F^*(L_\lambda/Q_\lambda)$ and $Y_\delta = \bigcap_{\mu\in\Delta(\delta)} Z_\mu$. Clearly $[Y_\delta, Q_\mu] = 1$ for any $\mu \in \Delta(\delta)$ and hence, as $\langle Q_\mu \mid \mu \in \Delta(\delta)\rangle \geq O^2(L_\delta)$, $[Y_\delta, O^2(L_\delta)] = 1$.

3.2.1. The following hold:

(i) $L_\alpha/C_{L_\alpha}(Z_\alpha) \cong AutM_{22}$ and $V_{\alpha'} \cap Q_\beta \not\leq F_\alpha$.

(ii) $\eta(L_\beta, V_\beta) = 2$.

(iii) $\mid Z_\alpha Q_{\alpha'}/Q_{\alpha'} \mid = 2^2$.

(iv) Let T be such that $S_{\alpha\beta} \geq T \geq Q_\alpha$ with $T/Q_\alpha = Z(S_{\alpha\beta}/Q_\alpha)$. Then $C_{Z_\alpha}(T)[Z_\alpha, T] \leq Y_\beta$ with $|[Z_\alpha, T]| = 2^4$. Also we have $Y_\beta = [Z_\alpha, V_{\alpha'} \cap Q_\beta]$.

(v) If $X \leq S_{\alpha\beta}$ and $[X, Y_\beta] = 1$, then $\mid XQ_\alpha/Q_\alpha \mid \leq 2^2$.

Parts (iv) and (v) are consequences of pinning down the structure of Z_α as a $GF(2)(L_\alpha/C_{L_\alpha}(Z_\alpha))$-module. Notice that part (iii) implies that $[S_{\alpha\beta} : Q_\alpha Q_\beta] \leq 2$. This is because $Z_\alpha Q_{\alpha'} \leq Q_{\alpha+2}Q_{\alpha'}$ and $Q_{\alpha+2}Q_{\alpha'}$ is G-conjugate to $Q_\alpha Q_\beta$.

One fact about $GF(2)S_5$-modules we shall use shortly is that if V is a non-trivial irreducible $GF(2)S_5$-module and A a fours subgroup of S_5, then $[V : C_V(A)] \geq 2^2$.

3.2.2. $\eta(L_{\alpha'}, Q_{\alpha'}) \leq 3$.

Let $\alpha' + 1 \in \Delta(\alpha')\backslash\{\alpha+2\}$, and put $R_{\alpha'} = \langle [U_{\alpha'+1}, Q_{\alpha'}]^{L_{\alpha'}}\rangle$. By Lemma 1.5(iii) and the definition of $U_{\alpha'+1}$, $U_{\alpha'+1} \leq L_{\alpha'}$. Also we observe that $U_{\alpha'+1} \not\leq Q_{\alpha'}$ for otherwise the transitivity of L_α on $\Delta(\alpha')$ gives $U_{\alpha+2} \leq Q_{\alpha'}$, whence $Z_\alpha \leq U_{\alpha+2} \leq Q_{\alpha'}$ contrary to $(\alpha, \alpha') \in \mathcal{C}$. Clearly $[U_{\alpha'+1}, Q_{\alpha'}] \leq R_{\alpha'}$, so as $Q_{\alpha'}\langle U_{\alpha'+1}{}^{L_{\alpha'}}\rangle \geq O^2(L_{\alpha'})$ (by the structure of $L_{\alpha'}/Q_{\alpha'} \cong S_5$) we see that $\eta(L_{\alpha'}, Q_{\alpha'}) = \eta(L_{\alpha'}, R_{\alpha'})$. Since $Y_{\alpha'} \leq Z_{\alpha'+1}$, $[U_{\alpha'+1}, Y_{\alpha'}] = 1$ (by Lemma 1.5(iii)) and so $[R_{\alpha'}, Y_{\alpha'}] = 1$. Hence, as $R_{\alpha'} \leq S_{\alpha'\alpha+2}$, $\mid R_{\alpha'}Q_{\alpha+2}/Q_{\alpha+2} \mid \leq 2^2$ by (3.2.1)(v). From $L_\beta/Q_\beta \cong S_5$ we infer that $|(R_{\alpha'} \cap Q_{\alpha+2})Q_\beta/Q_\beta| \leq 2^3$. Now $R_{\alpha'} \cap Q_{\alpha+2} \cap Q_\beta$ centralizes Y_β (as $Y_\beta \leq Z_{\alpha+2} \leq \Omega_1(Z(Q_{\alpha+2})))$ and so, using (3.2.1)(v) again, we obtain $|(R_{\alpha'} \cap Q_{\alpha+2} \cap Q_\beta)Q_\alpha/Q_\alpha| \leq 2^2$. Consequently,

$$[R_{\alpha'} : C_{R_{\alpha'}}(Z_\alpha)] \leq [R_{\alpha'} : R_{\alpha'} \cap Q_\alpha] \leq 2^7.$$

Combining this with (3.2.1)(iii) and Lemma 1.6 (together with the aforementioned facts about $GF(2)S_5$-modules) yields $\eta(L_{\alpha'}, R_{\alpha'}) \leq 3$. Therefore $\eta(L_{\alpha'}, Q_{\alpha'}) = \eta(L_{\alpha'}, R_{\alpha'}) \leq 3$.

As is clear from the arguments in (3.2.2) $\eta(G_{\alpha'}, Q_{\alpha'})$ can, in general, be bounded in terms of b and certain data concerning 2-subgroups and $GF(2)$-representations of L_λ/Q_λ ($\lambda \in V(\Gamma)$). Such bounds are usually of little value, except when b is small. Then the close proximity of β and α' allows us to produce much tighter bounds for $\eta(G_{\alpha'}, Q_{\alpha'})$.

For $\delta \in V(\Gamma)$ we define $K_\delta = \bigcap_{\lambda \in \Delta(\delta)} Q_\lambda$; note that $K_\delta \trianglelefteq L_\delta$, $K_\delta \leq Q_\delta$ and $K_\delta = C_{Q_\delta}(V_\delta)$. Also, as $b = 3$, $V_\delta \leq K_\delta$.

3.2.3. Let $X_{\alpha'}$ be a normal subgroup of $L_{\alpha'}$ with $K_{\alpha'} \leq X_{\alpha'} \leq Q_{\alpha'}$. Then $[X_{\alpha'} : K_{\alpha'}] \neq 2$.

Suppose (3.2.3) is false. So we have $[X_\alpha : K_{\alpha'}] = 2$ and (hence) $X_{\alpha'}/K_{\alpha'} \leq Z(L_{\alpha'}/K_{\alpha'})$. Since $X_{\alpha'} \not\leq K_{\alpha'}, X_{\alpha'} \not\leq Q_{\alpha+2}$ and therefore, by (3.2.1)(i) and [CCN], $X_{\alpha'}Q_{\alpha+2}/Q_{\alpha+2} = Z(S_{\alpha+2\,\alpha'}/Q_{\alpha+2})$. So, by (3.2.1)(iv), $[Z_{\alpha+2}, X_{\alpha'}] \leq Y_{\alpha'}$ with $|[Z_{\alpha+2}, X_{\alpha'}]| = 2^4$. Hence $[Z_{\alpha+2}, X_{\alpha'}]$ is centralized by $O^2(L_{\alpha'})$. Therefore

$$[Z_{\alpha+2}, X_{\alpha'}] \trianglelefteq \langle O^2(L_{\alpha'}), S_{\alpha'\alpha+2}\rangle = L_{\alpha'}.$$

Thus

$$[V_{\alpha'}, X_{\alpha'}] = [\langle Z_{\alpha+2}^{L_{\alpha'}}\rangle, X_{\alpha'}] = \langle [Z_{\alpha+2}, X_{\alpha'}]^{L_{\alpha'}}\rangle = [Z_{\alpha+2}, X_{\alpha'}].$$

Since $[X_{\alpha'} : K_{\alpha'}] = 2$, $X_{\alpha'}$ acts as an involution upon $V_{\alpha'}$ and so $[V_{\alpha'} : C_{V_{\alpha'}}(X_{\alpha'})] = |[V_{\alpha'}, X_{\alpha'}]| = |[Z_{\alpha+2}, X_{\alpha'}]| = 2^4$. From $[Z_{\alpha+2} : C_{Z_{\alpha+2}}(X_{\alpha'})] = |[Z_{\alpha+2}, X_{\alpha'}]| = 2^4$ we conclude that $V_{\alpha'} = Z_{\alpha+2}C_{V_{\alpha'}}(X_{\alpha'})$. This implies, using (3.2.1)(ii) (and $[Q_{\alpha+2}, Z_{\alpha+2}] = 1$, $Q_{\alpha+2} \not\leq Q_{\alpha'}$), that $\eta(L_{\alpha'}, C_{V_{\alpha'}}(X_{\alpha'})) = \eta(L_{\alpha'}, V_{\alpha'}) = 2$.

Next we consider the action of Z_α on $C_{V_{\alpha'}}(X_{\alpha'})$ (and it is here that we again exploit the closeness of β and α'). Employing (3.2.1)(iv) we obtain

$$[Z_\alpha, C_{V_{\alpha'}}(X_{\alpha'}) \cap Q_\beta] \leq Y_\beta \cap C_{V_{\alpha'}}(X_{\alpha'}) \leq Z_{\alpha+2} \cap C_{V_{\alpha'}}(X_{\alpha'}) \leq C_{Z_{\alpha+2}}(X_{\alpha'}) \leq Y_{\alpha'}.$$

But then

$$[C_{V_{\alpha'}}(X_{\alpha'})/Y_{\alpha'} : C_{C_{V_{\alpha'}}(X_{\alpha'})/Y_{\alpha'}}(Z_\alpha)] \leq 2^2,$$

which, by (3.2.1)(iii) and properties of $GF(2)S_5$-modules, forces

$$\eta(L_{\alpha'}, C_{V_{\alpha'}}(X_{\alpha'})/Y_{\alpha'}) \leq 1.$$

Since $\eta(L_{\alpha'}, Y_{\alpha'}) = 0$, this then contradicts $\eta(L_{\alpha'}, C_{V_{\alpha'}}(X_{\alpha'})) = 2$ and thus completes the proof of (3.2.3).

From (3.2.1)(i) $V_{\alpha'} \cap Q_\beta \not\leq F_\alpha$. Since $V_{\alpha'} \cap Q_\beta \leq U_{\alpha+2} \cap Q_\beta$ and $(U_\alpha \cap Q_\beta)[Q_\beta, Q_\beta]K_\beta \leq F_\alpha$, this yields

$$(U_{\alpha+2} \cap Q_\beta)[Q_\beta, Q_\beta]K_\beta \neq (U_\alpha \cap Q_\beta)[Q_\beta, Q_\beta]K_\beta.$$

Consequently, $\eta(L_\beta, Q_\beta/[Q_\beta, Q_\beta]K_\beta) \neq 0$ (note that the closeness of β and α' has been used again). Combining this with (3.2.1)(ii) and (3.2.2) we deduce that $\eta(L_\beta, [Q_\beta, Q_\beta]K_\beta/K_\beta) = 0$. Applying (3.2.3) (to L_β) forces $[Q_\beta, Q_\beta] \leq K_\beta$ ($\leq Q_\alpha$). But then $Q_\beta Q_\alpha/Q_\alpha$ is an abelian subgroup of $S_{\alpha\beta}/Q_\alpha$ of index at most 2, which is contrary to the structure of $S_{\alpha\beta}/Q_\alpha$, as $L_\alpha/C_{L_\alpha}(Z_\alpha) \cong AutM_{22}$. This contradiction concludes the proof of Lemma 3.2.

References

[CCN] J.H. Conway, R.T. Curtis, S.P. Norton, R.A. Parker and R.A. Wilson, *An Atlas of Finite Groups* (Oxford Univ. Press, London, 1985).

[FT] W. Feit and J.G. Thompson, Solvability of groups of odd order, *Pacific J. Math.* **13**(1963), 775–1029.

[Gol] D.M. Goldschmidt, Automorphisms of trivalent graphs, *Ann. Math.* **111** (1980), 377–404.

[Gom] K. Gomi, On the 2-local structure of groups of characteristic 2-type, *J. Algebra* **108**(1987), 492–502.

[MS] U. Meierfrankenfeld and G. Stroth, Quadratic GF(2)-modules for sporadic simple groups and alternating groups, *Comm. Algebra* **18**(7)(1990), 2099–2139.

[PR] C. Parker and P. Rowley, Amalgams which involve sporadic simple groups, in preparation.

[Se] J.P. Serre, *Trees* (Springer, New York, 1980).

[S1] B. Stellmacher, An application of the amalgam method: the 2-local structure of N-groups of characteristic 2 type, preprint, Kiel, 1992.

[S2] B. Stellmacher, On the 2-local structure of finite groups, in *Groups, Combinatorics and Geometry* (M. Liebeck and J. Saxl (eds.), Cambridge University Press, 1992).

[T1] J.G. Thompson, Ph.D thesis, University of Chicago, 1959.

[T2] J.G. Thompson, Normal p-complements for finite groups, *Math. Z.* **72**(1960), 332–354.

[T3] J.G. Thompson, Normal p-complements for finite groups, *J. Algebra* **1**(1964), 43–46.

VANISHING ORBIT SUMS IN GROUP ALGEBRAS OF p-GROUPS

M. ANWAR RAO and ROBERT SANDLING

Mathematics Department, The University, Manchester M13 9PL, England
E-mail: rsandling@manchester.ac.uk

Abstract

For G a nonabelian finite p-group, the non-fixed-point orbits of the unit group $U(\mathbb{F}_p G)$ in its conjugation action on the group algebra $\mathbb{F}_p G$ sum to 0 in $\mathbb{F}_p G$. Also shown are the related facts that $U(\mathbb{F}_p G)$ has no conjugacy class of order p and no abelian maximal subgroup.

1991 *Mathematics Subject Classification:* Primary 20C05; Secondary 16S34, 16U60, 20C20, 20D15, 20J05

The sums of the orbits of a group in its action on a module are encountered in the representation theory of groups and in the cohomology of groups. Most notable are the class sums, the sums of certain orbits of the conjugation action of a group on its group ring; these form a basis for the centre of a group ring of a finite group. Another significant instance is the norm element, the sum of the orbit of the identity in the right regular action of a finite group on its group ring. An orbit whose sum is equal to 0 seems less worthy of attention. Notwithstanding, this paper reports on such vanishing orbit sums. Its interest lies in the fact that, in the case considered, the vanishing is universal: all the non-fixed-point orbit sums vanish. Concepts from the cohomology of groups provide a context for our results. Let M be a module for a finite group P over some ring of coefficients. The homomorphism $M \to M^P$: $m \to m\sum P$, is of great significance in the case of integer coefficients, its cokernel being the Tate cohomology group $\hat{H}^0(P, M)$ (here M^P denotes the submodule of fixed points of P on M and $\sum P$ denotes the norm element of P). The element $m\sum P$ is not what we mean by the orbit sum of m as, in this expression, a given element of the orbit of m may enter into the sum repeatedly. In fact, $m\sum P$ is frequently 0; for example, it is 0 when P is a p-group for some prime p, the characteristic is p and $1 \neq C_P(m)$, the subgroup of all elements of P which fix m. The orbit sum considered here is, as the name indicates, simply the sum of the elements of the orbit mP of m; $\sum mP$ is thus the appropriate notation. There may be some interest in the mapping $M \to M^P : m \to \sum mP$, but it is not in general a homomorphism.

In this paper, the focus of attention is on the following situation. Let G be a finite p-group and F a field of characteristic p. In this case the augmentation ideal $I = I(FG)$ is nilpotent. The group $V = V(FG) = 1 + I$, the group of normalised units, is itself a p-group, finite if F is finite; we view it as the relevant object of study in preference to the full unit group $U(FG)$ to which our theorems also apply.

The inclusion of V in the full group of units of FG induces a ring homomorphism $FV \to FG$. By change of rings, every FG-module is then an FV-module. Our concern lies with FG as FV-module but not induced in this sense; instead we view FG as a right FV-module under the conjugation action of V: $\alpha \in FG$, $v \in V$ give $\alpha^v = v^{-1}\alpha v$. Our main theorem illuminates the case $P = V$ and $M = FG$ when $F = \mathbb{F}_p$, the field of p elements.

Theorem 1. *Let G be a finite p-group and $V = V(\mathbb{F}_pG)$. Let $\mathcal{O}$ be a non-trivial orbit of V in its action on $\mathbb{F}_pG$ by conjugation. Then $\sum \mathcal{O} = 0$.*

In particular, the sums of the non-central conjugacy classes of V, taken in $\mathbb{F}_pG$, vanish. It is then immediate that V has no conjugacy class of order p when $p = 2$. That this statement holds for any p is the second theorem of the paper; its proof is the major step in the proof of the first.

Theorem 2. *Let G be a finite nonabelian p-group and $V = V(\mathbb{F}_pG)$. Then V has no conjugacy class of order p, no centraliser of index p and no abelian maximal subgroup.*

This theorem makes a modest contribution to relieving the obscurity of the modular unit groups. That V is highly nonabelian has been known for some time [2]. The last point of the theorem, an immediate consequence of the second, is a further manifestation of this noncommutativity. It is not an aspect which follows from the statement of the result of Coleman and Passman (V has the wreath product $C_p \wr C_p$ as a section), but it can be deduced in some circumstances from other known results. For example, the refinements of [2] given by Shalev in [8] have this consequence. He showed that $C_p \wr C_{p^2}$ is a section of V under certain weak hypotheses on G; this wreath product does not have an abelian maximal subgroup: indeed, its Frattini subgroup is nonabelian. Coleman points out that another of Shalev's results, that V is not metabelian if G is nonabelian and $p \geq 5$ [9], also has this consequence.

Similar considerations about subgroups permit the following deduction.

Corollary. *Let G be a finite nonabelian p-group and let R be any ring with unit and of characteristic a power of p. Then the p-group $V(RG)$ has no abelian maximal subgroup.*

The proofs of our theorems begin with a lemma which gives a familiar and handy description of the annihilator of the augmentation ideal of a subgroup of G.

Lemma. *Let H be a subgroup of the finite group G and R any ring of coefficients, commutative with 1. The right annihilator of the R-submodule $I(RH)$ of RG is the right ideal $(\sum H)RG$. An element $\alpha \in RG$ belongs to this ideal if and only if it is constant on the right cosets of H in G.*

Note that, in the functional notation for the coefficients of a group ring element, the last condition means that $\alpha(hg) = \alpha(g)\ \forall h \in H\ \forall g \in G$.

The next assertion establishes Theorem 2 (of which it is also a consequence). The form of its statement is fashioned for its role as the main step of the proof of Theorem 1.

Proposition 1. *Let G be a finite p-group and $V = V(\mathbb{F}_pG)$. Let $\alpha \in \mathbb{F}_pG$. If α is not central, then* $\mathrm{codim} C_{\mathbb{F}_pG}(\alpha) \geq 2$ *and* $|V : C_V(\alpha)| \geq p^2$.

PROOF. The two conclusions are equivalent. This is because $C_{\mathbb{F}_pG}(\alpha) = \mathbb{F}_p + C_V(\alpha)$ so that $|V : C_V(\alpha)| = |\mathbb{F}_pG : C_{\mathbb{F}_pG}(\alpha)| = p^k$ where k is the codimension of $C_{\mathbb{F}_pG}(\alpha)$ in $\mathbb{F}_pG$.

By way of contradiction suppose that $k = 1$ so that $|V : C_V(\alpha)| = p$. It follows that

(a) if $W \leq V$, then $|W : C_W(\alpha)|$ divides p, and

(b) if J is a subspace of $\mathbb{F}_pG$, then the codimension of $C_J(\alpha)$ in J is ≤ 1.

For (a), note that if $W \not\leq C_V(\alpha)$, then $V = WC_V(\alpha)$ so that $W/C_W(\alpha)$ is isomorphic to $V/C_V(\alpha)$, a cyclic group of order p by hypothesis. Statement (b) is proved in a similar manner.

Let H denote the centraliser $C_G(\alpha)$ so that $\mathbb{F}_pH \subseteq C_{\mathbb{F}_pG}(\alpha)$. It follows that

(c) $|G : H| = p$, and

(d) for $g \in G$, $\mathbb{F}_pHg \cap C_{\mathbb{F}_pG}(\alpha) = \begin{cases} \mathbb{F}_pH & \text{if } g \in H, \\ I(H)g & \text{otherwise.} \end{cases}$

For (c), observe that statement (a) implies that, if $|G : H| \neq p$, then $H = G$ whence $\mathbb{F}_pG$ centralises α, a contradiction. For (d), note that $\mathbb{F}_pHg \cap C_{\mathbb{F}_pG}(\alpha) = Kg$ for some left ideal K of $\mathbb{F}_pH$. By (b), $|\mathbb{F}_pH : K| = |\mathbb{F}_pHg : C_{\mathbb{F}_pH_g}(\alpha)|$ divides p. It follows that $K = \mathbb{F}_pH$ or else $K = I(H)$, the only left ideal of $\mathbb{F}_pH$ having codimension 1. The case $K = \mathbb{F}_pH$ is equivalent to the condition $g \in C_{\mathbb{F}_pG}(\alpha) \cap G$ which is H.

To finish the proof, choose $g \in G \backslash H$. Then the Lie bracket $(\alpha, g) = \alpha g - g\alpha \neq 0$. By (c), G is the disjoint union of the right cosets Hg^i, $0 \leq i < p$. Now (α, g) is in the right annihilator of $I(H)$ since, by (d), $(h-1)g \in C_{\mathbb{F}_pG}(\alpha)$ for all $h \in H$ so that $(h-1)(\alpha, g) = (\alpha, (h-1)g) = 0$. Also, for any i, $(\alpha, g)(g^i) = (\alpha g)(g^i) - (g\alpha)(g^i) = \alpha(g^{i-1}) - \alpha(g^{i-1}) = 0$. Thus, (α, g) vanishes on at least one element of each right coset of H in G. But, by the Lemma, (α, g) is constant on each such coset. It follows that $(\alpha, g) = 0$ which provides the concluding contradiction. □

The final proposition completes the proof of Theorem 1. It is stated in a more general context than required in order to prompt the recognition elsewhere of the phenomenon of vanishing orbit sums.

Proposition 2. *Let P be a p-group, F a field of characteristic p and M a right FP-module. Suppose that $|P : C_P(m)| \geq p^2$ for any element m of M not in M^P. Let $\mathcal{O}$ be a finite orbit of P outside M^P. Then $\sum \mathcal{O} = 0$.*

PROOF. For $m \in \mathcal{O}$, the hypothesis implies that $\infty > |\mathcal{O}| = |P : C_P(m)| \geq p^2$. Let Q be a maximal subgroup of P containing $C_P(m)$ so that $|P : Q| = p$. As the orbit mQ of m under the action of Q is finite, we may form the sum $m' = \sum mQ$. That m' is a fixed point for P follows from the hypothesis and the fact that $C_P(m')$ contains Q so that $|P : C_P(m')| \leq p$.

Let T be a right transversal for $C_P(m)$ in Q; thus, $mQ = mT$. Let S be a right transversal for Q in P so that $\mathcal{O} = mP = mTS$. Then $\sum \mathcal{O} = \sum\{mts | t \in T, s \in S\} = \sum_{s \in S} m's = pm' = 0$. □

The provenance of the results reported here is instructive and deserves some comment. They came not from a well-founded conjecture or special insight; they emerged from study of examples. This seems unexceptional but consideration of the size of possible examples shows that such calculations could be done reliably only by computer. With the implementation in [7] achieved using CAYLEY [1] and SOGOS [3], centralisers in the modular group algebras of the groups of order 8 and 16 over the field of 2 elements could be examined. Examination of orbits was made possible through the work described in [5]. This implementation, done in CAYLEY and relying on Newman and O'Brien's provision of 2-groups [4], makes available abstract presentations of the normalised unit groups of the group algebras $\mathbb{F}_2 G$, where G has order dividing 32; the presentations obtained are power-commutator presentations. Rao's suite of programs also permits the user to pass between the unit group in this form and in its original form as $V(\mathbb{F}_p G)$. This enables one to calculate orbits and their sums. For example, to calculate the orbit sum corresponding to an element α of $\mathbb{F}_p G$ under the conjugation action of V, write $\alpha = a + v$, $a \in \mathbb{F}_p, v \in V$; then only the sum of the conjugacy class of v in V need be formed. For a p-group of moderate size given by a power-commutator presentation, calculation of a conjugacy class can be effected readily in CAYLEY. With the conjugacy class to hand, its elements are converted to vectors ($\mathbb{F}_p G$ is realised in CAYLEY as a vector space [6]) and their sum formed.

References

[1] Cannon, J.J., An introduction to the group theory language, Cayley, in *Computational group theory* (Academic Press, London, 1984), 145–183.

[2] Coleman, D.B. and Passman, D.S., Units in modular group rings, *Proc. Amer. Math. Soc.* **25**(1970), 510–512.

[3] Laue, R., Neübuser J. and Schoenwaelder, U., Algorithms for finite soluble groups and the SOGOS system, in *Computational group theory* (Academic Press, London, 1984), 105–135.

[4] Newman, M.F. and O'Brien, E.A., A CAYLEY library for the groups of order dividing 128, in *Group theory (Singapore, 1987)* (de Gruyter, Berlin, 1989), 337–342.

[5] Rao, M.A., *Computer calculations of presentations of the unit groups of the modular group algebras of the groups of order dividing 32* (Ph.D. thesis, Manchester Univ., 1993).

[6] Sandling, R., A group ring package for Cayley, *SIGSAM Bull.* **25**(1991), 60–64.

[7] Sandling, R., Presentations for unit groups of modular group algebras of groups of order 16, *Math. Comp.* **59**(1992), 689–701.

[8] Shalev, A., The nilpotency class of the unit group of a modular group algebra I, *Israel J. Math.* **70**(1990), 257–266.

[9] Shalev, A., Meta-abelian unit groups of group algebras are usually abelian, *J. Pure Appl. Algebra* **72**(1991), 295–302.

FROM STABLE EQUIVALENCES TO RICKARD EQUIVALENCES FOR BLOCKS WITH CYCLIC DEFECT

RAPHAËL ROUQUIER

DMI-ENS (CNRS UA 762), 45 Rue d'Ulm, 75005 Paris, France
E-mail: Raphael.Rouquier@ens.fr

1. Introduction

Let G and H be two finite groups, p a prime number. Let $\mathcal{O}$ be a complete discrete valuation ring with residue field k of characteristic p and with field of fractions K of characteristic 0, "big enough" for G and H. Let A and B be two blocks of G and H over $\mathcal{O}$.

Let M be a $(A \otimes B^{\circ})$-module, projective as A-module and as B°-module, where B° denotes the opposite algebra of B. We denote by M^* the $(B \otimes A^{\circ})$-module $\mathrm{Hom}_{\mathcal{O}}(M, \mathcal{O})$.

We say that M induces a *stable equivalence* between A and B if

$$M \otimes_B M^* \simeq A \oplus \text{projectives as } (A \otimes A^{\circ}) - \text{modules and}$$
$$M^* \otimes_A M \simeq B \oplus \text{projectives as } (B \otimes B^{\circ}) - \text{modules.}$$

Let C be a complex of $(A \otimes B^{\circ})$-modules, all of which are projective as A-modules and as B°-modules.

Denoting by C^* the $\mathcal{O}$-dual of C, we say that C induces a *Rickard equivalence* between A and B if $C \otimes_B C^*$ is homotopy equivalent to A as complexes of $(A \otimes A^{\circ})$-modules and $C^* \otimes_A C$ is homotopy equivalent to B as complexes of $(B \otimes B^{\circ})$-modules.

By [Ri4, 5.5], from a complex C inducing a Rickard equivalence between A and B, one can construct a module M inducing a stable equivalence between A and B as follows : In the derived bounded category of $A \otimes B^{\circ}$, the complex C is isomorphic to a complex with only one term which is not projective as $(A \otimes B^{\circ})$-module, V in degree $-n$ and then the n-th Heller translate (syzygy) $M = \Omega^n(V)$ induces a stable equivalence between A and B.

The main result of this note is a partial converse under very special assumptions (Theorem 6). Since there are well-known situations where a module M induces a stable equivalence between two blocks (Remark 9), for example when the Sylow p-subgroups of G are TI, H is the normalizer of a Sylow p-subgroup of G and A, B are principal blocks, it is tempting to try to construct a complex with two terms, M in degree 0 and a projective module in degree -1, inducing a Rickard equivalence between A and B. Using Theorem

6, we prove that it is indeed possible when the Sylow p-subgroups of G are cyclic or when $G = A_5$ or $SL_2(8)$ and $p = 2$.

2. A criterion for derived equivalences between blocks

2.1. Some lemmas

Let A' be an $\mathcal{O}$-free $\mathcal{O}$-algebra, finitely generated as an $\mathcal{O}$-module.

If V is an A'-module, let P_V be an A'-module which is a projective cover of V. We will denote by $\mathrm{Rad}(V)$ the radical of V and by $\mathrm{hd}(V)$ the head $V/\mathrm{Rad}(V)$ of V, *i.e.*, its largest semi-simple quotient.

If M and N are two A'-modules, we say that M and N are *disjoint* if they have no non-zero isomorphic direct summands. If M and N are projective, they are disjoint if and only if $\mathrm{Hom}_{A'}(M, \mathrm{hd}(N)) = 0$ or equivalently, $\mathrm{Hom}_{A'}(N, \mathrm{hd}(M)) = 0$.

If X is an $\mathcal{O}$-module, we define $\bar{X} = X \otimes k$.

Lemma 1. *Let P, Q and R be three projective A'-modules and $\varphi : P \oplus Q \twoheadrightarrow R$ a surjective morphism. Assume that Q and R are disjoint. Then, the restriction $\varphi_{|P}$ of φ to P is surjective.*

Let U,V and W be three injective $\bar{A}'$-modules and $\varphi : W \hookrightarrow U \oplus V$ an injective morphism. Assume that V and W are disjoint. Then, denoting by p_U the projection of $U \oplus V$ onto U, the map $p_U\varphi$ is injective.

PROOF. Let $h : R \to \mathrm{hd}(R)$ be the canonical projection. Since $h\varphi : P \oplus Q \twoheadrightarrow \mathrm{hd}(R)$ is surjective and $\mathrm{Hom}_{A'}(Q, \mathrm{hd}(R)) = 0$ by assumption, $h\varphi_{|P}$ is surjective. Hence, $\varphi(P) + \mathrm{Rad}(R) = R$ and by Nakayama's lemma, $\varphi(P) = R$. The second assertion follows immediately by duality since V and W are disjoint implies that V^* and W^* are disjoint. □

Lemma 2. *Let M be an $(A \otimes B^\circ)$-module, projective as A-module and as B°-module. A projective cover of M is*

$$\bigoplus_W P_{M \otimes_B W} \otimes P_W^*$$

where W runs over a complete set of representatives of isomorphism classes of simple B-modules. This module is isomorphic to

$$\bigoplus_V P_V \otimes P_{M^* \otimes_A V}^*$$

where V runs over a complete set of representatives of isomorphism classes of simple A-modules.

PROOF. Let V be an $\bar{A}$-module and W a $\bar{B}$-module. We have

$$\mathrm{Hom}_{\bar{B}^\circ}(\bar{M}, V \otimes W^*) \simeq \mathrm{Hom}_{\bar{B}^\circ}(\bar{M}, V \otimes W^*) \simeq \bar{M}^* \otimes_{\bar{B}^\circ} (V \otimes W^*)$$

since $\bar{M}$ is projective as $\bar{B}^\circ$-module. Hence,

$$\mathrm{Hom}_{\bar{B}^\circ}(\bar{M}, V \otimes W^*) \simeq (\bar{M} \otimes_{\bar{B}} W)^* \otimes V \simeq \mathrm{Hom}_k(\bar{M} \otimes_{\bar{B}} W, V)$$

and finally

$$\mathrm{Hom}_{\bar{A}\otimes\bar{B}^\circ}(\bar{M}, V \otimes W^*) \simeq \mathrm{Hom}_{\bar{A}}(\bar{M} \otimes_{\bar{B}} W, V).$$

Now, we have

$$\begin{aligned} \mathrm{hd}(M) &\simeq \bigoplus_{V,W} \dim \mathrm{Hom}_{A\otimes B^\circ}(M, V \otimes W^*)(V \otimes W^*) \\ &\simeq \bigoplus_{W} \left(\bigoplus_{V} \dim \mathrm{Hom}_A(M \otimes_B W, V) V \right) \otimes W^* \end{aligned}$$

where V (resp. W) runs over the simple A-modules (resp. B-modules) up to isomorphism, hence

$$\mathrm{hd}(M) \simeq \bigoplus_{W} \mathrm{hd}(M \otimes_B W) \otimes W^*,$$

so a projective cover of M is $\bigoplus_W P_{M\otimes_B W} \otimes P^*_W$.

To get the second description, replace A by B° and B by A° in the first description : a projective cover of M as $B^\circ \otimes (A^\circ)^\circ$-module is $\oplus_V P_{M\otimes_{A^\circ} V} \otimes P_{V^*}$ where V runs over the simple A°-modules. This module is isomorphic to $\oplus_V P_{M\otimes_{A^\circ} V^*} \otimes P_V$ where V runs over the simple A-modules, hence a projective cover of M as $(A \otimes B^\circ)$-module is $\oplus_V P_V \otimes P_{V^* \otimes_A M}$ where V runs over the simple A-modules. □

Lemma 3. (Linckelmann, [Li2, 6.8]) *Let M be an $(A\otimes B^\circ)$-module inducing a stable equivalence between A and B. Then, M has a unique non-projective direct summand, up to isomorphism.*

PROOF. Let $M = M_1 \oplus M_2$. Since $M^* \otimes_A M \simeq B \oplus$ projectives, we have $M^* \otimes_A M_1 \oplus M^* \otimes_A M_2 \simeq B \oplus$ projectives as $(B \otimes B^\circ)$-modules. As B is indecomposable as $(B \otimes B^\circ)$-module, there exists $i \in \{1, 2\}$ such that $M^* \otimes_A M_i$ is projective as $(B \otimes B^\circ)$-module, so $M \otimes_B M^* \otimes_A M_i$ is projective as $(A\otimes B^\circ)$-module. Now, $(M \otimes_B M^*) \otimes_A M_i \simeq M_i \oplus$ projectives as $(A\otimes B^\circ)$-modules, hence M_i is projective as $(A \otimes B^\circ)$-module. □

Remark 4. A similar proof shows that a complex of $(A \otimes B^\circ)$-modules C inducing a Rickard equivalence between A and B has a unique non-homotopy equivalent to zero direct summand, up to isomorphism.

Lemma 5. (Linckelmann, [Li2, 6.3]) *Let M be an indecomposable $(A \otimes B^\circ)$-module inducing a stable equivalence between A and B. For any simple B-module V, the A-module $M \otimes_B V$ is indecomposable.*

PROOF. (Linckelmann) Denote by $\mathrm{soc}(\bar{A})$ the largest semi-simple $\bar{A}$-submodule of $\bar{A}$. Recall that an $\bar{A}$-module V has no projective direct summand if and only if $\mathrm{soc}(\bar{A})V = 0$. We have $\mathrm{soc}(\bar{A} \otimes \bar{B}^\circ) = \mathrm{soc}(\bar{A}) \otimes \mathrm{soc}(\bar{B}^\circ)$. Since M has no projective direct summand, $\mathrm{soc}(\bar{A} \otimes \bar{B}^\circ)M = 0$, hence $\mathrm{soc}(\bar{A})(M \otimes_B \mathrm{soc}(\bar{B})) = 0$, which means that $M \otimes_B \mathrm{soc}(\bar{B})$ has no projective direct summand. But, if V is a simple B-module, it is a direct summand of $\mathrm{soc}(\bar{B})$, so $M \otimes_B V$ has no projective direct summand : as M induces a stable equivalence, $M \otimes_B V$ has a unique indecomposable non projective direct summand and the lemma follows. □

2.2. The criterion

We denote by $R_K(A)$ (resp. $R_K(B)$) the group of characters of $KA = K \otimes A$ (resp. KB).

Let us now state the main result:

Theorem 6. *Let M be an $(A \otimes B^\circ)$-module, projective as A-module and as B°-module. Let $\delta' : P' \twoheadrightarrow M$ be a projective cover of M. Let P be a direct summand of P', $\delta = \delta'_{|P}$ and $C = (0 \longrightarrow P \xrightarrow{\delta} M \longrightarrow 0)$ (M is in degree 0). Assume*

(a_1) $M^* \otimes_A M \simeq B \oplus Q$ where Q is a projective $(B \otimes B^\circ)$-module,

(a_2) $M \otimes_B M^* \simeq A \oplus R$ where R is a projective $(A \otimes A^\circ)$-module,

(b_1) $\mathrm{Res}^{A\otimes B^\circ}_{B^\circ} \bar{P}$ and $\mathrm{Res}^{A\otimes B^\circ}_{B^\circ} \bar{P}'/\bar{P}$ are disjoint,

(b_2) $\mathrm{Res}^{A\otimes B^\circ}_{A} \bar{P}$ and $\mathrm{Res}^{A\otimes B^\circ}_{A} \bar{P}'/\bar{P}$ are disjoint,

(c) KC induces an isometry between $R_K(A)$ and $R_K(B)$.

Then, C induces a Rickard equivalence between A and B.

PROOF. [1] Remark first that (b_1) implies that

(b′) $\mathrm{Res}^{B\otimes A^\circ}_{B} \bar{P}^*$ and $\mathrm{Res}^{B\otimes A^\circ}_{B} \left(\bar{P}'/\bar{P}\right)^*$ are disjoint.

We have

$$C^* \otimes_A C$$

$$= (0 \to M^* \otimes_A P \xrightarrow{(\delta^*\otimes id,\, id\otimes\delta)} P^* \otimes_A P \oplus M^* \otimes_A M \xrightarrow{\begin{pmatrix} id\otimes\delta \\ -\delta^*\otimes id\end{pmatrix}} P^* \otimes_A M \to 0).$$

[1] Using an unpublished result of J. Rickard, one can actually prove the theorem without the assumptions (a_2) and (b_2).

Since KC induces an isometry between $R_K(A)$ and $R_K(B)$, the character of $K(C^* \otimes_A C)$ as $(B \otimes B^\circ)$-module is equal to the character of B. Hence,

$$K(M^* \otimes_A P \oplus P^* \otimes_A M) \simeq K(P^* \otimes_A P \oplus Q).$$

We know that P is a projective $(A \otimes B^\circ)$-module and $\mathrm{Res}_B^{B\otimes A^\circ} M^*$ is projective, so $M^* \otimes_A P$ is projective as $(B \otimes B^\circ)$-module. Similarly, $P^* \otimes_A M$, $P^* \otimes_A P$ and Q are projective $(B \otimes B^\circ)$-modules. Hence

$$M^* \otimes_A P \oplus P^* \otimes_A M \simeq P^* \otimes_A P \oplus Q, \text{ and}$$

$$\bar{M}^* \otimes_A \bar{P} \oplus \bar{P}^* \otimes_A \bar{M} \simeq \bar{P}^* \otimes_A \bar{P} \oplus \bar{Q}. \tag{1}$$

Let $\bar{Q} = \bar{Q}_1 \oplus \bar{Q}_2$ where

$$\mathrm{Res}_{B^\circ}^{B\otimes B^\circ} \bar{Q}_2 \text{ and } \mathrm{Res}_{B^\circ}^{A\otimes B^\circ} \bar{P} \text{ are disjoint,} \tag{2}$$

$$\mathrm{Res}_{B^\circ}^{B\otimes B^\circ} \bar{Q}_1 \text{ and } \mathrm{Res}_{B^\circ}^{A\otimes B^\circ} \bar{P}'/\bar{P} \text{ are disjoint.} \tag{3}$$

(Since the map $p_{\bar{Q}}(id \otimes \bar{\delta}') : \bar{M}^* \otimes_A \bar{P}' \to \bar{Q}$ is surjective, every indecomposable direct summand of $\mathrm{Res}_{B^\circ}^{B\otimes B^\circ} \bar{Q}$ is isomorphic to a direct summand of $\mathrm{Res}_{B^\circ}^{B\otimes B^\circ} \bar{P}'$, so $\bar{Q}_1$ and $\bar{Q}_2$ are unique up to isomorphism).

The map $p_{\bar{Q}_1}(id \otimes \bar{\delta}') : \bar{M}^* \otimes_A \bar{P}' \to \bar{Q}_1$ is surjective and using (3),

$$\mathrm{Res}_{B^\circ}^{B\otimes B^\circ} \bar{Q}_1 \text{ and } \mathrm{Res}_{B^\circ}^{B\otimes B^\circ} \left(\bar{M}^* \otimes_A \bar{P}'\right) / \left(\bar{M}^* \otimes_A \bar{P}\right) \text{ are disjoint,}$$

$$\text{hence } \bar{Q}_1 \text{ and } \left(\bar{M}^* \otimes_A \bar{P}'\right) / \left(\bar{M}^* \otimes_A \bar{P}\right) \text{ are disjoint,}$$

and it follows from Lemma 1 that the map

$$p_{\bar{Q}_1}(id \otimes \bar{\delta}) : \bar{M}^* \otimes_A \bar{P} \to \bar{Q}_1$$

is surjective.

From (1), $\bar{Q}$ is isomorphic to a direct summand of $\bar{M}^* \otimes_A \bar{P} \oplus \bar{P}^* \otimes_A \bar{M}$, hence $\bar{Q}_2$ is isomorphic to a direct summand of $\bar{P}^* \otimes_A \bar{M}$ using (2). By (b'),

$$\mathrm{Res}_B^{B\otimes B^\circ} \left(\bar{P}^* \otimes \bar{M}\right) \text{ and } \mathrm{Res}_B^{B\otimes B^\circ} \left(\bar{P}'^* \otimes \bar{M} / \bar{P}^* \otimes \bar{M}\right) \text{ are disjoint,}$$

$$\text{hence } \mathrm{Res}_B^{B\otimes B^\circ} \bar{Q}_2 \text{ and } \mathrm{Res}_B^{B\otimes B^\circ} \left(\bar{P}'^* \otimes \bar{M} / \bar{P}^* \otimes \bar{M}\right) \text{ are disjoint.}$$

Now, since $(\bar{\delta}'^* \otimes id)_{|\bar{Q}_2} : \bar{Q}_2 \to \bar{P}'^* \otimes_A \bar{M}$ is injective, Lemma 1 implies that

$$(\bar{\delta}^* \otimes id)_{|\bar{Q}_2} : \bar{Q}_2 \to \bar{P}^* \otimes_A \bar{M}$$

is injective.

Let $\bar{R}_2$ be a submodule of $\bar{P}^* \otimes_A \bar{M}$ such that $\bar{P}^* \otimes_A \bar{M} = \bar{R}_2 \oplus \mathrm{Im}(\bar{\delta}^* \otimes id)_{|\bar{Q}_2}$. We introduce

$$f_2 = p_{\bar{R}_2}(id \otimes \bar{\delta}) : \bar{P}^* \otimes_A \bar{P} \to \bar{R}_2$$

$$\text{and } f_2' = p_{\bar{R}_2}(id \otimes \bar{\delta}') : \bar{P}^* \otimes_A \bar{P}' \to \bar{R}_2$$

We have $\bar{P}^* \otimes_A \bar{M} \simeq \bar{Q}_2 \oplus \bar{R}_2$, so, by (1), as $\bar{Q}_1$ is isomorphic to a direct summand of $\bar{M}^* \otimes_A \bar{P}$, the module $\bar{R}_1$ is a direct summand of $\bar{P}^* \otimes_A \bar{P}$, hence, by (b_1),

$$\mathrm{Res}_{B^\circ}^{B\otimes B^\circ} \bar{R}_2 \text{ and } \mathrm{Res}_{B^\circ}^{B\otimes B^\circ} \left((\bar{P}^* \otimes_A \bar{P}')/(\bar{P}^* \otimes_A \bar{P})\right) \text{ are disjoint.}$$

Since f_2' is surjective, Lemma 1 implies that f_2 is surjective.

It follows that the map $id \otimes \bar{\delta} - \bar{\delta}^* \otimes id$ is surjective. By Nakayama's lemma, the map $id \otimes \delta - \delta^* \otimes id$ is also surjective and hence splits. By duality, the map $\delta^* \otimes id + id \otimes \delta$ is injective and splits. Hence, the complex $C^* \otimes_A C$ is homotopy equivalent to B.

Similarly, the complex $C \otimes_B C^*$ is homotopy equivalent to A. Hence, the complex C induces a Rickard equivalence between A and B. □

2.3. An application

Let M be an indecomposable $(A \otimes B)^\circ$-module inducing a stable equivalence between A and B.

Assume that for every simple A-module V, the head of $M^* \otimes_A V$ is simple.

Theorem 7. *If there exists a direct summand P of*

$$\bigoplus_V P_V \otimes P^*_{M^*\otimes_A V} \simeq \bigoplus_W P_{M\otimes_B W} \otimes P^*_W$$

(V runs over the simple A-modules and W over the simple B-modules) such that $0 \to P \xrightarrow{0} M \to 0$ induces an isometry between $R_K(KA)$ and $R_K(KB)$, then there is a complex $C = 0 \to P \to M \to 0$ inducing a Rickard equivalence between A and B.

PROOF. The modules P and M being projective as A-modules and as B°-modules, the isometry induced by $0 \to P \xrightarrow{0} M \to 0$ is perfect [Br3, 1.2] and it follows that the algebras A and B have the same number s of isomorphism classes of simple modules [Br3, 1.5].

If V is a simple A-module, the modules P_V and $P^*_{M^*\otimes_A V}$ are indecomposable. Hence, a projective cover of M is a sum of s indecomposable $(A \otimes B^\circ)$-modules, which are mutually non-isomorphic when restricted to A or when restricted to B°. Hence, if P is a direct summand of P' then $\mathrm{Res}_{B^\circ}^{A\otimes B^\circ} P$ and $\mathrm{Res}_{B^\circ}^{A\otimes B^\circ} P'/P$ are disjoint and $\mathrm{Res}_{A}^{A\otimes B^\circ} P$ and $\mathrm{Res}_{A}^{A\otimes B^\circ} P'/P$ are disjoint. Now, Theorem 6 gives the conclusion. □

Let us denote by $\mathrm{CF}(G, K)$ the space of class functions $G \to K$, by $\mathrm{CF}(A, K)$ the subspace generated by $R_K(A)$. We denote by $\mathrm{CF}_p(G, K)$ (resp. $\mathrm{CF}_{p'}(G, K)$) the subspace of $\mathrm{CF}(G, K)$ consisting of class functions which vanish on p-regular (resp. p-singular) elements and $\mathrm{CF}_p(A,K)$ (resp. $\mathrm{CF}_{p'}(A,K)$) the intersection $\mathrm{CF}_p(G,K)\cap\mathrm{CF}(A,K)$ (resp. $\mathrm{CF}_{p'}(G,K)\cap\mathrm{CF}(A,K)$).

As the next lemma shows, in the situation of Theorem 7, if the map induced by $0\to P\xrightarrow{0} M\to 0$ is an isometry on a subspace of $\mathrm{CF}(A,K)$ which contains a complement of $\mathrm{CF}_p(A,K)$, then it is an isometry:

Lemma 8. *Let P_1, P_2 be two projective $(A\otimes B^\circ)$-modules and $C=0\to P_1\xrightarrow{0} M\oplus P_2\to 0$. Let I be the map between $R_K(A)$ and $R_K(B)$ induced by C. Let X be a subspace of $CF(A,K)$ such that $CF(A,K)=X+CF_p(A,K)$.*

If the restriction of I to X is an isometry, then I is an isometry.

PROOF. Let $f,g\in\mathrm{CF}(A,K)$. We decompose f and g as $f=f_p+f_{p'}$ and $g=g_p+g_{p'}$ where $f_p,g_p\in\mathrm{CF}_p(A,K)$ and $f_{p'},g_{p'}\in\mathrm{CF}_{p'}(A,K)$. Since I is perfect [Br3, 1.2], $I(f_p),I(g_p)\in\mathrm{CF}_p(B,K)$ and $I(f_{p'}),I(g_{p'})\in\mathrm{CF}_{p'}(B,K)$. Hence, the scalar product of $I(f)$ and $I(g)$ is $<I(f),I(g)>=<I(f_p),I(g_p)> + <I(f_{p'}),I(g_{p'})>$.

Furthemore, the restriction of I to $\mathrm{CF}_p(A,K)$ is an isometry because M induces a stable equivalence between A and B and as P_1 and P_2 are projective, the map induced by M between $R_K(A)$ and $R_K(B)$ is equal to I on $\mathrm{CF}_p(A,K)$ [Br2, 5.3]. It follows that $<I(f_p),I(g_p)>=<f_p,g_p>$ and we have now to prove that $<I(f_{p'}),I(g_{p'})>=<f_{p'},g_{p'}>$. But, as $\mathrm{CF}(A,K)=X+\mathrm{CF}_p(A,K)$, we can decompose $f_{p'}$ and $g_{p'}$ as $f_{p'}=f_1+f_2$ and $g_{p'}=g_1+g_2$ where $f_1,g_1\in X$ and $f_2,g_2\in\mathrm{CF}_p(A,K)$. Now, $<f_1,g_1>=<f_{p'},g_{p'}>-<f_2,g_2>$ and $<I(f_1),I(g_1)>=<I(f_{p'}),I(g_{p'})>-<I(f_2),I(g_2)>$. Finally, we know that $<I(f_1),I(g_1)>=<f_1,g_1>$ and $<I(f_2),I(g_2)>=<f_2,g_2>$, hence $<I(f_{p'}),I(g_{p'})>=<f_{p'},g_{p'}>$. □

Remark 9. Stable equivalences induced by bimodules arise for example in the following situation [Br2, 6.4]:

Assume that H is a subgroup of G with index prime to p and e, f are the units of A and B. Following Broué, let us assume that for every non trivial p-subgroup P of H, we have $N_G(P)=N_H(P)O_{p'}C_G(P)$. Then, the $(A\otimes B^\circ)$-module $e\mathcal{O}Gf$ induces a stable equivalence between A and B. Let M be an indecomposable non-projective direct summand of $e\mathcal{O}Gf$; by Lemma 3, such a module is unique up to isomorphism ; we have $e\mathcal{O}Gf=M\oplus$ projectives, so M induces a stable equivalence between A and B.

Example 1. Let $G=SL_2(4)=A_5$ and $H=A_4=2^2\rtimes 3$ a Borel subgroup, $p=2$. The principal block ekG of G has three simple modules : k, S_1 and S_2 of dimension 2. The module $\mathrm{Res}_H^G(S_1)$ is a non-split extension of V_2 by V_1, where V_1 and V_2 are the two non-trivial non-isomorphic simple kH-modules and $\mathrm{Res}_H^G(S_2)$ is a non-split extension of V_1 by V_2. An immediate character calculation shows that

$$0 \to P_{S_1} \otimes P_{V_1}^* \oplus P_{S_2} \otimes P_{V_2}^* \xrightarrow{0} e\mathcal{O}G \to 0$$

induces an isometry between $R_K(eKG)$ and $R_K(KH)$. Hence, by Remark 9 and Theorem 7, there exists a complex $0 \to P_{S_1} \otimes P_{V_1} \oplus P_{S_2} \otimes P_{V_2} \to e\mathcal{O}G \to 0$ inducing a Rickard equivalence between the principal blocks of G and H, a result due to J. Rickard [Ri3].

Example 2. Let $G = SL_2(8)$ and $H = 2^3 \rtimes 7$ a Borel subgroup, $p = 2$. Then, Theorem 7 applies also to construct a complex inducing a Rickard equivalence between the principal blocks of G and H : The $(A \otimes B^\circ)$-bimodule $e\mathcal{O}G$ is indecomposable. We leave to the reader to check that a projective cover of this module is :

$$P_1 \otimes Q_1^* \oplus P_{2_1} \otimes Q_{2_1}^* \oplus P_{2_2} \otimes Q_{2_2}^* \oplus P_{2_3} \otimes Q_{2_3}^* \oplus P_{4_1} \otimes Q_{4_1}^* \oplus P_{4_2} \otimes Q_{4_2}^* \oplus P_{4_3} \otimes Q_{4_3}^*$$

(where P_1 (resp. Q_1) is a projective cover of the trivial A-module (resp. B-module), P_{2_1}, P_{2_2} and P_{2_3} (resp. P_{4_1}, P_{4_2} and P_{4_3}) are projective covers of the three non-isomorphic 2-dimensional (resp. 4-dimensional) simple A-modules and $Q_{2_1}, Q_{2_2}, Q_{2_3}, Q_{4_1}, Q_{4_2}, Q_{4_3}$ are projective covers of the six non-isomorphic non-trivial simple B-modules) and that the complex

$$0 \to \oplus P_{4_1} \otimes Q_{4_1}^* \oplus P_{4_2} \otimes Q_{4_2}^* \oplus P_{4_3} \otimes Q_{4_3}^* \xrightarrow{0} e\mathcal{O}G \to 0$$

induces an isometry between $R_K(A)$ and $R_K(B)$, so that by Remark 9 and Theorem 7, there exists a complex $0 \to \oplus P_{4_1} \otimes Q_{4_1}^* \oplus P_{4_2} \otimes Q_{4_2}^* \oplus P_{4_3} \otimes Q_{4_3}^* \to e\mathcal{O}G \to 0$ inducing a Rickard equivalence between the principal blocks of G and H.

3. Application to principal blocks with cyclic defect

Let G be a finite group with a cyclic Sylow p-subgroup P and let $H = N_G(P)$. As before, $A = \mathcal{O}Ge$ and $B = \mathcal{O}Hf$ are the principal blocks of G and H, where e and f are primitive idempotents of the centers of $\mathcal{O}G$ and $\mathcal{O}H$.

The functor $erm\,\mathrm{Ind}_H^G$ induces a stable equivalence between A and B with inverse functor $f\mathrm{Res}_H^G$ (Remark 9).

As conjectured by J. Rickard (cf [Ri2]), a slight modification of these functors leads to a derived equivalence, and this proves in particular the conjecture of Broué and Rickard on abelian defect, for principal blocks with cyclic defect (cf [Br1]) :

Theorem 10. *There exists a projective $(A \otimes B^{\circ})$-module Y and a map $\phi : Y \to eOGf$ such that, if $C = 0 \longrightarrow Y \xrightarrow{\phi} eOGf \longrightarrow 0$, then C induces a Rickard equivalence between A and B. In particular, C is a Rickard tilting complex of p-permutation modules.*

Note that the fact that A and B are derived-equivalent was already known by the work of Rickard and Linckelman (cf [Ri1] and [Li1]).

3.1. Construction of C

Let us quote some classical results about A (cf [Gr]).

The set of irreducible characters of KA is $\mathrm{Irr}(A) = \{\chi_1, \ldots, \chi_e\} \cup \{\chi_\lambda\}_{\lambda \in \Lambda}$ where $\chi_1, \ldots, \chi_e$ are the non-exceptional characters and the $\chi_\lambda, \lambda \in \Lambda$, are the exceptional characters. (In the case there is only one exceptional character, one can choose it different from the character 1_G.)

Define $\chi_{e+1} = \sum_{\lambda \in \Lambda} \chi_\lambda$ and $\Gamma = \{\chi_1, \ldots, \chi_{e+1}\}$.

The Brauer tree $\mathcal{T}_A$ is then defined as follows :

- the set of its vertices is Γ,
- two vertices v and v' are incident if and only if $v + v'$ is the character of a projective indecomposable A-module. We denote by $\{v, v'\}$ the corresponding edge.

The vertex χ_{e+1} is called the exceptional vertex of $\mathcal{T}_A$. Every character of a projective indecomposable A-module is an edge of $\mathcal{T}_A$ and we have a bijection beween the set of edges of $\mathcal{T}_A$ and the set of characters of projective indecomposable A-modules. If v and v' are two vertices of $\mathcal{T}_A$, we denote by $d(v, v')$ the distance between v and v'.

There is a "walk" on $\mathcal{T}_A$ starting from 1_G, the trivial character of G, *i.e.*, a sequence $v_0 = 1_G, v_1, \ldots, v_{2e}$ of vertices of $\mathcal{T}_A$ such that v_i is incident with v_{i+1} for $0 \leq i \leq 2e - 1$, with the following properties :

- Each edge is traversed twice, *i.e.*, denoting by l_i the edge $\{v_i, v_{i+1}\}$, then for every edge l of $\mathcal{T}_A$, there exists i and j two distinct integers, $0 \leq i, j \leq 2e - 1$, such that $l = l_i = l_j$;
- denote by P_i a projective indecomposable module with character l_i. Then, we have a minimal projective resolution of the A-module $\mathcal{O}$, periodic of period $2e$:

$$\cdots \to P_0 \to P_{2e-1} \to \cdots \to P_1 \to P_0 \to \mathcal{O} \to 0. \qquad (4)$$

We have $v_{2e} = v_0$. Given three vertices $v, v', v"$ of $\mathcal{T}_A$, we have $d(v, v') + d(v', v") \equiv d(v, v") \pmod 2$, hence $d(v_i, v_0) \equiv i \pmod 2$. Suppose $l_i = l_j$.

Since $\mathcal{T}_A$ is a tree, we have $v_i = v_{j+1}$ and $v_j = v_{i+1}$, hence $i \equiv j+1 \pmod 2$. It follows that $\{l_{2i}\}_{0\leq i\leq e-1}$ is the set of all edges of $\mathcal{T}_A$.

If X is an A-module (resp. a B-module) and i an integer, we define $\Omega^i_A(X)$ (resp. $\Omega^i_B(X)$) to be the i-th Heller translate of X.

The character of $\Omega^i_A\mathcal{O}$ is v_i.

The block B has a similar description which is a particular case of the previous one :

The Brauer tree of B, $\mathcal{T}_B$, is a star whose center is the exceptional vertex, *i.e.*, every edge of $\mathcal{T}_B$ is of the form $\{w, w'\}$ where w' is the exceptional vertex. There is a walk $w_0 = 1_H, w_1, \ldots, w_{2e}$ on $\mathcal{T}_B$ such that :

- Every edge is traversed twice ;
- denote by Q_i a projective indecomposable module with character $w_i + w_{i+1}$. Then, we have a minimal projective resolution of the B-module $\mathcal{O}$, periodic of period $2e$:

$$\cdots \to Q_0 \to Q_{2e-1} \to \cdots \to Q_1 \to Q_0 \to \mathcal{O} \to 0.$$

Note that for any i, $0 \leq i \leq e-1$, w_{2i+1} is the exceptional vertex and $\{w_{2i}\}_{0\leq i\leq e-1}$ is the set of all non-exceptional characters of KB. The module $\Omega^{2i}_B\mathcal{O}$ remains irreducible modulo p and its character is w_{2i}.

Since $e\mathcal{O}Gf$ induces a stable equivalence between A and B, we have $e\mathcal{O}Gf = M \oplus U$ as $(A \otimes B^\circ)$-modules, where M is indecomposable – and then $\bar{M}$ is also indecomposable since M is a p-permutation module – and U is projective (cf Lemma 3). We still have

$$M \otimes_B M^* \simeq A \oplus \text{projectives} \quad \text{and} \quad M^* \otimes_A M \simeq B \oplus \text{projectives}.$$

Since M induces a stable equivalence between A and B, tensoring by M commutes with Heller translates, up to projectives, hence $M \otimes_B \Omega^{2i}_B\mathcal{O} \simeq \Omega^{2i}_A\mathcal{O} \oplus$projectives. Since $\bar{M}$ is indecomposable and $\Omega^{2i}_B k$ is simple, $\bar{M} \otimes_B \Omega^{2i}_B k$ is indecomposable (cf Lemma 5), so that

$$M \otimes_B \Omega^{2i}_B\mathcal{O} \simeq \Omega^{2i}_A\mathcal{O}. \tag{5}$$

Now, since a projective cover of $\Omega^{2i}_A\mathcal{O}$ is P_{2i} and a projective cover of $\Omega^{2i}_B\mathcal{O}$ is Q_{2i}, it follows from Lemma 2 that a projective cover of M is :

$$\bigoplus_{0\leq i\leq e-1} P_{2i} \otimes Q^*_{2i} \xrightarrow{\psi} M.$$

For $l = \{v', v"\}$ an edge and v a vertex of $\mathcal{T}_A$, define $\delta(l, v) = \inf(d(v', v), d(v", v))$. Let x be an integer, $0 \leq x \leq 2e$, such that v_x is the exceptional vertex of $\mathcal{T}_A$. Let

$$X = \bigoplus_{\delta(l_{2i}, v_x)\equiv x \pmod 2} P_{2i} \otimes Q^*_{2i}$$

and ϕ be the restriction of ψ to X. We then define D to be $0 \longrightarrow X \xrightarrow{\phi} M \longrightarrow 0$ (where M is in degree 0).

3.2. Proof of Theorem 10

Let i and j be two integers, $0 \le i, j \le e-1$. We have $Q_{2j}^* \otimes_B \Omega_B^{2i}\mathcal{O} \simeq \mathrm{Hom}_B(Q_{2j}, \Omega_B^{2i}\mathcal{O})$. Since Q_{2j} is a projective cover of $\Omega_B^{2i}\mathcal{O}$ if and only if $i = j$, we have

$$Q_{2j}^* \otimes_B \Omega_B^{2i}\mathcal{O} \simeq \begin{cases} \mathcal{O} & \text{if } i = j, \\ 0 & \text{otherwise.} \end{cases}$$

Hence, we have

$$X \otimes_B \Omega_B^{2i}\mathcal{O} \simeq \begin{cases} P_{2i} & \text{if } \delta(l_{2i}, v_x) \equiv x \pmod 2, \\ 0 & \text{otherwise.} \end{cases}$$

It follows from (5) that

$$D \otimes_B \Omega_B^{2i}\mathcal{O} \simeq \begin{array}{ll} 0 \to 0 \to \Omega_A^{2i}\mathcal{O} \to 0 & \text{if } \delta(l_{2i}, v_x) \not\equiv x \pmod 2, \\ 0 \to P_{2i} \to \Omega_A^{2i}\mathcal{O} \to 0 & \text{if } \delta(l_{2i}, v_x) \equiv x \pmod 2 \end{array}$$

where in both cases, $\Omega_A^{2i}\mathcal{O}$ is in degree 0. Let I be the map between the group of characters of B, $R_K(B)$, and the ring of characters of A, $R_K(A)$, induced by D. By (4), we have:

$$I(w_{2i}) = \begin{array}{ll} v_{2i} & \text{if } \delta(l_{2i}, v_x) \not\equiv x \pmod 2, \\ -v_{2i+1} & \text{if } \delta(l_{2i}, v_x) \equiv x \pmod 2. \end{array}$$

Lemma 11. *The restriction of the map I to the submodule of $R_K(B)$ with basis $\{w_0, w_2, \ldots, w_{2(e-1)}\}$ is an isometry.*

PROOF. We have $\delta(l_{2i}, v_x) \equiv x \pmod 2$ if and only if $\delta(l_{2i}, v_x) = d(v_{2i}, v_x)$, since $d(v_{2i+1}, v_x) \equiv x + 1 \pmod 2$. Hence, $\delta(l_{2i}, v_x) \equiv x \pmod 2$ if and only if $d(v_{2i}, v_x) < d(v_{2i+1}, v_x)$. So, $I(w_{2i})$ is, up to sign, the furthest vertex of l_{2i} from v_x. Since $\mathcal{T}_A$ is a tree, the vertices corresponding to $I(w_{2i})$ and $I(w_{2j})$ are equal if and only if $w_{2i} = w_{2j}$. Note furthermore that $I(w_{2i})$ is, up to sign, an irreducible character. Hence, the lemma follows. □

Corollary 12. *The map I is an isometry.*

PROOF. Indeed, we have $CF(B, K) = K < w_0, w_2, \ldots, w_{2(e-1)} > \oplus CF_p(B, K)$ and the result is given by Lemma 8 and Lemma 11. □

The following is now a direct consequence of Theorem 7:

Theorem 13. *The complex D induces a Rickard equivalence between A and B.*

We obtain the exact formulation of Theorem 10 by replacing D by $0 \longrightarrow X \oplus U \xrightarrow{\delta + id} M \oplus U \longrightarrow 0$, which is homotopy equivalent to D.

References

[Br1] M. Broué, Rickard equivalences and block theory, these Proceedings.

[Br2] M. Broué, Equivalences of blocks of group algebras, preprint LMENS (1993); to appear in *Proceedings of the International Conference on Representations of Algebras* (Ottawa, August 1992).

[Br3] M. Broué, Isométries parfaites, types de blocs, catégories dérivées, *Astérisque* **181-182**(1990), 61–92.

[Gr] J.A. Green, Walking around the Brauer tree, *J. Austral. Math. Soc.* **17**(1974), 197–213.

[Li1] M. Linckelmann, Derived equivalences for cyclic blocks over a p-adic ring, *Math. Z.* **207**(1991), 293–304.

[Li2] M. Linckelmann, The isomorphism problem for blocks with cyclic defect groups, preprint , 1993.

[Ri1] J. Rickard, Derived categories and stable equivalence, *J. Pure Appl. Algebra* **61**(1989), 303–317.

[Ri2] J. Rickard, Conjectural calculation of a two-sided tilting complex for cyclic blocks, preprint, 1990.

[Ri3] J. Rickard, Derived equivalences for principal blocks of A_4 and A_5, preprint, 1990.

[Ri4] J. Rickard, Derived equivalences as derived functors, *J. London Math. Soc.* **43**(1991), 37–48.

FACTORIZATIONS IN WHICH THE FACTORS HAVE RELATIVELY PRIME ORDERS

A.D. SANDS

Department of Mathematics and Computer Science, The University, Dundee DD1 4HN, Scotland

We shall assume throughout that G is a finite abelian group. We say that G is the direct sum of subsets A, B if each $g \in G$ can be expressed uniquely as $g = a + b, a \in A, b \in B$. We write $G = A + B$ and call this a factorization of G. Clearly $|G| = |A||B|$, where $|X|$ denotes the order of a subset X of G.

If $|A|, |B|$ are relatively prime then G possesses unique subgroups H, K with $|A| = |H|$ and $|B| = |K|$. G is the direct sum of these subgroups. In [3] the question was raised whether, under these conditions, $G = A + B$ implies either that $G = H + B$ or that $G = A + K$. It is shown there that if A has prime power order then $G = A + K$. In [4] it is shown that if G is cyclic and A has prime power order then also $G = H + B$. It is also shown in [4] that if A has order 6 then $G = H + B$.

In this note we generalise these last two results. In the first case we show that the result holds whenever the p-component of G is cyclic where the order of A is a power of the prime p. The case $|A| = 6$ is a special case of the following result. Let $|A| = m$ and $|B| = n$, where m and n are relatively prime. Let p, q be the two least prime factors of n. Then if $m < p + q$ and $G = A + B$ it follows that $G = H + B$.

We shall use characters of G. If χ is such a character we shall denote the sum of all the complex numbers $\chi(a), a \in A$, by $\chi(A)$.

Theorem 1. *Let A, B be subsets of G such that $G = A + B$. Let $|A| = p^e$, $|B| = n$, where p is a prime not dividing n, and let G have cyclic p-component. Then, if H, K are the subgroups of G of orders p^e, n, we have $G = H + B = A + K$.*

PROOF. It has already been proved in [3] that $G = A + K$. The proof given there is rather inelegant and may be replaced by the following proof based on later results of Redei [2] on the replacement of factors. Redei has shown that if B, C are subsets of G with $|B| = |C|$ such that, for each character χ of G, $\chi(B) = 0$ implies $\chi(C) = 0$ then, in any factorization of G involving B, B may be replaced by C.

If possible let χ be a character of G such that $\chi(B) = 0$ and $\chi(K) \neq 0$. Since K is a subgroup we must have $\chi(k) = 1$ for all $k \in K$. So for each $g \in G$ it follows that $\chi(g)$ is a p^e-th root of unity. Let ρ be a primitive p^e-th root of unity. Then $0 = \chi(B) = \sum \rho^{u_i}$, for some non-negative integers u_i. From this it follows that the cyclotomic polynomial $F_{p^e}(x)$ divides $\sum x^{u_i}$ and so, replacing x by 1, that p divides n. As this is false it follows that B may be replaced by K and so that $G = A + K$.

We did not use the assumption that the p-component of G was cyclic in the above proof, but it is required now. We proceed by induction on the order of G. If n is prime the result that $G = H + B$ follows from the above result. We assume this result holds for groups of lower order than $|G|$. It is shown in [5] that if $G = A + B$ and the given conditions hold then either A or B is periodic. Suppose first that B is periodic and so there is a non-zero subgroup M such that $B = M + C$. Then $G = A + M + C$ leads to a factorization of the quotient group, giving $G/M = (A+M)/M + (C+M)/M$. By the induction assumption we have that $G/M = (H + M)/M + B/M$. This gives $G = H + B$, as required. If A is periodic there is a non-zero subgroup L such that $A = L+D$. This leads to a factorization of the quotient group as $G/L = A/L + (B + L)/L$. The inductive assumption implies that $G/L = H/L + (B + L)/L$. This implies that $G = H + B$.

This completes the proof of the theorem by induction. □

Theorem 2. *Let A, B be subsets of G such that $G = A+B$, $|A| = m, |B| = n$, where m, n are relatively prime. If $m < p+q$, where p, q are the two least prime factors of n, then $G = H + B$, where H is the subgroup of G of order m.*

PROOF. Since the case of prime power order is covered by Theorem 1, there is no loss of generality in assuming that n has at least two distinct prime factors. We assume that $p < q$ and so q is odd. If possible let χ be a character of G such that $\chi(A) = 0$ but $\chi(H) \neq 0$. As H is a subgroup we must have $\chi(h) = 1$ for all $h \in H$. It follows that $\chi(g)$ is an n-th root of unity for all $g \in G$. Let ρ be an n–th primitive root of unity. Let $n = p^e q^f k$, where p, q do not divide k. Then $\rho = \alpha\beta\gamma$ where α, β, γ are respectively p^e–th, q^f–th, k–th primitive roots of unity.

From $\chi(A) = 0$ it follows that $\sum_{i=1}^{m} \rho^{t_i} = 0$ for some non–negative integers t_i. Let $t_i \equiv u_i (\text{mod } p^e)$, $t_i \equiv v_i (\text{mod } q^f)$, $t_i \equiv w_i (\text{mod } k)$, where $0 \leq u_i < p^e$, $0 \leq v_i < q^f$, $0 \leq w_i < k$. Then $\sum \alpha^{u_i}\beta^{v_i}\gamma^{w_i} = 0$. Since $F_{p^e}(x)$ is irreducible over the field of $q^f k$–th roots of unity it follows that it divides $\sum \beta^{v_i}\gamma^{w_i}x^{u_i}$ and then that the degree of the quotient is less than p^{e-1}. Thus in this polynomial we have

$$\text{coefft}(x^c) = \text{coefft}(x^{c+p^{e-1}}) = \ldots = \text{coefft}(x^{c+(p-1)p^{e-1}}),$$

for $c = 0, 1, \ldots, p^{e-1} - 1$. We need to consider only those values of c for which at least one of these coefficients is not the empty sum. If one of these p terms corresponding to a value of c is not the empty sum but another sum here is empty then each of these coefficients is zero. Let r be the number of terms occurring in such a sum. Then $0 < r \leq m < p+q < 2q$. Since $\beta\gamma$ is a $q^f k$–th root of unity and q is its least prime factor it follows by [6, Theorem 1] that r divides $q^f k$ and that $r \geq q$. Since $m < 2q$ at most one such term can exist and $r = q$. If, for some value of c, all p coefficients involve non-empty sums then at least p terms are contributed to B. Since $m < p + q$ we cannot have two values of c, one contributing q terms and one at least p. Since $m = q$ is not possible we cannot have only the first case arising. So we may assume that only the second case occurs.

If, for each c, the sums in the p coefficients are identical then p divides m. This is false. So, for some c, we must have two coefficients consisting of sums which are not identical. After cancellation of identical terms it follows by [1, Theorem 4] that one of these sums contains at least q terms. The remaining $p - 1$ coefficients must contribute at least one term each to B. This gives at least $q + p - 1$ terms in B, which is the maximum possible order for B. So we must have one coefficient which is a sum of q terms and each of the other $p - 1$ coefficients consists of one term. This gives

$$\sum_{i=1}^{q} \beta^{v_i} \gamma^{w_i} = \beta^v \gamma^w.$$

As before from $F_{q^f}(x)$ dividing $(\sum_{i=1}^{q} \gamma^{w_i} x^{v_i}) - \gamma^w x^v$ we deduce equality of coefficients of powers of x in sets of q. If $k = 1$ then $\gamma = 1$ and this leads, as before, to q dividing $q - 1$. Thus we have $k > 1$. A non-empty sum of powers of γ equal to zero contains at least r terms, where r is the least prime factor of k. Since q is odd, $r > q$ implies $r > q + 1$. This gives too many terms in the sum. All q coefficients non-zero leads as before to at least $q + r - 1$ coefficients, which again exceeds $q + 1$.

We have obtained a contradiction in all cases. Hence A can be replaced by H and so $G = H + B$, as required. □

The method of proof used does not extend to the case $m = p + q$. It is possible in this case to have $\chi(A) = 0$ but $\chi(H) \neq 0$. This does not necessarily lead to a negative answer to the question. In order to deduce that $G = A + B$ implies $G = H + B$ one needs to show that $\chi(B) \neq 0$ implies $\chi(H) = 0$ for all non-identity characters χ. This does not follow as it is possible to have both $\chi(A) = 0$ and $\chi(B) = 0$.

We give one example where this occurs. Let G be the additive cyclic group of integers modulo 30. Let $m = 5$, $n = 6$. Then $p = 2$, $q = 3$ and $m = p + q$. We take $A = \{0, 4, 8, 12, 21\}$ and $B = \{0, 5, 10, 15, 20, 25\} = K$. Since A is a complete set of residues modulo 5 we have $G = A + B$. Let ρ be a primitive 6-th root of unity and let χ be the character defined by $\chi(n) = \rho^n$. It is clear that $\chi(B) = 0$ and also $\chi(A) = 1 + \rho^4 + \rho^8 + \rho^{12} + \rho^{21} = (1 + \rho^4 + \rho^2) + (1 + \rho^3) = 0$. We also have $\chi(h) = 1$ for all $h \in H$.

References

[1] K.Corradi, A.D.Sands and S.Szabo, Simulated Factorizations, *J. Algebra* **151** (1992), 12–25.

[2] L.Redei, Die neue Theorie der endlichen Abelschen Gruppen und Verallgemeinerung des Hauptsatzes von Hajos, *Acta Math.Acad.Sci.Hungar.* **16**(1965), 329–373.

[3] A.D.Sands, On a Problem of L. Fuchs, *J. London Math. Soc.* **37**(1962), 277–284.

[4] A.D.Sands, On the Factorization of Finite Groups, *J. London Math. Soc.(2)* **7**(1974), 627–631.

[5] A.D.Sands, On the Factorisation of Finite Abelian Groups III, *Acta Math. Acad. Sci. Hungar.* **25**(1974), 279–284.

[6] A.D.Sands, Simulated Factorizations II, *Aequationes Math.* **44**(1992), 48–59.

SOME PROBLEMS AND RESULTS IN THE THEORY OF PRO-P GROUPS

ANER SHALEV

Institute of Mathematics, The Hebrew University, Jerusalem 91904, Israel

1. Introduction

The theory of finite p-groups and their inverse limits, known as pro-p groups, seems to attract increasing attention in the past few years. Various new results and methods have been introduced; moreover, progress in this field, especially in p-adic analytic groups, is proving to have an impact on several other areas (see [DDMS] and the reference list therein).

The purpose of this paper is to outline some recent developments in the study of pro-p groups, and – more importantly – to present some problems (or conjectures) which I consider natural. Unlike several recent survey papers dealing with p-adic analytic groups and their applications to discrete groups (see [M1],[Se]), this paper is devoted solely to pro-p groups. Moreover, p-adic analytic groups, though mentioned, are not the focus of this paper; they are viewed from a somewhat wider perspective. For recent remarkable results on pro-p groups which are not reported here see Zelmanov [Z1],[Z2].

Roughly speaking, we shall try to draw a 'map' of the universe of pro-p groups, starting with relatively tame groups, and passing gradually to wilder ones. More specifically, we consider the following families of pro-p groups:

1. Groups of finite coclass.
2. p-adic analytic groups.
3. Groups with analytic structure over a general pro-p ring Λ (e.g. $\Lambda = \mathbb{F}_p[[t]]$).
4. The Nottingham group and other monsters.
5. Groups of finite width.

Basic properties of groups in each class will be presented (with some emphasis on groups of finite coclass); it will become clear that, in a certain sense, there is a continuous transition from one class to the next. While our understanding of the first two classes is quite satisfactory (though several problems still remain open; see below), the knowledge of the last classes is somewhat scattered (and so the questions there are more fundamental). This applies, for example, to the third class, that of Λ-analytic groups, which is particularly challenging. The investigation of these groups, carried out in [LS], enables us to settle a problem in p-adic analytic groups posed by Lubotzky and Mann in [LM1] (and reposed in the last St Andrews Conference [M1, Problem 3]).

The 'map' drawn here is by no means exhaustive: indeed, most pro-p groups do not lie in any of these classes; however, I believe that this map has some methodological advantages and hope that it will inspire further research.

2. Groups of finite coclass

In 1980 C.R. Leedham-Green and M.F. Newman proposed a program for classifying finite p-groups using coclass as the primary invariant [LGN]. The coclass of a group of order p^n and nilpotency class c is defined to be $n - c$; a pro-p group of coclass r is an inverse limit of finite p-groups of coclass r. Thus, groups of coclass 1 are exactly the groups of maximal class studied by N. Blackburn [Bl] and many others.

We can now state the coclass conjectures made in [LGN], ordered by increasing strength (that is, each conjecture implies the previous ones).

Conjecture E. Given p and r, there are only finitely many isomorphism types of infinite soluble pro-p groups of coclass r.

Conjecture D. Given p and r, there are only finitely many isomorphism types of infinite pro-p groups of coclass r.

Conjecture C. Pro-p groups of finite coclass are soluble.

Conjecture B. For some function g, every p-group of coclass r has derived length at most $g(p,r)$.

Conjecture A. For some function f, every p-group of coclass r has a normal subgroup N of class at most 2 (1, if $p = 2$) and index at most $f(p,r)$.

Some of these conjectures may be reformulated as combinatorial statements on the structure of certain graphs. Thus, let $T_{p,r}$ be a (directed) tree whose vertices are all p-groups of coclass r, where (H,G) is an edge if there is an epimorphism from H to G whose kernel has order p. Then Conjecture D is tantamount to saying that there are only finitely many infinite chains in $T_{p,r}$.

There is a certain counter-intuitive aspect in some of the conjectures, in that they assert that groups of small coclass, that is, of very large nilpotency class, are in a sense almost abelian.

From a methodological (but not historical!) viewpoint the work on the coclass conjectures may be divided into two phases: the first, which does not use Lie algebras and Lie methods, and the second, which does use them.

To the first phase belong a proof of Conjecture E by Leedham-Green, S. McKay and W. Plesken [LMP1], [LMP2], partial proofs of Conjecture A by McKay [MK1],[MK2] and Mann [M2] (the uncovered case), and also some work by other authors. Here we focus on the second phase, which applies Lie methods; it includes works by S. Donkin [D], Leedham-Green [LG1],[LG2], Zelmanov and myself [SZ], [Sh4], as well as Alperin's early paper [A] which confirms Conjecture B in the case $r = 1$ about 18 years before it was formulated.

An important discovery of Leedham-Green paved the way for several Lie-theoretic attacks on the coclass conjectures. He showed that pro-p groups of finite coclass are p-adic analytic [LG1] (see also Mann [M2] for a shorter, quantitative proof of this fact). In particular it follows that, with each pro-p group G of finite coclass one can associate a p-adic Lie algebra L, which proves quite useful.

Let me now describe Donkin's approach (with some unavoidable simplifications), which, combined with Leedham-Green's above-mentioned result, yields a proof of Conjecture C for $p > 3$. Let G be a counter-example to Conjecture C and let L be its associated p-adic Lie algebra. Since G is non-soluble (by assumption) it easily follows that L is non-soluble. The fact that G has finite coclass implies that the adjoint action of G on L is *uniserial* (i.e. the $\mathbb{Z}_pG$-submodules of L are linearly ordered by inclusion). We have thus obtained a non-soluble p-adic Lie algebra L acted on uniserially by a pro-p group. The next stage is to reduce to the case where L (tensored with the p-adic field) is simple. This leads to the following question: *can a pro-p group act uniserially on a simple p-adic Lie algebra?* By the above discussion, a negative answer would imply Conjecture C. Using the classification of simple p-adic Lie algebras and Iwahori and Matsumoto's theory of p-adic Chevalley groups, Donkin was able to obtain a negative answer for $p > 3$, thus confirming Conjecture C for these primes. Donkin's proof is rather complicated and highly non-elementary. Another non-elementary approach, which applies the theory of buildings, has been suggested by J. Tits.

Recently, using a different (and more elementary) approach, Zelmanov and I have obtained a short proof of Conjecture C for all primes p [SZ]. We first observed that a p-adic analytic pro-p group with an element whose centralizer is finite is soluble. This can be deduced from Kreknin's result [Kr] on Lie rings with fixed-point-free automorphisms (or alternatively from results of Steinberg on endomorphisms of algebraic groups). At this stage one is led to the following question: *does every pro-p group of finite coclass have an element whose centralizer is finite?* Clearly, a positive answer would imply Conjecture C. By making additional reductions to questions on modular Lie algebras we gave an affirmative answer to that question, and deduced Conjecture C for all primes p (the case $p = 2$ in fact being the simplest). Note that, if $x \in G$ and the centralizer $C_G(x)$ is finite, then x is automatically of finite order. As a by-product it follows that pro-p groups of finite coclass cannot be torsion-free. We are not aware of any direct proof of this somewhat curious fact.

Let us now discuss the strongest of the coclass conjectures, namely Conjecture A. This conjecture was deduced from Conjecture C by Leedham-Green in [LG2]. The deduction makes use of various limit arguments; consequently, though the existence of a function $f(p, r)$ with the required property is established, no explicit bounds are obtained.

It turns out that certain bounds on f may be obtained by using the following method (which may be regarded as an extension of Alperin's approach). First, one proves that the derived length of a finite p-group G is effectively bounded in terms of the order of any centralizer $C_G(x)$ of an element $x \in G$. This result, which is established in [Sh3], extends a classical theorem of Alperin [A] dealing with the case where x has order p. Secondly, one proves that every finite p-group G of coclass r has an element x satisfying $|C_G(x)| \leq k(p, r)$ for some function k which may be effectively bounded. It then follows that the derived length of a p-group of coclass r is effectively bounded in terms of p and r (this confirms an effective version of Conjecture B). In the final stage the bound on the derived length is used in order to find an explicit bound on the index of a class 2 subgroup. This type of argument (which in practice is very long and tedious) gives rise to the inequality

$$f(p, r) \leq p^{2^{2^{p^{p^{2r}}}}},$$

a bound which, though explicit, seems a bit exaggerated, compared with existing examples.

The final approach which I would like to discuss is related to the notion of *non-singular derivations*. A derivation D of a finite-dimensional Lie algebra L is called non-singular if it is non-singular as a linear transformation. Note that 'natural' (e.g. inner) derivations are singular, though some Lie algebras (e.g. abelian ones) do admit non-singular derivations. In general it seems a strong restriction on a Lie algebra to admit a non-singular derivation. For example, Jacobson has shown that a finite-dimensional Lie algebra over a field of characteristic zero which admits a non-singular derivation is nilpotent [J].

The relevance of all this to the current discussion emerges from the following fact: assuming a certain effective version of Conjecture A is violated, it is possible to construct a Lie algebra L having the following properties:

1. $L \neq 0$ is finite-dimensional over some global field of characteristic p.
2. L admits a derivation D satisfying $D^{p-1} = 1$.
3. L is perfect, namely $L' = L$.

The construction of L is long and very technical; it makes essential use of the theory of powerful p-groups [LM1], and other tools. However, once it is constructed, Conjecture A (with explicit bounds) is reduced to the following question: *does a Lie algebra satisfying conditions 1–3 exist?*

Clearly the derivation D above is non-singular. However, this alone is not enough to derive a contradiction: it can be shown that perfect – and even simple – modular Lie algebras admitting a non-singular derivation do exist (the split 3-dimensional Lie algebra of characteristic 2 is the smallest example). Thus we have to use the equation $D^{p-1} = 1$, which implies that D is semisimple and that all its eigenvalues are in $\mathbb{F}_p^*$. Decomposing $L = \sum_{\alpha \in \mathbb{F}_p} L_\alpha$ as a direct sum of eigenspaces for D we obtain a grading of L over the additive group of $\mathbb{F}_p$, such that $L_0 = 0$. It is now straightforward to verify that L satisfies the $(p-1)$ Engel condition for homogeneous elements, and is thus nilpotent (by a variation of Jacobson on Engel's theorem [J]); alternatively, the nilpotency of L may be deduced from a Lie-theoretic result of G. Higman, implicitly contained in [H]. In any event, having shown that L is nilpotent, it cannot be perfect. This shows that there is no Lie algebra L satisfying conditions 1–3, thus providing the required answer to the question posed above.

This line of argument gives rise to the first part of the following theorem.

Theorem 1. *Let G be a finite p-group of coclass r.*

1. *Suppose $p > 2$. Then $\gamma_{2(p^r - p^{r-1} - 1)}$ has class at most 2.*
2. *Suppose $p = 2$ and $|G| \geq 2^{2^{2r+5}}$. Then $\gamma_{7.2^r - 2}$ is abelian.*

For odd primes this confirms Conjecture A with

$$f(p,r) \leq p^{2(p^r - p^{r-1} - 1) + r - 1} < p^{2p^r},$$

a bound which is quite realistic. The case $p = 2$ requires some extra arguments, and the bound obtained is probably less accurate.

Let us now turn back to infinite pro-p groups. It follows from Conjecture C, combined with early results of Leedham-Green and Newman [LGN], that just-infinite pro-p groups of finite coclass have the structure of (uniserial) *p-adic space groups*: they are formed by extending a finitely generated free p-adic module T (the translation group) by a finite p-group P (the point group) acting faithfully and uniserially. The simplest such example is the split extension of $\mathbb{Z}_2$ by an involution acting as multiplication by -1. Thus there are interesting relations between the coclass theory and certain (p-adic) crystallographic groups.

Moreover, by a result of Leedham-Green, there are rather surprising relations between the p-adic crystallographic groups mentioned above and the structure of finite p-groups. Indeed, it is shown in [LG2] that there exists a function h such that every finite p-group G of coclass r has a normal subgroup N whose order is at most $h(p,r)$ such that G/N is obtained from a p-adic space group of coclass at most r via a series of 'elementary operations' (such finite p-groups are called *constructible*). While the precise result for odd p is not easily formulated, for $p = 2$ this simply means that G/N is a *quotient* of some 2-adic space group of coclass at most r.

Problem 1. Find explicit bounds on the function $h(p,r)$.

No such bounds are known up to now. This problem has a geometrical interpretation concerning the twig lengths in the tree $T_{2,r}$. Recently M.F. Newman and E.A. O'Brien have made several delicate conjectures on the structure of that tree, based on a thorough analysis of the case $r = 3$.

We close this section with another problem. For an infinite pro-p group G define a series $c_n = c_n(G) \leq \infty$ $(n \geq 1)$ by

$$|G : \gamma_n| = p^{c_n}.$$

Note that G has coclass r if and only if $c_n(G) = n + r - 1$ for all sufficiently large n; thus G has finite coclass if and only if the series $\{c_n(G) - n\}$ is bounded. However, one may try to impose weaker conditions on the series $\{c_n\}$. For example, in [Sh2] it is shown that the inequality $c_n < n + [\log_p(n/2)]$ for a single value of n already implies that G has finite coclass. We therefore ask:

Problem 2. Given n, find the maximal integer $f(n)$ such that $c_n(G) \leq f(n)$ implies that G has finite coclass.

It can be shown, using the example of the Nottingham group (see section 5), that $f(n) < n + [(n-2)/(p-1)]$. This still leaves open a fairly large gap.

3. p-adic analytic groups

There are various characterizations of p-adic analytic pro-p groups, such as being finitely generated and virtually powerful (Lazard [La]), or being of finite rank (Lubotzky and Mann [LM1]). By combining these characterizations with Zelmanov's recent solution to the Restricted Burnside Problem [Z1] it is possible to derive the following characterization:

Theorem 2. *Let G be a finitely generated pro-p group. Then G is p-adic analytic if and only if, for some n, G has no open sections isomorphic to the wreath product $C_p \wr C_{p^n}$.*

This result, proved in [Sh1], is actually equivalent to Zelmanov's theorem.

Problem 3. Let G be a finitely generated pro-p group and suppose G does not involve $C_p \wr \mathbb{Z}_p$ as a closed section; does it follow that G is p-adic analytic?

Note that $C_p \wr \mathbb{Z}_p$ may be identified with the inverse limit of the finite groups $C_p \wr C_{p^n}$, which are all involved in G.

An affirmative solution would present $C_p \wr \mathbb{Z}_p$ as the *minimal* non p-adic analytic finitely generated pro-p group. This would imply the following powerful procedure: let P be any property of groups which is inherited by closed subgroups and quotients; then, in order to show that the property P implies p-adic analyticity for finitely generated pro-p groups, it is enough to verify that $C_p \wr \mathbb{Z}_p$ does not satisfy P.

The next problem is of a similar flavour. By an old result of Serre, p-adic analytic pro-p groups are Noetherian, i.e. closed subgroups are finitely generated (as topological groups).

Problem 4. Is every Noetherian pro-p group p-adic analytic?

This problem is posed in [LM1] (and subsequently [M1]). A positive solution to problem 3 yields a positive solution to problem 4 (as the wreath product $C_p \wr \mathbb{Z}_p$ is not Noetherian).

The next question, which arose in a conversation with Avinoam Mann, is somewhat related.

Problem 5. Let G be a pro-p group, and suppose every closed subgroup $H \subset G$ of infinite index is p-adic analytic; does it follow that G is p-adic analytic?

It is easy to see that, if G is as above, then G is Noetherian. Hence an affirmative solution to problem 4 implies an affirmative solution to problem 5.

Note that, if G is p-adic analytic, then there is a bound on the length of chains of closed subgroups $1 = G_0 \subset G_1 \subset \ldots \subset G_n = G$ with the property that the indices $|G_{i+1} : G_i|$ are all infinite (as G_i are all p-adic analytic and $\dim G_i$ is strictly increasing). A positive answer to problem 5 would imply the converse: namely, that the existence of such a bound is equivalent to G being p-adic analytic.

Let us now consider the power structure of pro-p groups. Denote by G^p the *set* of all pth powers in a pro-p group G. It is known that p-adic analytic pro-p groups have many pth powers, in the sense that G^p contains an open subgroup. Indeed, such a group G has a powerful open subgroup H, and the set H^p is an open subgroup contained in G^p.

It is interesting to know whether the converse is true.

Problem 6. Let G be a pro-p group and suppose G^p contains an open subgroup; does it follow that G is p-adic analytic?

It is not even clear if the assumption that G^p is a subgroup implies analyticity (see [Sh1] for some related results). As for discrete groups, it has recently been shown in [HKLS] that for a finitely generated linear (or soluble) group G, the set G^p contains a finite index subgroup if and only if G is nilpotent-by-finite.

Finally, consider the subgroup growth of pro-p groups. As usual, let $a_n = a_n(G)$ denote the number of open subgroups of index n in G. It was shown by Lubotzky and Mann that a pro-p group has polynomial subgroup growth if and only if it is p-adic analytic [LM2, Theorem 3.1]. This result is sharpened in [Sh2], where it is shown that, if c is any constant less than $1/8$, then the condition $a_n(G) \leq n^{c \log_p n}$ for all sufficiently large n already implies that G is p-adic analytic.

Problem 7. Find the best constant $c = c(p)$ such that, if G is a pro-p group satisfying $a_n(G) \leq n^{c \log_p n}$ for all sufficiently large n, then G is p-adic analytic.

It is known that $\frac{1}{8} - \epsilon < c(p) < 1 + \frac{2}{p-1} + \epsilon$ for all $\epsilon > 0$ and $p \geq 5$ (see Section 5).

4. Λ-analytic groups

While the theory of Lie groups is highly developed in characteristic zero, very little seems to be known about Lie groups over local fields K of characteristic p, say $K = \mathbb{F}_p((t))$. Some basic information on these analytic groups may be found in Serre [S] and Bourbaki [B]. In particular Serre shows that these group have open subgroups, called *standard*, which are obtained from a single (say, d-dimensional) formal group law F defined over the valuation ring $\mathbb{F}_p[[t]]$, when applied to the appropriate free module (of rank d) over the maximal ideal $t\mathbb{F}_p[[t]]$. This reduces various questions on K-analytic groups to the case of standard groups. A typical example of a standard group over $\mathbb{F}_p[[t]]$ is the first congruence subgroup of $SL_m(\mathbb{F}_p[[t]])$, which we denote by $SL_m^1(\mathbb{F}_p[[t]])$.

In [LS] Lubotzky and I began a systematic study of these $\mathbb{F}_p[[t]]$-standard groups. At a certain stage it became clear that these groups share many properties with Λ-standard groups, where Λ is any commutative Noetherian complete local ring whose residue class field is finite. Thus one may take Λ to be any epimorphic image of the power series ring $\mathbb{Z}_p[[X_1, \ldots, X_r]]$ in any (finite) number of variables. Of course, if Λ is finitely generated as a p-adic module, then the resulting groups are p-adic analytic. It turns out that the converse is also true; in other words, if $|\Lambda/p\Lambda| = \infty$ then Λ-standard groups are *not* p-adic analytic.

The additive group of $\mathbb{F}_p[[t]]$ and other 1-dimensional groups demonstrate that Λ-standard groups need not be finitely generated (as pro-p groups).

From certain points of view these examples may be considered 'degenerate'. However, there is a natural class of Λ-standard groups, which we call Λ-*perfect*, whose structure seems much more rigid (in particular, they are finitely generated). Roughly speaking, a Λ-standard group G is said to be Λ-perfect if a certain Lie algebra associated to G is perfect. The groups $SL^1_m(\mathbb{F}_p[[t]])$ are $\mathbb{F}_p[[t]]$-perfect unless $p = m = 2$.

The following theorem summarizes some of the main results on Λ-perfect groups obtained in [LS].

Theorem 3. *Let G be a Λ-perfect group.*

1. *For some constant c we have $a_n(G) \leq n^{c \log n}$ for all n.*
2. *Suppose the formal group associated to G is defined over the prime subring of Λ. Then there exists an integer d and a series of open subgroups $\{H_n\}$ forming a neighbourhood base to the identity in G, such that each H_n is d-generated.*
3. *G satisfies the Golod-Shafarevich inequality; namely, if $\langle X; R\rangle$ is a minimal presentation of G (as a pro-p group), then $|R| \geq |X|^2/4$.*

The minimal integer d satisfying condition 2 above is called the *lower rank* of G (it is defined as ∞ if there is no such integer). Thus part 2 asserts that many Λ-perfect groups (including $SL^1_m(\mathbb{F}_p[[t]])$ for $(p, m) \neq (2, 2)$) have finite lower rank. This settles a problem posed in [LM1] as to whether all pro-p groups of finite lower rank are p-adic analytic; see also [M1, Problem 3]. The simplest counter-example is $SL^1_2(\mathbb{F}_p[[t]])$ ($p > 2$), whose exact lower rank is – however – unknown (it is either 2 or 3).

Theorem 3 shows that, in some sense, there is a continuous transition between p-adic analytic pro-p groups and (certain) analytic groups over general pro-p rings Λ. For example, a polynomial subgroup growth (in the p-adic case) is replaced by a growth of the type $n^{c \log n}$ (in the general case), which (up to a constant) is the minimal type of growth of non-PSG pro-p groups. Similarly, the property of having finite rank (in the p-adic analytic case) is replaced by that of having finite lower rank. And the Golod-Shafarevich inequality (established by Koch and Lubotzky for p-adic analytic groups) remains valid in the more general situation. We remark that important results on the Golod-Shafarevich inequality for other types of pro-p groups have recently been established by Wilson and Zelmanov [W],[WZ].

Part 1 of Theorem 3 has recently been established by Inga Levich for $\mathbb{F}_p[[t]]$-standard groups of Chevalley type which are not $\mathbb{F}_p[[t]]$-perfect.

We close this section with some fundamental problems which are still very much open.

Problem 8. Find a group-theoretic characterization of Λ-analytic (Λ-standard, or Λ-perfect) pro-p groups.

The important case $\Lambda = \mathbb{F}_p[[t]]$ may be regarded as Hilbert's fifth problem in characteristic p. Unfortunately, we are not even close to formulating a reasonable conjecture (a characterization in the language of groups with a given operator – correponding to the action of t – is more likely to be found).

Problem 9. Are Λ-analytic pro-p groups always linear?

The answer is affirmative for $\Lambda = \mathbb{Z}_p$. We conjecture that it is positive in general. Progress in these directions is likely to have profound applications to the characterization of linear groups in characteristic p (see Lubotzky [Lu] for the zero characteristic case).

Next, we call attention to the fact that, while closed subgroups of p-adic analytic groups are themselves p-adic analytic, this is no longer the case for Λ-analytic groups. For example, $\mathbb{Z}_p$ can usually be embedded in analytic groups over $\mathbb{F}_p[[t]]$, though it is not itself $\mathbb{F}_p[[t]]$-analytic. It is therefore natural to ask:

Problem 10. Which pro-p groups can be embedded as closed subgroups of Λ-analytic pro-p groups? In particular, can a (non-abelian) free pro-p group be obtained in this way?

A negative answer to the last question would imply that every analytic pro-p group satisfies some pro-p identity, and may help lay the foundation for a possible theory of groups with *topological* identities (which might be related to the classical theory of PI rings). Preliminary results in this direction were obtained by Zubkov [Zu]. Questions of this type also occur in number theory, in the study of pro-p groups as Galois groups (see Boston [Bo], as well as [LS, Section 6] where some conjectures from [Bo] are settled).

5. Pro-p groups associated with non-classical Lie algebras

One of the reasons why problem 8 above may be very difficult is the existence of the so called *Nottingham group*. This group, which was studied by D. Johnson and his student I. York in Nottingham, may be described as the group of normalized automorphisms of the ring $\mathbb{F}_p[[t]]$, namely, those automorphisms acting trivially on $t\mathbb{F}_p[[t]]/t^2\mathbb{F}_p[[t]]$. Alternatively, this group may be described as the group of all power series of the form $t + a_2t^2 + a_3t^3 + \dots$ $(a_i \in \mathbb{F}_p)$ under substitution. See [Y] for basic properties of the Nottingham group (in particular, the determination of its lower central series).

As the following result shows, the Nottingham group looks rather tame from the top, and rather wild from the bottom.

Theorem 4. *Let G be the Nottingham group for $p \geq 5$. Then*

1. $a_n(G) \leq 2n^{c \log_p n}$ *where* $c = 1 + \frac{2}{p-1}$.
2. *G has lower rank 2.*
3. *G satisfies the Golod-Shafarevich inequality.*
4. *Every finite p-group can be embedded in G.*
5. *G is not linear over any field.*

Part 5 above is an immediate consequence of part 4 (as the derived length of soluble linear groups in a given dimension is bounded).

Parts 1-3 indicate that, from a certain perspective, the Nottingham group G looks very much like an $\mathbb{F}_p[[t]]$-analytic pro-p group (being of infinite rank, G is certainly not p-adic analytic). However, it is very likely that G is *not* $\mathbb{F}_p[[t]]$-analytic (though a rigorous argument is still required; the fact that G is not $\mathbb{F}_p[[t]]$-standard can be proved rather easily). Of course, since the Nottingham group is not linear, a positive solution to problem 9 would immediately imply that it is not analytic.

For the proofs of the above result, see [LGSW]. It is interesting to note that the proof of part 4 (which is due to Leedham-Green and A. Weiss) is Galois-theoretic. Weiss has also shown that the Nottingham group has infinite elementary abelian subgroups.

Problem 11. Does the Nottingham group have a closed subgroup (or section) isomorphic to $C_p \wr \mathbb{Z}_p$?

A positive answer would imply a negative answer to problem 3.

Problem 12. Does G have (non-abelian) free (discrete) subgroups?

The answer ought to be 'yes'.

There is an interesting relation between the Nottingham group and the first Witt algebra W_1 in characteristic p. The Lie algebra W_1 has an $\mathbb{F}_p$-basis $e_0, e_1, \ldots, e_{p-1}$ with $[e_i, e_j] = (i-j)e_{i+j}$ (operations on indices are mod p). It turns out that G has a natural 'congruence' filtration $\{G_n\}$ which gives rise to a graded Lie algebra L, which is isomorphic to a finite index $\mathbb{F}_p$-subalgebra of $W_1 \otimes \mathbb{F}_p[[t]]$. Now, using the solubility of $\mathrm{Aut} W_1$, it can be shown that there is no $\mathbb{F}_p[[t]]$-standard group whose Lie algebra is W_1; so in a sense the group corresponding to W_1 is the Nottingham group, which is not analytic in the classical sense, but still shares some common properties with analytic groups.

However, W_1 is only one example of a non-classical simple modular Lie algebra. It can be shown that other non-classical simple Lie algebras (of Cartan type; see, e.g., [SW]) give rise to some associated pro-p groups, of which the Nottingham group is a special case.

Problem 13. Construct and study pro-p groups corresponding to arbitrary simple modular Lie algebras of Cartan type.

Roughly speaking, while Lie algebras of Cartan type arise as collections of derivations annihilating a given differential form, their corresponding groups should consist of automorphisms preserving that form.

6. Groups of finite width

We have started this tour with pro-p groups of finite coclass, which in a sense are particularly narrow. Attempting to generalize the coclass theory, Leedham-Green has proposed the study of pro-p groups of *finite width.* Here the width may be defined as the supremum of the orders of the lower central factors of the group. Thus finite coclass implies finite width, but not the converse. Some other narrowness conditions have been introduced by other authors (such as C.M. Scoppola).

Classification of pro-p groups according to width is undoubtedly one of the main challenges in the field. The difficulty of the subject lies in the fact that, unlike groups of finite coclass, pro-p groups of finite width need be neither soluble, nor p-adic analytic.

A typical example of a non-soluble pro-p group of finite width is $SL_2^1(\mathbb{Z}_p)$; it is of course p-adic analytic. However, various $\mathbb{F}_p[[t]]$-analytic groups (such as $SL_2^1(\mathbb{F}_p[[t]])$) have finite width, and they are *not* p-adic analytic. Moreover, the Nottingham group discussed above has finite width (which equals p^2), and it is not even linear.

We see that the investigation of pro-p groups of finite width involves the study of (at least) three interesting types of pro-p groups discussed in previous sections: the p-adic analytic ones, the groups which are analytic over a local field of characteristic p, and the Nottingham group.

Problem 14. Does every just-infinite pro-p group of finite width belong to one of the above three classes?

7. Subgroup growth revisited

Some results on subgroup growth were already mentioned briefly in previous sections. Here I would like to pursue the matter a bit further. Let us say that two sequences $\{a_n\}$ and $\{b_n\}$ of natural numbers are *equivalent* if there are constants $c_2 > c_1 > 0$ such that $a_n^{c_1} \leq b_n \leq a_n^{c_2}$ for all n. Given a group G, let $s_n(G)$ denote the number of subgroups of index *at most* n in G. We say that the groups G and H have the same *growth type* if the sequences $\{s_n(G)\}$ and $\{s_n(H)\}$ are equivalent. If, for some $c > 0$ we have $s_n(G) \geq s_n(H)^c$ for all n we say that the growth type of G is greater than or equal to that of H.

It would be interesting to investigate the ordered set $\mathcal{G}$ of growth types of finitely generated pro-p groups. It can be shown that all PSG pro-p groups which are not virtually cyclic have the same growth type (namely n). The next type of subgroup growth for pro-p groups is $n^{\log n}$ [Sh2], and the maximal growth type is 2^n [I]. It is not clear whether the growth type $n^{\log n}$ has a consecutive element in $\mathcal{G}$, or even whether $\mathcal{G}$ is linearly ordered.

The fact that $\mathcal{G}$ is not well-ordered follows from a recent work of Dan Segal and myself [SS], realizing certain new growth types for pro-p (and discrete) groups.

Theorem 5. *For every $d \geq 2$ and a prime p there is a finitely presented pro-p group whose growth type is $2^{n^{1/d}}$.*

Our examples are metabelian, and the proof applies some algebraic geometry and geometry of numbers. It can also be shown (using [W]) that the growth type of a finitely presented soluble pro-p group is at most $2^{\sqrt{n}}$ (this holds for all f.p. pro-p groups which satisfy the Golod-Shafarevich inequality). We have not been able to settle the following:

Problem 15. For which real numbers $0 < \alpha < 1$ are there pro-p groups whose growth type is 2^{n^α}?

The methods of [SS] might cope with rational values of α, but certainly not with the general case.

References

[A] J.L. Alperin, Automorphisms of solvable groups, *Proc. Amer. Math. Soc.* **13**(1962), 175–180.

[Bl] N. Blackburn, On a special class of p-groups, *Acta Math.* **100**(1958), 49–92.

[Bo] N. Boston, Explicit deformation of Galois representations, *Invent. Math.* **103**(1991), 181–196.

[B] N. Bourbaki, *Lie groups and Lie algebras, Chapters 1-3* (Springer, Berlin, 1980).

[DDMS] J. Dixon, M.P.F. Du Sautoy, A. Mann and D. Segal, *Analytic Pro-p Groups* (London Math. Soc. Lecture Note Series **157**, Cambridge University Press, Cambridge, 1991).

[D] S. Donkin, Space groups and groups of prime power order. VIII. Pro-p groups of finite coclass and p-adic Lie algebras, *J. Algebra* **111**(1987), 316–342.

[H] G. Higman, Groups and Lie rings having automorphisms without non-trivial fixed points, *J. London Math. Soc.* **32**(1957), 321–334.

[HKLS] E. Hrushovski, P.H. Kropholler, A. Lubotzky and A. Shalev, Powers in finitely generated groups, Preprint, 1993.

[I] I. Ilani, Counting finite index subgroups and the P. Hall enumeration principle, *Israel J. Math.* **68**(1989), 18–26.

[J] N. Jacobson, A note on automorphisms and derivations of Lie algebras, *Proc. Amer. Math. Soc.* **6**(1955), 281–283.

[Kr] V.A. Kreknin, Solvability of Lie algebras with a regular automorphism of finite period, *Soviet. Math. Dokl.* **4**(1963), 683–685.

[La] M. Lazard, Groupes analytiques p-adiques, *Publ. Math. I.H.E.S.* **26**(1965), 389–603.

[LG1] C.R. Leedham-Green, Pro-p groups of finite coclass, to appear.

[LG2] C.R. Leedham-Green, The structure of finite p-groups, to appear.

[LMP1] C.R. Leedham-Green, S. McKay and W. Plesken, Space groups and groups of prime power order. V. A bound to the dimension of space groups with fixed coclass, *Proc. London Math. Soc.* **52**(1986), 73–94.

[LMP2] C.R. Leedham-Green, S. McKay and W. Plesken, Space groups and groups of prime power order. VI. A bound to the dimension of a 2-adic group with fixed coclass, *J. London Math. Soc.* **34**(1986), 417–425.

[LGN] C.R. Leedham-Green and M.F. Newman, Space groups and groups of prime power order I, *Arch. Math.* **35**(1980), 193–202.

[LGSW] C.R. Leedham-Green, A. Shalev and A. Weiss, Reflections on the Nottingham group, in preparation.

[Lu] A. Lubotzky, A group-theoretic characterization of linear groups, *J. Algebra* **113**(1988), 207–214.

[LM1] A. Lubotzky and A. Mann, Powerful p-groups. I,II. *J. Algebra* **105**(1987), 484–515.

[LM2] A. Lubotzky and A. Mann, On groups of polynomial subgroup growth, *Invent. Math.* **104**(1991), 521–533.

[LS] A. Lubotzky and A. Shalev, On some Λ-analytic pro-p groups, Preprint, 1992.

[M1] A. Mann, Some applications of powerful p-groups, in *Groups – St Andrews 1989*, Vol.2 (London Math. Soc. Lecture Note Series **160**, Cambridge University Press, Cambridge, 1990).

[M2] A. Mann, Space groups and groups of prime power order. VII. Powerful p-groups and uncovered p-groups, *Bull. London Math. Soc.* **24**(1992), 271–276.

[MK1] S. McKay, On a special class of p-groups, *Quart. J. Math. Oxford* (2) **38**(1987), 489–502.

[MK2] S. McKay, On a special class of p-groups II, *Quart. J. Math. Oxford* (2) **41**(1990), 431–448.

[Se] D. Segal, Residually finite groups, in *Groups – Canberra 1989* (Lecture

Notes in Math. **1456**, Springer, Berlin, 1990).

[SS] D. Segal and A. Shalev, Groups with fractionally exponential subgroup growth, *J. Pure Appl. Algebra* **88**(1993) (The Gruenberg Volume), 205–223.

[S] J.-P. Serre, *Lie groups and Lie algebras* (new edition) (Lecture Notes in Math. **1500**, Springer, Berlin, 1991).

[Sh1] A. Shalev, Characterization of p-adic analytic groups in terms of wreath products, *J. Algebra* **145**(1992), 204–208.

[Sh2] A. Shalev, Growth functions, p-adic analytic groups, and groups of finite coclass, *J. London Math. Soc.* **46**(1992), 111–122.

[Sh3] A. Shalev, On almost fixed point free automorphisms, *J. Algebra*, to appear.

[Sh4] A. Shalev, The structure of finite p-groups: effective proof of the coclass conjectures, *Invent. Math.*, to appear.

[SZ] A. Shalev and E.I. Zelmanov, Pro-p groups of finite coclass, *Math. Proc. Cambridge Philos. Soc.* **111**(1992), 417–421.

[SW] H. Strade and R.L. Wilson, Classification of simple Lie algebras over algebraically closed fields of prime characteristic, *Bull. Amer. Math. Soc.* **24**(1991), 357–362.

[W] J.S. Wilson, Finite presentations of pro-p groups and discrete groups, *Invent. Math.* **105**(1991), 177–183.

[WZ] J.S. Wilson and E.I. Zelmanov, Identities for Lie algebras of pro-p groups, *J. Pure Appl. Algebra* **81**(1992), 103–109.

[Y] I.O. York, *The group of formal power series under substitution*, Ph.D. Thesis, Nottingham, 1990.

[Z1] E.I. Zelmanov, *The solution of the restricted Burnside problem for groups of prime power exponent* (Yale University Notes, 1990).

[Z2] E.I. Zelmanov, On periodic compact groups, *Israel J. Math.* **77**(1992), 83–95.

[Zu] A.N. Zubkov, Non-abelian free pro-p groups cannot be represented by 2-by-2 matrices, *Siberian Math. J.* **28**(1987), 742–747.

ON EQUATIONS IN FINITE GROUPS AND INVARIANTS OF SUBGROUPS

S.P. STRUNKOV[1]

Department of Higher Mathematics, Moscow Engineering Physics Institute, Kashirskoe Shosse 31, 115409 Moscow, Russia

The main aim of this paper is to show connections between some questions in the representation theory of finite groups and in the theory of equations on groups. Most of the proofs of the results stated here can be found in [St1]-[St11].

1. On invariants of subgroups of a finite group

Let G be a finite group, H be a subgroup of G, χ be the character of an ordinary representation R of G. We consider the number $c_\chi(H) = \sum_{g \in H} \chi(g)$. This number has an interpretation in invariant theory, because it is equal to the dimension of the linear space of invariants of the first degree for H in the restriction of R on H. We are interested in relations between the numbers $c_\chi(H)$ for different subgroups H of G. This approach is fruitful. It gives possibility to generalise some arithmetical facts and to get new applications of relations for these numbers to the theory of finite groups.

We denote by $M(\chi)$ the set of natural numbers, each number a of which is relatively prime to $| G |$ and is representable in the form $p_1^{k_1} \dots p_s^{k_s}$ in which each factor $p_i^{k_i}$ is equal to the order of a finite field in which the representation R can be realised under its reduction to a field of prime characteristic p_i.

Theorem 1.1. *For any characters $\chi_1, \dots, \chi_n$ and $a_i \in M(\chi_i)$*

$$\sum_{T \supseteq H} \mu(H,T) a_1^{c_{\chi_1}(T)} \dots a_n^{c_{\chi_n}(T)} \equiv 0 \ (\mathrm{mod} \ | N_G(H)/H |),$$

where $\mu(H,T)$ is Möbius function of the set of all subgroups of G partially ordered by inclusion and the sum is produced over all subgroups $T \supseteq H$.

If $G = Z_m, H = 1, n = 1, \chi_1$ is the sum of all faithful irreducible characters of G, then this relation turns into the Euler's theorem $a^{\varphi(m)} \equiv 1 \ (\mathrm{mod} \ m)$.

Theorem 1.1 is proved by an algebraic method (it was proved for $n = 1$ in [St8]). Other relations for the numbers $c_\chi(H)$ can be obtained using geometric methods. With this method L.Scott found the following remarkable relation

$$\sum_{i=1}^{n} c_\chi(< a_i >) + c_\chi(< a_1 a_2 \dots a_n >) \leq (n-1)\chi(1)$$

[1]Supported by International Science Foundation.

for any finite group $G =< a_1, \ldots, a_n >$ and its character χ with $(\chi, 1) = 0$ [Sc]. This relation generalises the well-known "Brauer trick" and now it plays a key role in investigations of generators of simple finite groups. Using geometric method we can prove:

Theorem 1.2. ([St10]) *If $G =< a_1, \ldots, a_n >, a_i$ are involutions and χ does not contain one-dimensional components, then*

$$\sum_{i=1}^{n} c_\chi(< a_i, a_{i+1} >) + \chi(1) \leq \sum_{i=1}^{n} c_\chi(< a_i >),$$

where $a_{n+1} = a_1$.

This relation is similar but not equivalent to Scott's one. It was proved by calculating the first homology group of a two-dimensional compact surface, on which the group G acts. The following fact shows the reason for our interests in the numbers $c_\chi(H)$ for small subgroups H.

Theorem 1.3. ([St8]) *For any finite group G and its noncyclic sugroup H*

$$\sum_{T \subseteq H} (\mid T \mid / \mid N_G(H) \mid) c_\chi(T) \mu(T, H) = 0.$$

Thus, in a sense, the numbers $c_\chi(H)$ for noncyclic subgroups H are defined by the numbers $c_\chi(T)$ for cyclic subgroups T.

Now we want to demonstrate one application of numbers c_χ to problems about possibility to define a group, that is to problems of Burnside type.

Theorem 1.4. *Let n be a natural number, $\sum$ a set of non-isomorphic finite groups such that*

1) *any $G \in \sum$ can be generated by not greater than n elements;*
2) *$exp(G) \leq n$ for any $G \in \sum$;*
3) *every $G \in \sum$ has subgroups $H_1, \ldots, H_k$ $(k \leq n, \mid H_i \mid \leq n)$ such that for any $\chi_j \in Irr(G)$ and for some subgroup H_{s_j} at least one of the inequalities $c_{\chi_j}(H_{s_j}) \leq n$ or $c_{\chi_j}(H_{s_j}) \geq \chi(1) - n$ holds.*

Then the set $\sum$ is finite.

If we can prove that 3) is a consequence of 1) and 2) then we will have proved the restricted Burnside problem for all finite groups (but not only for p-groups).

Finally we demonstrate one arithmetical relation between the numbers $c_\chi(H)$ and some equations on finite groups.

Theorem 1.5. *Let G be a finite group, H a subgroup of G, $\chi_1, \ldots, \chi_r$ all complex irreducible characters of G, s_k the number of solutions of the equation $u_1^{x_1} u_2^{x_2} \ldots u_k^{x_k} = g$ $(x_i \in G, u_i \in H)$ for some $g \in G$. Then for any $k \geq 1$*

$$\sum_{i=1}^{r} \left(\frac{\mid G \mid}{\chi_i(1)} \right)^{k-1} c_{\chi_i}(H)^k \mid H \mid^k \chi_i(g) = s_k.$$

2. On p-blocks of defect 0

Let G be a finite group as before, p be a prime dividing the order $\mid G \mid$ of G, χ be a character of some ordinary absolutely irreducible representation of G. If $\mid G \mid = p^c m, \chi(1) = p^{c_1} m_1$ (m and m_1 are prime to p), then the non-negative number $c - c_1$ is called the p-defect of χ. The following problems of R.Brauer [Br1] are very important in the representation theory of finite groups:

1) What are necessary and sufficient conditions for the existence of characters of p-defect 0 ?
2) How many are there?

There is one-to-one correspondence between characters of p-defect 0 in G and its p-blocks of defect 0, therefore these questions are equivalent to questions of existence and the number of p-blocks of defect 0.

More general problems about necessary and sufficient conditions for the existence of p-blocks with a given defect group D and the number of such blocks are reduced via the Brauer correspondence [Br2] to problems 1), 2) above for smaller group, which, in a sense, are easier than the questions about characters of p-defect 0.

These problems were first posed thirty years ago, however, it is only recently that active investigation on them has been carried out. One example is a theorem of Y.Tsushima's [Ts] on idempotents of p-blocks of defect 0, which was reformulated in Karpilovski's book [Ka] as a wonderful criterion for the existence of such blocks in term of the equation $xy = g$ (x, y are p-regular). G.R.Robinson expressed the number of p-blocks with a given defect group by means of the rank of the product of two matrices whose entries are the cardinals of certain subsets of G [Ro]. Zhang Ji-Ping found a criterion for the existence of characters of p-defect 0 in groups with cyclic Sylow p-subgroup [Zh]. The following result gives an infinite collection of such criterions in terms of equations on groups.

Theorem 2.1. ([St9]) *Let G be a finite group, p a prime number and G_p a Sylow p-subgroup of G. Let $f(x_1, \ldots, x_k, u_1, \ldots, u_l)$ be a function of G, which is a product of "so called" elementary functions $[x_i, x_{i+1}]$ and $u_j^{x_s}$ with $x_i \in G, u_j \in G_p$. Moreover; assume that the variables in different elementary factors are distinct and that $k \geq 2$. Then G has a p-block of defect 0 if and only if for some $g \in G$ the number of solutions of the equation*

$$f(x_1, x_2, \ldots, x_k, u_1, u_2, \ldots, u_l) = g$$

is not divisible by $p \mid G_p \mid^l$.

Corollary 1. *The following conditions are equivalent:*

A) a finite group G has a p-block of defect 0;

B) *the number of solutions of the equation* $[x,y] = g$ *is prime to* p *for some* $g \in G$ $(x,y \in G)$;

C) *the number of solutions of the equation* $u^x v^y = g$ *is not divisible by* $p|G_p|^2$ *for some* $g \in G$ $(x,y \in G, u,v \in G_p)$;

D) *Let* k *be a natural number. Then the number of solutions of the equation* $[x_1,x_2][x_3,x_4]\dots[x_{2k-1},x_{2k}] = g$ *is prime to* p *for some* $g \in G$ $(x_1,\dots,x_{2k} \in G)$.

It was known that if G has a p-block of defect 0, then G has two Sylow p-subgroups U, V with $U \cap V = 1$. This fact is a special case of the well-known theorem of J.A.Green [Gr] about defect groups of p-blocks. Obviously, this was one of the reasons for the questions of R.Brauer. Therefore it is useful to get such criteria in structural terms, that is in terms of subgroups, cosets and their intersections, (see [CH]). These criteria must make the theorem of J.A. Green more precise. Theorem 2.1 gives such a possibility. Some of the conditions of this theorem can be restated in structural terms.

Definition. An ordered pair of Sylow p-subgroups U, V is called (g,p)-vector $(g \in G)$, if $U \cap V = 1$, $g \in UV$ (or in another words $Ug \cap V \neq \emptyset$).

We note that (g,p)-vector is a structural object.

Corollary 2. *G has a p-block of defect 0 if and only if the number of* (g,p)*-vectors in* G *is prime to* p *for some* $g \in G$.

Now we consider the question about the number of p-blocks of defect 0 in finite groups in terms of equations on groups.

Theorem 2.2. ([St9]) *Let* s_i *be the number of solutions of the equation* $[x_1,x_2][x_3,x_4]\dots[x_{2i+1},x_{2i+2}] = 1, \sigma_i = s_i/\mid G\mid$,

$$v_n = \sum \frac{(-1)^{i_1+i_2+\dots+i_n}}{1^{i_1}2^{i_2}\dots n^{i_n} i_1! i_2! \dots i_n!} \cdot \sigma_1^{i_1}\sigma_2^{i_2}\dots\sigma_n^{i_n}$$

$(n = 1,2,\dots,r; r = dim Z(CG)$; *the sum runs over all sets* $i_1,\dots,i_n$ *of non-negative integers such that*

$$i_1 + 2i_2 + \dots + ni_n = n).$$

Then

1) *all the numbers* v_n *are integers;*

2) *the number* k_0 *of p-blocks of defect 0 is equal to the largest* n *such that the number* v_n *is prime to* p.

We want to note in conclusion that there is parallel theory of real p-blocks of defect 0. It is easy to obtain criteria for the existence of real p-blocks of defect 0 in a finite group and characterisations of the number of such blocks in terms of equations on groups [St9].

References

[Br1] Brauer R., *Representations of finite groups* (Lect. in Math., v.1, Wiley, New York, 1963), 133-175.

[Br2] Brauer R., Zur Darstellungstheorie der Gruppen endlicher Ordnung, I, *Math. Z.* **63**(1956), 406–444.

[CH] Chillag D. and Herzog M., Defect groups, trivial intersections and character tables, *J. Algebra* **61**(1979), 152–160.

[Fe] Feit W., *The representation theory of finite groups* (North-Holland, 1982).

[Gr] Green J.A., Blocks of modular representations, *Math. Z.* **79**(1962), 100–115.

[Ka] Karpilovski G., *Structure of blocks of group algebras* (Longman, Essex, 1987).

[Ro] Robinson G.R., The number of blocks with a given defect group, *J. Algebra* **84**(1983), 493–502.

[Sc] Scott L.L., Matrices and cohomology, *Ann. Math.* **105**(1977), 473–492.

[St1] Strunkov S.P., On one problem of R.Brauer, *Dokl. Akad. Nauk SSSR* **310**(1990), 35–36.

[St2] Strunkov S.P., On some conditions of the existence of p-blocks of defect 0 in finite groups, *Mat. Sb.* **181**(1990), 1144–1149.

[St3] Strunkov S.P., On the spectrum of sums of generators of a finite groups, *Izv. Akad. Nauk SSSR*, Ser. Mat. **54**(1990), 1108–1111; Engl. transl. in *Math. USSR Izv.* **37**(1991), 461–463.

[St4] Strunkov S.P., On characters of minimal p-defect in finite groups, *Dokl. Akad. Nauk SSSR* **314**(1990), 1349–1352; Engl. transl. in *Soviet Math. Dokl.* **42**(1991), 689–692.

[St5] Strunkov S.P., Generalisations of Euler's divisibility theorem and other arithmetic properties of representations of finite permutation groups, *Dokl. Akad. Nauk SSSR* **316**(1991), 1323–1326; Engl. transl. in *Soviet Math. Dokl.* **43**(1991), 288–291.

[St6] Strunkov S.P., Some arithmetical properties of finite groups and their linear representations, *Contemp. Math.* **131**(1992), 375–382.

[St7] Strunkov S.P., On blocks of defect 0 in finite groups, *Izv. Akad. Nauk SSSR*, Ser. Mat. **53**(1989), 657–665; Engl. transl. in *Math. USSR Izv.* **34**(1990), 677-683.

[St8] Strunkov S.P., On the identity components of restictions of a representation of a finite group, *Algebra i Analiz* **3**(1991), 135–155; Engl. transl. in *St. Petersburg Math. J.* **3**(1992), 613–629.

[St9] Strunkov S.P., On the existence and the number of p-blocks of defect 0 in finite groups, *Algebra i Logika* **30**(1991), 655–668.

[St10] Strunkov S.P., On one representation of finite groups generated by involutions, *Ukrain. Mat. Zh.*, **43**(1991), 1013–1017.

[St11] Strunkov S.P., On one generalisation of Fermat's divisibility theorem, *Izv. Akad. Nauk SSSR*, Ser. Mat. **55**(1991), 218–220.

[Ts] Tsushima Y., On p-blocks of defect zero, *Nagoya Math. J.* **44**(1971), 57–59.

[Zh] Zhang Ji-Ping, A condition for the existence of p-blocks of defect 0, *Proc. Sympos. Pure Math.* **47**(1987).

GROUP PRESENTATIONS WHERE THE RELATORS ARE PROPER POWERS

RICHARD M. THOMAS

Department of Mathematics and Computer Science, University of Leicester, University Road, Leicester LE1 7RH, U.K.
E-mail: rmt@uk.ac.le

In this paper, we review some of what is known about groups defined by certain presentations in which the relators are all proper powers. The question that interests us here is that of asking when such groups are finite. In the particular cases we shall be looking at (the choice of which has been somewhat influenced by those presentations we have played with ourselves) the order of each generator will usually be mentioned explicitly among the relations, so that we will be considering presentations of the form

$$< a_1, a_2, \ldots\ldots, a_n : a_1^{m(1)} = a_2^{m(2)} = \ldots\ldots = a_n^{m(n)}$$
$$= \alpha_1^{r(1)} = \alpha_2^{r(2)} = \ldots\ldots = \alpha_k^{r(k)} = 1 >,$$

where each $m(i)$ and $r(i)$ is greater than one and each α_i is a cyclically-reduced word in the generators and their inverses which is not itself a proper power. We will assume that $n \geq 2$ (else we have a finite cyclic group) and then that $k \geq 1$ (else we have a free product of cyclic groups, and hence an infinite group). We will also often identify a word α with the corresponding element of the group without further comment.

The simplest examples of such groups are the *triangle groups* (m, n, k) defined by the presentations

$$< a, b : a^m = b^n = (ab)^k = 1 > .$$

These are known not to *collapse*, in that the elements a, b and ab really do have orders m, n and k respectively. Note that we are using the word *collapse* here in the sense of [Col] to mean that there is a relation of the form $\alpha^p = 1$ in our presentation (with $p > 1$) where the corresponding element α of the group has order less than p. The collapse is not necessarily *total*, in that the group is not necessarily trivial. Another way of putting all this here is to state that the cyclic groups of orders m, n and k embed naturally into (m, n, k). If we let $g = \frac{1}{m} + \frac{1}{n} + \frac{1}{k}$, then the group is infinite if $g \leq 1$ and finite if $g > 1$.

In general, this is the sort of thing we expect to happen: provided that it does not collapse, a group defined by such a presentation will tend to be infinite for large powers. To be more precise, we have the following:

Theorem 1. *Let G be the group defined by the presentation*

$$< x_1, x_2, \ldots\ldots, x_n : \alpha_1^{s(1)} = \alpha_2^{s(2)} = \ldots\ldots = \alpha_r^{s(r)} = 1 >,$$

where each α_i is a word in the generators and their inverses, and let $g = \frac{1}{s(1)} + \frac{1}{s(2)} + \dots + \frac{1}{s(r)}$. Then, provided that each element α_i really does have order $s(i)$ in G, we have that G is infinite if $g \leq n-1$.

This result has appeared in many different places. An elementary proof, couched in terms of the Cayley graph of the group, may be found in [Th1] and a cohomological version in [HoP]. Dave Johnson has informed us that Charles Sims presented a proof of this result using coset tables at the computational group theory conference at Durham in 1982, so that there are (at least) three different references to this theorem; the Bellman, at least, would approve! [Car]

The assumption in Theorem 1 that the group does not collapse is critical. We know that this is true for the triangle groups, and so Theorem 1 gives yet another proof that these groups are infinite when $g \leq 1$. In fact, even if we allow the word ab to be replaced by another word, this is still true, as we shall now see.

To be more precise, we define a *generalized triangle group* to be a group defined by a presentation of the form

$$< a, b : a^m = b^n = \alpha^k = 1 >,$$

where $m, n, k \geq 2$ and α is some cyclically reduced word involving both a and b which is not itself a proper power. It was shown in [BMS], and, independently, in [FHR], that a generalized triangle group G does not collapse, since there is always a representation ϕ of G in $PSL(2, \mathbf{C})$ such that $a\phi$, $b\phi$ and $\alpha\phi$ have orders m, n and k respectively. Given this, such groups are infinite for $g \leq 1$ by Theorem 1; a previous proof of this fact had been given in [BMS], and a slightly weaker form in [FiR].

In fact, the non-collapsing of the generalized triangle groups seems to be a special case of a more general result. If A and B are any two groups with finite presentations $< X : R >$ and $< Y : S >$ respectively, and if α is any word involving elements from both X and Y which is cyclically reduced as a word in $A * B$, then it is known that A and B both embed into the group with presentation

$$< X \cup Y : R \cup S \cup \{\alpha^k = 1\} >$$

provided that $k \geq 4$; see [Ho1, Ho2]. It seems likely that, as with the generalized triangle groups, this result should extend to $k \geq 2$.

Let us return to the generalized triangle groups. Unlike the triangle groups, a generalized triangle group with $g > 1$ need not be finite; so it seems natural to ask which of these groups are finite. If $m = n = 2$, we have a dihedral group; so we will assume that $(m, n) \neq (2, 2)$.

A complete classification of the presentations defining finite generalized triangle groups with $k \geq 3$ was given in [FL2]. There is essentially only one (apart from the triangle groups), namely the group with presentation

$$< a, b : a^2 = b^3 = (ababab^{-1})^3 = 1 >,$$

which has structure $C_2.(A_4 \times A_5)$, by which we mean an extension of the cyclic group C_2 of order 2 by a direct product of the alternating groups A_4 and A_5 of degrees 4 and 5, and hence the group is finite of order 1440; in fact, the extension does not split over either factor. We say *essentially* here in that there is certainly only one group, and the only presentations of this form defining it are obtained from this one by cyclic permutations of the last word, inverting the last word, or replacing b by b^{-1} throughout. In general, we regard a presentation for a generalized triangle group as being equivalent to the presentation

$$\wp \; = < a, b : a^m = b^n = \alpha^k = 1 >$$

if it is obtained from $\wp$ by a sequence of moves of the form:

1. cyclic permutations of α;
2. inversion of α;
3. automorphisms of $\mathbf{Z}_m$ or $\mathbf{Z}_n$;
4. interchange of the two free factors (if $m = n$).

Given the result for $k \geq 3$, let us now concentrate on the case $k = 2$, that is to say on groups defined by presentations of the form

$$< a, b : a^m = b^n = \alpha^2 = 1 > .$$

There are many interesting groups defined by such presentations. For example, if we consider our old friends the *Fibonacci groups* $F = F(2, 2m)$ defined by the presentations

$$< x_1, x_2, \ldots\ldots, x_{2m} : x_1x_2 = x_3, x_2x_3 = x_4, \ldots\ldots, x_{2m-2}x_{2m-1} = x_{2m},$$
$$x_{2m-1}x_{2m} = x_1, x_{2m}x_1 = x_2 >,$$

then it is known [Bru, Lyn] that F is infinite for $m \geq 4$; see [Th3] for a general survey of these groups. However, F has index $2m$ in the group $L(m)$ with presentation

$$< a, b : a^2 = b^m = (abababab^{-1}ab^{-1})^2 = 1 >,$$

and one can use elementary trace arguments to show that $L(m)$ has a representation $< A, B >$ in $PSL(2, \mathbf{C})$ with AB of infinite order for $m \geq 4$, thus giving another proof that F is infinite; see [Th2, Th3] for details. This matrix representation of F also occurs in [HKM], where it is shown that F is the fundamental group of a manifold with hyperbolic structure, so that F

has a faithful representation as a discrete subgroup of $PSL(2, \mathbf{C})$. In fact, Jim Howie pointed out to us that F is isomorphic to $\pi_1(M_m)$, where M_m is the m-fold cyclic branched cover of the figure-8 knot. On the other hand, Helling noted that $F(2, 2m+1)$ contains elements of order 2 for $m \geq 4$, and so cannot be the fundamental group of a manifold with hyperbolic structure. We haven't seen Helling's proof, but an account of this last fact is given in [Th5].

The Fibonacci groups are often said to be *cyclically presented* due to the cyclic nature of their relations. Another class of such groups are the groups $G(m)$ defined by the presentations

$$< x_1, x_2, \ldots\ldots, x_m : x_1x_3 = x_2, x_2x_4 = x_3, \ldots\ldots, x_{m-2}x_m = x_{m-1},$$
$$x_{m-1}x_1 = x_m, x_mx_2 = x_1 > .$$

Ann-Chi Kim and Jim Howie pointed out to us that these are also 3-manifold groups, although not, in this case, hyperbolic. In fact, $G(m)$ is the fundamental group of the m-fold cyclic branched cover of the 3-sphere, branched over the trefoil knot; see [Th5] for more details. As with the Fibonacci groups, the $G(m)$ are subgroups of finite index in generalized triangle groups, in this case of index $2m$ in the group $K(m)$ with presentation

$$< a, b : a^2 = b^m = (ababab^{-1})^2 = 1 > .$$

See [Th4, Th5] for details.

For small values of m, these groups are finite. We clearly have that $K(2)$ is dihedral of order 12 and that $G(2)$ is cyclic of order 3. If $m = 3$, we have that $K(3)$ is isomorphic to $GL(2,3)$, with $G(3)$ the quaternion group Q_8. The groups $G(4)$ and $G(5)$ are isomorphic to $SL(2,3)$ and $SL(2,5)$ (see [JoM] for example), so that $K(4)$ and $K(5)$ are finite of orders 192 and 1200 respectively. However, the picture changes if $m \geq 6$.

Using the trick (again) of representing our groups in $PSL(2, \mathbf{C})$, we can show that the group $K(m)$ is infinite if $m \geq 7$; indeed, $K(m)$, and hence $G(m)$, has a free subgroup of rank 2 in this case [Th4, Th5]. An alternative (unpublished) proof that the $G(m)$ are the fundamental groups of cyclic branched covers and are infinite had been obtained by Martin Dunwoody; see also [GiH], [HMV] and [LeR] for further information on these groups.

The group $K(6)$ is rather different. If we take the presentation

$$< x_1, x_2, \ldots\ldots, x_6 : x_1x_3 = x_2, x_2x_4 = x_3, \ldots\ldots, x_6x_2 = x_1 >,$$

and eliminate the generators x_6, x_4, x_2 and x_5 in turn, and then write x_1 as u and x_3 as v so we don't have to bother with subscripts, we get the presentation

$$< u, v : uv^2uv^{-1}u^{-2}v^{-1} = vuv^{-1}u^{-1}v^{-1}uvu^{-1} = 1 > .$$

We introduce a new generator $w = uvu^{-1}$; the second relation is then equivalent to $[v,w] = 1$ and the first relation to $w^2uw^{-1}u^{-1}v^{-1} = 1$, and hence to $uwu^{-1} = v^{-1}w^2$. So we have the presentation

$$< u, v, w : uvu^{-1} = w, uwu^{-1} = v^{-1}w^2, [v,w] = 1 > .$$

If we introduce a new generator $z = wv^{-1}$, and then delete $w = zv$, this becomes

$$< u, v, z : uvu^{-1} = zv, uzvu^{-1} = v^{-1}zvzv, [v,z] = 1 >,$$

which simplifies to

$$< u, v, z : [u,v] = z, [u,z] = [v,z] = 1 > .$$

This is a presentation for the *Heisenberg group* consisting of all matrices of the form $\begin{pmatrix} 1 & r & s \\ 0 & 1 & t \\ 0 & 0 & 1 \end{pmatrix}$ with r, s, $t \in \mathbf{Z}$; see Section 5.2 of [Joh] for example.

Returning to the general question of which generalized triangle groups are finite, it follows from the results in [Con] that the only presentations of the form

$$< a, b : a^2 = b^3 = \alpha^2 = 1 >$$

where α has length at most 12 (other than the presentations for the triangle groups and the presentation for $K(3)$ above) defining finite groups are (essentially)

$$< a, b : a^2 = b^3 = (abab^{-1})^2 = 1 >,$$
$$< a, b : a^2 = b^3 = (abababab^{-1})^2 = 1 >,$$
$$< a, b : a^2 = b^3 = (ababab^{-1}ab^{-1})^2 = 1 >,$$
$$< a, b : a^2 = b^3 = (ababab^{-1}abab^{-1})^2 = 1 >,$$
$$< a, b : a^2 = b^3 = (ababababab^{-1}ab^{-1})^2 = 1 >,$$

and the corresponding groups are

$$C_2 \times A_4,\ C_2 \times A_5,\ C_2.(S_4 \times A_4), C_2.(S_3 \times A_5), C_2.(S_4 \times A_5)$$

of orders 24, 120, 576, 720 and 2880 respectively. By [Ros], the only further presentations of the form

$$< a, b : a^m = b^n = (a^pb^qa^rb^s)^2 = 1 >$$

with $0 < p, r < m$ and $0 < q, s < n$ defining finite groups are

$$< a, b : a^3 = b^3 = (aba^{-1}b^{-1})^2 = 1 >,$$
$$< a, b : a^3 = b^3 = (abab^{-1})^2 = 1 >,$$
$$< a, b : a^2 = b^5 = (abab^{-1})^2 = 1 >,$$

where the groups have orders 288, 180 and 120 respectively, the third presentation being another presentation for $C_2 \times A_5$. It was then shown in [LeR] that the only further presentations of the form

$$< a, b : a^m = b^n = (a^p b^q a^r b^s a^t b^u)^2 = 1 >$$

with $0 < p, r, t < m$ and $0 < q, s, u < n$ defining finite groups are the presentations for $K(4)$ and $K(5)$ given above, with groups of order 192 and 1200, and the one further presentation

$$< a, b : a^2 = b^5 = (abab^2ab^4)^2 = 1 > .$$

Since $(abababab^4)^2$ and $(abab^2ab^4)^2$ have the same normal closure in the group defined by $< a, b : a^2 = b^5 = 1 >$, the group defined by this last presentation is also isomorphic to $K(5)$. It was also shown in [LeR] that there are no further finite groups defined by presentations of the form

$$< a, b : a^m = b^n = (a^p b^q a^r b^s a^t b^u a^v b^w)^2 = 1 > .$$

So far we have exhibited eleven finite generalized triangle groups (apart from the triangle groups). It was shown in [HMT] that, apart from these groups, there are at most two more finite cases, the relevant presentations being

$$< a, b : a^2 = b^3 = (abababab^{-1}abab^{-1}ab^{-1})^2 = 1 >,$$
$$< a, b : a^2 = b^3 = (abababab^{-1}ab^{-1}abab^{-1}ab^{-1})^2 = 1 > .$$

Bernd Souvignier, Levente Lévai and Gerhard Rosenberger [LRS] have shown that the first group is infinite, but, using the computer system GAP [Sch], that the second is finite of order $2^{20}3^4 5$. Given this, we have precisely twelve finite generalized triangle groups which are not triangle groups.

Moving up the scale, we could consider presentations with two generators a and b and four relators. If a and b both have order 2, then the group is dihedral, so that the smallest non-trivial case is where a has order 2 and b has order 3. In this case, ab is essentially the only possible word of length 2 in the free product of $< a >$ and $< b >$ involving both a and b, and $ab^{-1}ab$ essentially the only such word of length 4 which is not a proper power. If we add proper powers of these two words as relators, we have the group $(2, 3, p; q)$ with presentation

$$< a, b : a^2 = b^3 = (ab)^p = [a, b]^q = 1 > .$$

Several finite examples of such groups were discovered by Coxeter and others in the 1930's; see [Col] in particular. The finiteness or otherwise of these groups has now been almost completely determined:

Theorem 2. *With the possible exception of $(p,q) = (13,4)$, the group $(2,3,p;q)$ is finite if and only if p and q satisfy one of the following five conditions:*

(i) $p \leq 6$
(ii) $p = 7$ *and* $q \leq 8$;
(iii) $p = 8$ *or* 9 *and* $q \leq 5$;
(iv) $p = 10$ *or* 11 *and* $q \leq 4$;
(v) $p \geq 12$ *and* $q \leq 3$.

See [Co2, HoP, HT1, Ed1].

The status of $(2,3,13;4)$ is still unknown. Derek Holt has shown, using computational techniques, that it has a homomorphic image of the form $E.PSL(3,3) \times PSL(2,25)$, where E is elementary abelian of order 2^{12}, and he also showed that $PSL(2,25)$ and $PSL(3,3)$ are the only finite simple groups of order less than 10^6 which are homomorphic images of $(2,3,13;4)$. We also know that there is no other $PSL(2,q)$ which is a homomorphic image of $(2,3,13;4)$; see [HT2] for details.

We can generalize this problem by considering the groups $(m,n,p;q)$ defined by the presentations

$$< a, b : a^m = b^n = (ab)^p = [a,b]^q = 1 > .$$

The picture is largely completed by the following result of Martin Edjvet [Ed2]:

Theorem 3. *If $2 \leq m \leq n \leq p$, $2 \leq q$, $(m,n) \neq (2,2)$, $(m,n) \neq (2,3)$ and $(m,q) \neq (3,2)$, then $(m,n,p;q)$ is finite if and only if m, n, p and q satisfy one of the following conditions:*

(i) $m = 2, n = 4, p \geq 4, q = 2$;
(ii) $m = 2, n = 4, p = 4, q \geq 3$;
(iii) $m = 2, n = 4, p = 5, 3 \leq q \leq 4$;
(iv) $m = 2, n = 4, p = 7, q = 3$;
(v) $m = 2, n = 5, 5 \leq p \leq 9, q = 2$;
(vi) $m = 2, n = 6, p = 7, q = 2$;
(vii) $m = 3, n = 3, p = 3, q \geq 3$;
(viii) $m = 3, n = 3, p = 4, q = 3$.

Some similar results have been obtained in [Cha]. Subsequently, Martin Edjvet and Jim Howie have proved:

Theorem 4. *The group $(3,n,p;2)$ is infinite if either*

(i) $n = 4, p \geq 17$, *or*
(ii) $n \geq 5, p \geq 11$.

Given Theorems 2, 3 and 4, we are left with the cases $(3,4,p;2)$ with $4 \leq p \leq 16$ and $(3,5,p;2)$ with $5 \leq p \leq 10$. A lot of computing results have been obtained on the groups $(m,n,p;q)$ by several people, including Derek Holt, Mike Newman, Edmund Robertson, Berndt Souvignier, Geoff Smith and the author, and much of the information exchanged by means of the *Group Pub Forum.* This e-mail discussion group is a splendid invention of Geoff Smith's, the atmosphere being like that of a pub at a good group theory conference where one can ask questions and exchange information. (To join, e-mail Geoff at *gcs@uk.ac.bath.maths.*) In particular, the following results have been found:

$(3,4,p;2)$ is finite for $p \leq 7$; $(3,4,8;2)$ is infinite;
$(3,4,10;2)$ is infinite; $(3,4,12;2)$ is infinite;
$(3,4,14;2)$ is infinite; $(3,4,p;2)$ is infinite for $14 \leq p \leq 16$;
$(3,5,5;2)$ is infinite; $(3,5,p;2)$ is infinite for $8 \leq p \leq 10$;
$(3,n,p;2)$ is infinite for $6 \leq n \leq p \leq 10$.

The methods here are varied, but include techniques such as finding palpably infinite subgroups (such as those with infinite abelianization) and using versions of the Golod- Šafarevič theorem. The computer software used included Cayley [Can], GAP [Sch] and Quotpic [HoR]. The upshot of all this is that the remaining open cases are:

$$(2,3,13;4); \quad (3,4,9;2); \quad (3,4,11;2);$$
$$(3,4,13;2); \quad (3,5,6;2); \quad (3,5,7;2).$$

Of course, one can go on and on considering more and more complicated presentations by adding further generators and relators. One such interesting class of examples is that of the ***generalized tetrahedron groups*** defined by presentations of the form

$$< a,b,c : a^p = b^q = c^r = \alpha(a,b)^l = \beta(b,c)^m = \gamma(c,a)^n = 1 >;$$

see [FL1] for example, where it is shown that such a group admits a representation ϕ in $PSL(2,\mathrm{C})$ such that $a\phi$, $b\phi$, $c\phi$, $\alpha\phi$, $\beta\phi$ and $\gamma\phi$ have orders p, q, r, l, m and n respectively, so that Theorem 1 applies. In fact, it is shown in [FL1] that a generalized tetrahedron group is not only infinite, but has a subgroup of finite index mapping onto the free group of rank 2, if

$$\frac{1}{p} + \frac{1}{q} + \frac{1}{r} + \frac{1}{l} + \frac{1}{m} + \frac{1}{n} < 1,$$

and some further results are obtained there.

One general result in this area, however, is the following [BaS]:

Theorem 5. *Let p be a prime, $e > 0$, $q = p^e$, and let the group G be defined by the presentation*

$$< a_1, a_2, ..., a_n : \alpha_1^q = \alpha_2^q = ... = \alpha_r^q = 1 > .$$

If $r < (n-1)(q+1)$, then G is infinite; indeed, G contains an element of infinite order.

If we put $m(i) = q$ for all i in Theorem 1, the condition there becomes that $r \leq (n-1)q$, which is slightly more restrictive than the hypothesis that $r < (n-1)(q+1)$ from Theorem 5; moreover, we have the extra hypothesis in Theorem 1 that the group does not collapse. On the other hand, Theorem 1 copes with cases where the exponents are different, whereas in Theorem 5 we must have the same exponent throughout, which, furthermore, must be a prime power.

In [BaS], Theorem 5 is used to show that the group G_1 defined by the presentation

$$< a, b : a^4 = b^4 = (ab)^4 = (a^2b^2)^4 = 1 >$$

is infinite. This would also follow from Theorem 1 if we knew that there was no collapse. It is clear (from considering the abelianization) that a, b and ab all have order 4 in G_1, and that a^2b^2 has order 2 or 4. If a^2b^2 had order 2, we would have the group G_2 defined by the presentation

$$< a, b : a^4 = b^4 = (ab)^4 = (a^2b^2)^2 = 1 >,$$

and one can check that G_1 and G_2 are not isomorphic by applying the nilpotent quotient algorithm with the prime 2 (in that we find that the fourth factors in the corresponding series are distinct). So a^2b^2 has order 4 in G_1, thus giving another proof that G_1 is infinite.

In fact, even G_2 is infinite. To see this, we consider the group defined by the presentation

$$< a, b : a^r = b^r = (ab)^s = (a^2b^2)^t = 1 >,$$

and adjoin the automorphism c of order 2 interchanging a and b, to get the presentation

$$< a, c : a^r = c^2 = (ac)^{2s} = (a^2c)^{2t} = 1 > .$$

If we introduce $d = ca^{-1}$, and then delete $c = da = a^{-1}d^{-1}$, we get the presentation

$$< a, d : a^r = d^{2s} = (ad)^2 = (a^{-1}d)^{2t} >,$$

which defines the group $(r, 2s \mid 2, 2t)$, which is infinite if

$$2\ sin(\frac{\pi}{r})\ sin(\frac{\pi}{2s}) \ \leq \ cos(\frac{\pi}{2t})$$

by [Col]. Since $2sin(\frac{\pi}{8}) = \sqrt{2-\sqrt{2}} < 1$, we have that $(4, 8 \mid 2, 4)$ is infinite, and hence that G_2 is infinite.

Acknowledgements. The author is very grateful to Jim Howie for a great deal of helpful advice, encouragement and information, and also to Martin Edjvet, Derek Holt, Dave Johnson, Mike Newman, Gerhard Rosenberger and Geoff Smith for several helpful conversations in general and comments about the material in this paper in particular. He would also like to thank Hilary Craig for all her help and encouragement.

References

[BMS] G. Baumslag, J. W. Morgan and P. B. Shalen, Generalized triangle groups, *Math. Proc. Cambridge Philos. Soc.* **102**(1987), 25–31.

[BaS] G. Baumslag and P. B. Shalen, Affine sets and some infinite finitely presented groups, in *Essays in Group Theory* (S.M. Gersten (ed.), M.S.R.I. Publications **8**, Springer Verlag, Berlin, Heidelberg, New York, 1987), 1–14.

[Bru] A. M. Brunner, The determination of Fibonacci groups, *Bull. Australian Math. Soc.* **11**(1974), 11–14.

[Can] J. J. Cannon, An introduction to the group theory language Cayley, in *Computational Group Theory* (M.D. Atkinson (ed.), Academic Press, London, 1984), 145–183.

[Car] L. Carroll, *The Hunting of the Snark*, 1876.

[Cha] H. Chaltin, Among Coxeter's groups $((l, m, n; p))$, "many" are infinite and do not collapse, preprint.

[Con] M. D. E. Conder, Three-relator quotients of the modular group, *Quart. J. Math.* **38**(1987), 25–31.

[Co1] H. S. M. Coxeter, The abstract groups $G^{m,n,p}$, *Trans. Amer. Math. Soc.* **45**(1939), 73–150.

[Co2] H. S. M. Coxeter, Groups generated by unitary reflections of period two, *Canad. J. Math.* **9**(1957), 243–272.

[Ed1] M. Edjvet, An example of an infinite group, in *Discrete Groups and Geometry* (W. J. Harvey and C. Maclachlan (eds.), London Math. Soc. Lecture Note Series **173**, Cambridge University Press, Cambridge, 1992), 66–74.

[Ed2] M. Edjvet, On the abstract groups $(m, n, p; q)$, preprint.

[FHR] B. Fine, J. Howie and G. Rosenberger, One-relator quotients and free products of cyclics, *Proc. Amer. Math. Soc.* **102**(1988), 249–254.

[FL1] B. Fine, F. Levin, F. Roehl and G. Rosenberger, The generalized tetrahedron groups, preprint.

[FL2] B. Fine, F. Levin and G. Rosenberger, Free subgroups and decompositions of one-relator products of cyclics. Part 1: The Tits alternative, *Archiv. Math.* **50**(1988), 97–109.

[FiR] B. Fine and G. Rosenberger, A note on generalized triangle groups, *Abh. Math. Sem. Univ. Hamburg* **56**(1986), 233–244.

[GiH] N. D. Gilbert and J. Howie, LOG groups and cyclically presented groups, preprint.

[HKM] H. Helling, A. C. Kim and J. L. Mennicke, On Fibonacci groups, preprint.

[HMV] H. Helling, J. L. Mennicke and B. Vinberg, On some generalized triangle groups and tetrahedral orbifolds (in Russian), preprint.

[HoP] D. F. Holt and W. Plesken, A cohomological criterion for a finitely presented group to be infinite, *J. London Math. Soc.* **45**(1992), 469–480.

[HoR] D. F. Holt and S. Rees, A graphics system for displaying finite quotients of finitely presented groups, in *Groups and Computation* (L. Finkelstein and W. M. Kantor (eds.), DIMACS Series in Discrete Mathematics and Theoretical Computer Science **11**, American Mathematical Scoiety, Providence, Rhode Island, 1993), 113–126.

[Ho1] J. Howie, The quotient of a free product of groups by a single high-powered relator I. Pictures. Fifth and higher powers, *Proc. London Math. Soc.* **59**(1989), 507–540.

[Ho2] J. Howie, The quotient of a free product of groups by a single high-powered relator II. Fourth powers, *Proc. London Math. Soc.* **61**(1990), 33–62.

[HMT] J. Howie, V. Metaftsis and R. M. Thomas, *Finite generalized triangle groups* (Technical Report **1993/10**, Department of Mathematics and Computer Science, University of Leicester, July 1993).

[HT1] J. Howie and R. M. Thomas, The groups $(2, 3, p; q)$; asphericity and a conjecture of Coxeter, *J. Algebra* **154**(1993), 289–309.

[HT2] J. Howie and R. M. Thomas, Proving certain groups infinite, in *Geometric Group Theory, Volume 1* (G. A. Niblo and M. A. Roller (eds.), London Math. Soc. Lecture Note Series **181**, Cambridge University Press, Cambridge, 1993), 126–131.

[Joh] D. L. Johnson, *Presentations of Groups* (London Math. Soc. Student Texts **15**, Cambridge University Press, Cambridge, 1990).

[JoM] D. L. Johnson and H. Mawdesley, Some groups of Fibonacci type, *J. Austral. Math. Soc.* **20A**(1975), 199–204.

[LRS] L. Lévai, G. Rosenberger and B. Souvignier, All finite generalized triangle groups, preprint.

[LeR] F. Levin and G. Rosenberger, On free subgroups of generalized triangle groups, part II, in *Proceedings of the Ohio State meeting in 1992 in honour of H. Zassenhaus* (S. Seghal et al. (eds.), to appear).

[Lyn] R. Lyndon, On a family of infinite groups introduced by Conway, unpublished.

[Ros] G. Rosenberger, On free subgroups of generalized triangle groups, *Algebra i Logika* **28**(1989), 227–240.

[Sch] M. Schönert et al, *GAP - Groups, Algorithms and Programming* (Lehrstuhl D für Mathematik, RWTH Aachen, 1992).

[Th1] R. M. Thomas, Cayley graphs and group presentations, *Math. Proc. Cam-*

bridge Philos. Soc. **103**(1988), 385–387.

[Th2] R. M. Thomas, The Fibonacci groups $F(2, 2m)$, *Bull. London Math. Soc.* **21**(1989), 463–465.

[Th3] R. M. Thomas, The Fibonacci groups revisited, in *Groups St Andrews 1989, Volume 2* (C. M. Campbell and E. F. Robertson (eds.), London Math. Soc. Lecture Note Series **160**, Cambridge University Press, Cambridge, 1991), 445–454.

[Th4] R. M. Thomas, On a question of Kim concerning certain group presentations, *Bull. Korean Math. Soc.* **28**(1991), 219–224.

[Th5] R. M. Thomas, *Matrices, Messages and Manifolds* (Technical Report **53**, Department of Computing Studies, University of Leicester, May 1992).

A CONDENSING THEOREM

NALSEY B. TINBERG[1]

Department of Mathematics, Occidental College, Los Angeles, California 90041, U.S.A.

1. Introduction

Let k be an algebraically closed field, G a finite group. Let Y be any kG-module. Suppose that $Y = Y_1 \oplus \cdots \oplus Y_s$ is a decomposition of Y into indecomposable kG-modules, and that $1 = e_1 + \cdots + e_s$ is the corresponding decomposition of $1 \in E = End_{kG}(Y)$ into primitive idempotents. Let $\psi : E \to k$ be a multiplicative character of E. There is a unique e_i with $\psi(e_i) \neq 0$. Green ([3], p.5) considers $Y_i = e_i(Y)$ as the component of Y *related to* ψ. Write Y_ψ for this Y_i.

In this paper we let $Y = k_U^G$ for $U \leq G$, and we calculate and compare the multiplicities of Y_ψ in the decompositions of k_H^G and k_K^G, respectively, for subgroups $H \leq K \leq G$ using results from Green ([2], [3]). Results from the subgroup H case are "condensed" to yield those of K. We apply our results in the case of $U \leq K$ to find the multiplicity of Y_ψ in k_K^G. In particular, we consider the Scott module and the Steinberg component of a finite group with a split (B, N)-pair. We prove that the Steinberg component is a component of $k_{G_J}^G$, for a particular parabolic subgroup G_J, if and only if $J = \emptyset$. This result holds true in *any* characteristic.

2. Multiplicities

Let Y be any kG-module, $Y = Y_1 \oplus \cdots \oplus Y_s$, E and ψ as above. Let M be any kG-module. Green determined the multiplicity $[Y_\psi|M]$ of Y_ψ in M (*i.e.*, the number of terms $M_i \cong Y_\psi$ in a decomposition of $M = M_1 \oplus \cdots \oplus M_t$ of indecomposable submodules). Details of Green's work can be found in [3, Sections 1–4] or in [2, (3.5)]. This multiplicity is equal to the rank of the following bilinear form:

$$\Phi : (M, Y)_G \times (Y, M)_G \to k \tag{1}$$

given by $\Phi(fg) = \psi(fg)$ for all $f \in (M, Y)_G = \mathrm{Hom}_{kG}(M, Y)$ and for all $g \in (Y, M)_G$.

[1]The author acknowledges support from the NSA in the form of grant MDA 904-90-H-4030. The author is also grateful to J. A. Green for discussions on the subject matter of this paper.

Now assume $Y = k_U^G$ for some $U \leq G$. Let $y = 1_G \otimes 1_k$, and let S be a transversal of G/U so that Y has basis $\{sy = s \otimes 1_k | s \in S\}$. Then there exist natural k-isomorphisms $\alpha : (M,k)_U \to (M,Y)_G$ and $\beta : (k,M)_U \to (Y,M)_G$ as follows: For $\xi \in (M,k)_U$, $\alpha(\xi) : x \to \sum_{s \in S} s \otimes \xi(s^{-1}x) = \sum_{s \in S} \xi(s^{-1}x)sy$; for $\eta \in (k,M)_U$, $\beta(\eta) : gy \to g\eta(1_k)$ for all $g \in G$ (see, for example [1, p.243]). We may "lift" Φ via α and β and define

$$\Phi'(\xi,\eta) = \Phi(\alpha(\xi),\beta(\eta)) = \psi(\alpha(\xi)\beta(\eta)). \tag{2}$$

Now $\alpha(\xi)\beta(\eta) \in E$ takes $y = 1_G \otimes 1_k$ to $\alpha(\xi)(\eta(1_k)) = \sum_{s \in S} \xi(s^{-1}\eta(1_k))sy$. Since ξ, η are U-maps, $\xi(s^{-1}\eta(1_k))$ is unchanged when s is replaced by any other element of UsU. Let N be a transversal of $U\backslash G/U$, and we break up the sum above collecting for each $n \in N$ all the terms with $s \in UnU$. So $\alpha(\xi)\beta(\eta)$ is the kG-endomorphism of Y taking y to

$$\sum_{n \in N} \xi(n^{-1}\eta(1_k))p_n y \quad \text{where} \quad p_n = \sum_{s \in UnU} s. \tag{3}$$

Remember that E has basis $\{A_n | n \in N\}$ where $A_n(y) = p_n y$ ([4, p.717]). Hence $\alpha(\xi)\beta(\eta) = \sum_{n \in N} \xi(n^{-1}\eta)A_n$ and

$$\Phi'(\xi,\eta) = \sum_{n \in N} \xi(n^{-1}\eta)\psi(A_n). \tag{4}$$

Because of the isomorphisms α and β, Φ and Φ' have the same rank so that $[Y_\psi|M]$ is the rank of Φ'. We replace $(k,M)_U$ by $I_U(M)$, the set of U-invariant elements of M, by means of the k-isomorphism $(k,M)_U \to I_U(M)$ where $\gamma \in (k,M)_U$ is mapped to $\gamma(1_k)$. As $(M,k)_U = I_U(M^*)$ for M^* the dual of M, we replace Φ' by Λ where

$$\begin{aligned} &\Lambda : I_U(M^*) \times I_U(M) \to k \text{ and} \\ &\Lambda(\xi,x) = \textstyle\sum_{n \in N} \xi(n^{-1}x)\psi(A_n), \quad \forall \xi \in I_U(M^*),\ x \in I_U(M). \end{aligned} \tag{5}$$

3. Condensing

For any subset T of G let $[T] = \sum_{t \in T} t$. Hence $Y \cong kG[U]$. Take $H \leq G$, $M = kG[H] \cong k_H^G$. Let X_H be a transversal for the U-H double cosets, *i.e.*, $G = \bigcup_{a \in X_H} UaH$. Then $I_U(M)$ has basis $\{x_a = [UaH] | a \in X_H\}$. Let U_a be a subset of U satisfying $UaH = \bigcup_{u \in U_a} uaH$. Then M has basis $\{ua[H] | a \in X_H, u \in U_a\}$, and $I_U(M^*)$ has basis $\{\xi_a | a \in X_H\}$, where $\xi_a(ub[H]) = \delta_{a,b}$. Let $\Lambda_H(\psi)$ be the matrix of the bilinear form (5) with respect to these bases of $I_U(M)$ and $I_U(M^*)$. So $\Lambda_H(\psi) = (\Lambda_H(\xi_a, x_b))_{a,b \in X_H}$, and the rank of this matrix gives the multiplicity of Y_ψ in any decomposition of $kG[H]$. So

$$\text{The } (a,b) \text{ entry of } \Lambda_H(\psi) \text{ is } \Lambda_H(\xi_a, x_b) = \textstyle\sum_{n \in N} \xi_a(n^{-1}x_b)\psi(A_n). \tag{6}$$

We write $\xi_a(nx_b) = \theta(nx_b) \cdot 1_k$, where $\theta_a(nx_b)$ is the number of cosets gH in $nUbH \cap UaH$. Left-multiplication by n^{-1} gives $\xi_a(nx_b) = \xi_b(n^{-1}x_a)$. (7)

Now consider two subgroups $H \leq K \leq G$. We wish to relate the multiplicity of Y_ψ in $kG[H]$ to its multiplicity in $kG[K]$ for any ψ. Write $G = \bigcup_{a \in X_K} UaK$, $x_a^K = [UaK]$ for $a \in X_K$. Denote the factor in each summand of the (a,b) entry of $\Lambda_K(\psi)$ by $\xi_a^K(nx_b^K)$ for $a, b \in X_K$. Hence $\xi_a^K(nx_b^K) = \theta_a^K(nx_b^K) \cdot 1_k$ where $\theta_a^K(nx_b^K) = \#\{\text{cosets } gK \text{ in } nUbK \cap UaK\}$, as in (7). Write $K = \bigcup_{v \in K/H} vH$ where K/H is a transversal of K by H. So $G = \bigcup_{\substack{a \in X_K \\ v \in K/H}} UavH$ so that $X_H = \{av | a \in X_K, v \in K/H\}$. Now

$$\begin{aligned} nUbK \cap UaK &= n\left(\bigcup_{w \in K/H} UbwH\right) \bigcap \left(\bigcup_{v \in K/H} UavH\right) \\ &= \bigcup_{v,w \in K/H} (nUbwH \cap UavH). \end{aligned}$$

Counting the number of H cosets gH in each side yields

$$[K : H]\, \theta_a^K(nx_b^K) = \sum_{v,w \in K/H} \theta_{av}^H(nx_{bw}^H) \quad \text{for all } a, b \in X_K. \tag{8}$$

Define $\Sigma_a^H(nx_b^K)$ to be the integer $\sum_{v,w \in K/H} \theta_{av}^H(nx_{bw}^H)$ for any $a, b \in X_K$. By (8) $\Sigma_a^H(nx_b^K)$ is an integer divisible by $[K : H]$ and we have:

Lemma 1. $\xi_a^K(nx_b^K) = \left(\frac{1}{[K:H]}\Sigma_a^H(nx_b^K)\right) \cdot 1_k$. *This equation holds in any characteristic.*

Hence the entries of $\Lambda_K(\psi)$ can be obtained by a "condensing" of the entries of $\Lambda_H(\psi)$. In fact,

Theorem 1. $\Lambda_K(\psi)_{a,b} = \frac{1}{[K:H]} \sum_{v,w \in K/H} \Lambda_H(\psi)_{av,bw}$ *for any* $a, b \in X_K$.

We see that the (a,b) entry of $\Lambda_K(\psi)$ is obtained by summing the entries of the (a,b) "coset block" of $\Lambda_H(\psi)$. We now generalize Theorem 1 to include the following case:

Hypothesis (P). Consider the case $k = \mathbb{C}$ where $E = E_\mathbb{C}$ is the endomorphism algebra of $\mathbb{C}G[U]$. Say there exists a character $\psi_0 : E_\mathbb{C} \to \mathbb{C}$ such that $\psi_0(A_n) \in \mathbb{Z}$ for all $n \in N$, and that for any k, the character $\psi : E \to k$ is given by $\psi(A_n) = \psi_0(A_n) \cdot 1_k$ (for all $n \in N$).

We know from (6) that the (a,b) entry in $\Lambda_H(\psi_0)$ is $\sum_{n\in N}\theta_a^H(nx_b^H)\psi_0(A_{n^{-1}})$. Since $\psi(A_n)=\psi_0(A_n)\cdot 1_k$ gives $\Lambda_H(\psi)=\Lambda_H(\psi_0)\cdot 1_k$ for any k, we may generalize our theorem as follows:

Theorem 2. *Let ψ,ψ_0 satisfy (P), $H\le K$. Then*

$$\Lambda_K(\psi)_{a,b}=\left(\frac{1}{[K:H]}\sum_{v,w\in K/H}\Lambda_H(\psi_0)_{av,bw}\right)\cdot 1_k$$

for all $a,b\in X_K$. The condensing works for all k.

We now consider the special case of $H=U$ and relate the entries of $\Lambda_U(\psi)$ to the values of ψ. We know (see, for example [1, p.291]) that if $a,b\in N$, then $A_aA_b=\sum_{n\in N}\mu_{abn}A_n$, where the coefficient μ_{abn} is the coefficient given by $\#\{\text{cosets } tU \text{ in } nUa^{-1}U\cap UbU\}=\xi_b^U(nx_{a^{-1}}^U)$. So $A_aA_b=\sum_{n\in N}\xi_b^U(nx_{a^{-1}}^U)A_n=\sum_{n\in N}\xi_{a^{-1}}^U(n^{-1}x_b^U)A_n$ by (7). Hence

$$A_{a^{-1}}A_b=\sum_{n\in N}\xi_a^U(n^{-1}x_b^U)A_n.$$

Apply ψ to both sides to yield

$$\psi(A_{a^{-1}})\psi(A_b)=\sum_{n\in N}\xi_a^U(n^{-1}x_b^U)\psi(A_n) \tag{9}$$

for all $a,b\in N$, so that

Lemma 2. $\Lambda_U(\psi)_{a,b}=\psi(A_{a^{-1}})\psi(A_b)$, *for all $a,b\in N$.*

The matrix $\Lambda_U(\psi)$ is then of rank one, as expected, since we have calculated the multiplicity of Y_ψ in Y.

Now let $K\ge U$. We aim to get information on $\Lambda_K(\psi)$ from condensing $\Lambda_U(\psi)$. By Theorem 1 and Lemma 2,

$$\begin{aligned}\Lambda_K(\psi)_{a,b} &= \frac{1}{[K:U]}\sum_{v,w\in K/U}\psi(A_{v^{-1}a^{-1}})\psi(A_{bw}), &&\forall a,b\in X_K.\\ &= \frac{1}{[K:U]}\sum_{v\in K/U}\psi(A_{v^{-1}a^{-1}})\sum_{w\in K/U}\psi(A_{bw}), &&\forall a,b\in X_K.\end{aligned} \tag{10}$$

We may state our general result as follows:

Theorem 3. *Let $U\le K$, ψ, ψ_0 satisfy (P). Then*

$$\Lambda_K(\psi)_{a,b}=\left(\frac{1}{[K:U]}\sum_{v\in K/U}\psi(A_{v^{-1}a^{-1}})\sum_{w\in K/U}\psi(A_{bw})\right)\cdot 1_k.$$

This expression is of the form $\left(\frac{1}{[K:U]}\Phi_1(a)\cdot\Phi_2(b)\right)\cdot 1_k$ so that the rank of $\Lambda_K(\psi)\le 1$ (in any characteristic).

4. Applications to split (B, N)-pairs

Let G be a finite group with an unsaturated split (B, N)-pair with Weyl group W, fundamental reflection set R, and fundamental root system Δ. We now take $U = B$ in our discussions above, $Y = kG[B]$ so that E has basis $\{T_w | w \in W\}$. Then the character $\psi : E \to k$ given by $\psi(T_w) = (-1)^{l(w)} \cdot 1_k$, where $l(w)$ is the length of w as a product of fundamental reflections, satisfies (P). The component of $kG[B]$ corresponding to ψ is called the Steinberg Component, St_B (see [5]). Let $J \subseteq R$ and $K = G_J$ be the corresponding parabolic subgroup with Weyl group W_J. Write X_J for X_{G_J}, the transversal of B-G_J double cosets. Now $X_J = \{w \in W | w(\Delta_J) > 0\}$ where Δ_J is the set of fundamental roots for reflections in J. If $v \in W_J$ and $a \in X_J$, $l(av) = l(a) + l(v)$. We now prove:

Theorem 4. *Let k be a field of any characteristic. Then St_B is a component of $k_{G_J}^G$ if and only if $J = \emptyset$.*

PROOF. We know that the rank of $\Lambda_K(\psi)$ is at most one by Theorem 3. Now, the (a, b) entry of $\Lambda_K(\psi)$ is of the form $\frac{1}{[G_J:B]} \sum_{v,w \in W_J} \psi(T_{v^{-1}a^{-1}})\psi(T_{bw})$ for $a, b \in X_J$ by (10). But

$$\begin{aligned}
\sum_{v,w \in W_J} \psi(T_{v^{-1}a^{-1}})\psi(T_{bw}) &= \sum_{v,w \in W_J} \psi(T_{av})\psi(T_{bw}) \\
&= \sum_{v \in W_J} \psi(T_{av}) \sum_{w \in W_J} \psi(T_{bw}) \\
&= \left(\sum_{v \in W_J} (-1)^{l(av)} \sum_{w \in W_J} (-1)^{l(bw)} \right) \cdot 1_k \\
&= \left(\sum_{v \in W_J} (-1)^{l(a)+l(v)} \sum_{w \in W_J} (-1)^{l(b)+l(w)} \right) \cdot 1_k \\
&= (-1)^{l(a)+l(b)} \left(\sum_{v \in W_J} (-1)^{l(v)} \right)^2 \cdot 1_k.
\end{aligned}$$

But $\sum_{v \in W_J} (-1)^{l(v)} = 0$ if $J \neq \emptyset$ because elements in W_J can be broken up into a disjoint union of cosets $\{w\} \cup \{ww_i\}$ where $l(ww_i) = l(w) - 1$ for $w_i \in J$ so that lengths cancel in pairs. $\square$

Similarly we may study the multiplicity of the Scott module in this same context. If G is any group, $U \leq G$, the character Φ given by $\Phi(A_n) = [U : U \cap nUn^{-1}] \cdot 1_k$ satisfies (P). The component in Y corresponding to Φ is called the Scott module (see [3, p.9]). When G is a Chevalley group over a finite field of q elements (taking $U = B$ in our discussion above), the character Φ is given by $\Phi(T_w) = q^{l(w)} \cdot 1_k$ and satisfies (P). We shall call the module in $kG[B]$ corresponding to Φ the Scott B-module, Sc_B. Hence, we prove:

Lemma 3. *Let k be any characteristic. The multiplicity of Sc_B in $k_{G_J}^G$ is 1, unless $\sum_{v\in W_J} q^{l(v)} = 0$.*

PROOF. The multiplicity of Sc_B is determined by the matrix entries given in (10). As above

$$\sum_{v,w\in W_J} \Phi(T_{v^{-1}a^{-1}})\Phi(T_{bw}) = q^{l(a)}q^{l(b)}\left(\sum_{v\in W_J} q^{l(v)}\right)^2 \quad \text{for any } a,b \in X_K.$$

References

[1] C.W. Curtis and I. Reiner, *Methods of representation theory, with applications to finite groups and orders I*, (Wiley-Interscience, New York,1981).

[2] J.A. Green, Multiplicities, Scott modules and lower defect groups, *J. London Math. Soc. (2)* **28**(1983), 282–292.

[3] J.A. Green, Functor categories and group representations, *Portugal. Math.* **43**(1985-1986), 1–16.

[4] N.B. Tinberg, Modular representation of finite groups with unsaturated split (B,N)-pairs, *Canad. J. Math.* **31**(1980), 714–733.

[5] N.B. Tinberg, The Steinberg component of a finite group with a split (B,N)-pair, *J. Algebra* **104**(1986), 126–134.

LIE RING METHODS IN THE THEORY OF NILPOTENT GROUPS

E.I. ZELMANOV

Mathematics Department, University of Wisconsin–Madison, 480 Lincoln Drive, Madison, Wisconsin 53706, U.S.A.

0.1. The title of this survey has already appeared in the literature at least twice (see G. Higman [14] and G. E. Wall [35]). As in [14, 35] we will not try to develop some general theory but rather will concentrate on particular group-theoretic problems in which Lie algebra methods proved to be useful.

Our main objects will be finite p-groups and their relations: pro-p groups and residually-p groups.

In §1 we consider residually-p groups whose Lie algebras satisfy polynomial identities. To show that this class is well behaved we sketch the proof that a finitely generated periodic group with this property is finite.

The §2 is dedicated to another "ring theoretic" problem in p-groups: the famous Golod-Shafarevich inequalities.

0.2. As we have already mentioned above our main object of interest is a finite p-group. However, since this is too difficult an object to be studied individually, we will study arrays of finite p-groups. More precisely, let G be a group. A system of homomorphisms $\varphi_i : G \rightarrow G_i$, $i \in I$, is said to approximate G if for any arbitrary element $1 \neq a \in G$ there exists a homomorphism φ_i such that $\varphi_i(a) \neq 1$. Let $H_i = \text{Ker } \varphi_i$. The definition above says that the system of homomorphisms $\{\varphi_i, i \in I\}$ approximates G if and only if $\cap_{i\in I} H_i = (1)$. In this case we also say that G can be approximated by groups G_i, $i \in I$.

Let p be a prime number. A group G is said to be a residually-p group if it can be approximated by finite p-groups.

In a residually-p group G the kernels $H_i = \text{Ker } \varphi_i$ of approximating homomorphisms φ_i can be taken for a basis of neighborhoods of 1 in G. A residually-p group is said to be a pro-p group if it is complete in this topology. It is easy to see that a pro-p group is just an inverse limit of finite p-groups G/H_i.

Every residually-p group G gives rise to a pro-p group $G_{\hat{p}}$ which is the completion of G.

It is easy to see that any pro-p group is compact.

0.3. Let G be a group. For elements $x, y \in G$ let $(x, y) = x^{-1}y^{-1}xy$ denote their group commutator. If $x_1, \ldots, x_r \in G$ then we define $(x_1, \ldots, x_r) = (\cdots(x_1, x_2), x_3), \ldots, x_r)$. For arbitrary elements $x, y, z \in G$ we have

$$(x, y)^{-1} = (y, x)$$

$$(xy, z) = (x, z)(x, z, y)(y, z)$$
$$(x, yz) = (x, z)(x, y)(x, y, z)$$
$$(x, y^{-1}, z)^y (y, z^{-1}, x)^z (z, x^{-1}, y)^x = 1,$$

where $x^y = y^{-1}xy$. These laws can be either directly verified or found in M. Hall [11] or in any other text-book in Group Theory.

0.4. Let $G = G_1 \geq G_2 \geq \ldots$ be a descending chain of normal subgroups. This chain is said to be a *central series* if

$$(G_i, G_j) \subseteq G_{i+j}$$

for any $i, j \geq 1$. Any descending central series gives rise to a Lie ring.

Consider abelian factors G_i/G_{i+1} and their direct sum

$$L(G) = \bigoplus G_i/G_{i+1}, \quad i \geq 1.$$

For arbitrary cosets $a_i G_{i+1} \in G_i/G_{i+1}$, $b_j G_{j+1} \in G_j/G_{j+1}$ we define

$$[a_i G_{i+1}, b_j G_{j+1}] = (a_i, b_j) G_{i+j+1}.$$

The right hand side does not depend on the choice of representatives a_i, b_j and thus is well defined. From (0.3) it follows that the bracket [,] defines a structure of a Lie ring on $L(G)$.

0.5. Now we shall introduce two important central series. Let C_k be the subgroup of G generated by all commutators $(x_1, \ldots, x_k)$, $x_i \in G$. From the formulae (0.3) it follows that

$$G = C_1 \geq C_2 \geq \cdots$$

is a central series which is called the lower central series. We shall denote

$$L(G) = \bigoplus_{i=1}^{\infty} C_i/C_{i+1}.$$

As above, by p we denote a fixed prime number. Let D_k be the subgroup of G generated by all elements $(x_1, \ldots, x_i)^{p^j}$, where $i \cdot p^j \geq k$ and $x_1, x_2, \ldots$ are arbitrary elements from G. Clearly,

$$G = D_1 \geq D_2 \geq \cdots$$

is a descending chain of normal subgroups every factor of which is an elementary abelian p-group. Thus, D_i/D_{i+1} can be viewed as a vector space over the field F_p of residues modulo p.

The following lemma is due to M. Lazard [23].

Lemma. *Consider the free group in two generators x and y. Then for an arbitrary prime number p we have*

$$(y, x^p) = (y, \underbrace{x, \ldots, x}_{p})\rho_1 \cdots \rho_n \sigma_1^p \cdots \sigma_m^p \ ,$$

where σ_i are commutators of length ≥ 2 and ρ_j are commutators in x, y that either contain $> p$ elements x or p elements x and ≥ 2 elements y.

From this lemma it follows that

$$(D_i, D_j) \subseteq D_{i+j}.$$

Hence, $D_1 \geq D_2 \geq \cdots$ is a central series which is called the lower central p-series of G or the Lazard central series. The Lie ring $\bigoplus_{i\geq 1} D_i/D_{i+1}$ is actually an algebra over the field F_p. We shall denote the subalgebra of $\bigoplus_{i\geq 1} D_i/D_{i+1}$ generated by D_1/D_2 as $L_p(G)$.

There is another way to introduce the subgroups D_k. Consider the group algebra F_pG and its augmentation ideal Δ. Then

$$D_k = \{g \in G \mid 1 - g \in \Delta^k\}.$$

1. Infinitesimally PI residually-p groups

1.1. Let L be a Lie algebra over a ground field F and let $f(x_1, \ldots, x_n)$ be a nonzero element of the free Lie algebra on the free generators $x_1, \ldots, x_n$.

A Lie algebra L is said to satisfy the identity $f(x_1, \ldots, x_n) = 0$ if for arbitrary elements $a_1, \ldots, a_n \in A$ we have $f(a_1, \ldots, a_n) = 0$. In this case we say that L is a Lie algebra with a polynomial identity (PI).

1.2. Let G be a residually-p group. We say that G is infinitesimally PI if the Lie algebra $L_p(G)$ is PI.

Example 1. Let G be a pro-p group which is (topologically) finitely generated. M. Lazard proved (see [23, 4]) that G is p-adic analytic if and only if the Lie algebra $L_p(G)$ is nilpotent. Thus, p-adic analytic pro-p groups are infinitesimally PI.

Example 2. Let (Λ, M) be a complete commutative Noetherian local ring whose residue field Λ/M is finite, char $\Lambda/M = p$. Apart from the ring of p-adic integers the archetypal example of such a Λ is $F_p[[t]]$, the ring of infinite series over the field F_p, $M = t\Lambda$. In [25] A. Lubotzky and A. Shalev considered pro-p groups that are analytic over Λ. Let G be a perfect Λ-analytic group (for the definition of such a group see [25]). We shall only mention that the main congruence subgroups $SL_m^1(\Lambda)$ are perfect unless $p = m = 2$.

Suppose that $p\Lambda = (0)$. Then $L_p(G) \cong L_0 \otimes gr(M)$, where L_0 is a finite-dimensional Lie algebra over Λ/M and $gr(M) = \bigoplus_{n=1}^{\infty} M^n/M^{n+1}$ is the graded algebra of M.

Thus, the algebra $L_p(G)$ is PI.

Example 3. The Nottingham group Nott(p) is the group of automorphisms of the algebra $F_p[[t]]$ acting trivially on $tF_p[[t]]/t^2F_p[[t]]$. We have $L_p(\text{Nott}(p)) \cong W_1 \otimes tF_p[t]$, where W_1 is the Witt algebra with the basis $e_0, \ldots e_{p-1}$ over F_p and the multiplication $[e_i, e_j] = (i-j)e_k$, $k = i+j \mod p$.

Example 4. Let $w(x,y)$ be a nonidentical element of the free group on the free generators x, y. We say that a group G satisfies the law $w = 1$ if $w(a,b) = 1$ for arbitrary elements $a, b \in G$.

Any group law implies a polynomial identity on the Lie algebra $L_p(G)$ (see Lemma (1.3) below). So, a group that satisfies a law is infinitesimally-PI.

For a pro-p group G it was proved in [37] that either G contains a free abstract subgroup of rank 2 or G is infinitesimally PI.

Let p be a prime number. The law $x^p = 1$ implies the Engel's identity

$$E_{p-1}: \; [x, \underbrace{y, \ldots, y}_{p}] = 0$$

on the Lie algebra $L(G)$ (Magnus [26]).

The law $x^{p^k} = 1$, where $k > 1$, implies only the linearized Engel's identity

$$\tilde{E}_{p^k-1}: \; \sum_{\sigma \in S_{p^k-1}} [x, y_{\sigma(1)}, y_{\sigma(2)}, \ldots, y_{\sigma(p^k-1)}] = 0$$

(G. Higman, [14]). These facts are of crucial importance for the Burnside Problem on periodic groups (see the discussion of the Restricted Burnside Problem in [21, 31] and also in (1.16) below).

1.3. Let $F(n)$ be the free group generated by the free generators $x_1, \ldots, x_n$. For an arbitrary prime number p the group $F(n)$ is residually-p. Consider the topology on $F(n)$ generated by all normal subgroups of $F(n)$ of p-power index. The closure of $F(n)$ with respect to this topology is called the free pro-p group on the set of free generators $x_1, \ldots, x_n$. We shall denote this closure by $F(n)_{\hat{p}}$. For any pro-p group G an arbitrary mapping $x_i \to G$, $1 \le i \le n$, can be uniquely extended to the continuous homomorphism $F(n)_{\hat{p}} \to G$.

Elements of $F(n)_{\hat{p}}$ can be throught of as infinite products

$$c = c_1 c_2 \ldots, \qquad (*)$$

where each factor c_i is either 1 or a product of p-powers of commutators $(x_{i_1}, \ldots, x_{i_k})^{p^t}$ such that $k \cdot p^t = i$.

We say that a nonidentical element $c \in F(n)_{\hat{p}}$ is a pro-p identity on a pro-p group G (or, G satisfies $c = 1$) if any continuous homomorphism $F(n)_{\hat{p}} \to G$ maps c to 1.

Lemma. *If a pro-p group G satisfies a pro-p identity then the Lie algebra $L_p(G)$ is PI.*

PROOF. Let c be a pro-p identity satisfied by G and let $c = c_1c_2\ldots$ be a decomposition $(*)$. Let k be the minimal number such that $c_k \neq 1$. Then $c_k = \rho_1^{p^{k_1}} \cdots \rho_s^{p^{k_s}}$, where ρ_j are commutators in $x_1, \ldots, x_n$ of length $\ell(\rho_j)$ and for each j we have $\ell(\rho_j)p^{k_j} = k$. In the free group $F(n+1)$ we have

$$(x_{n+1}, c_k) = (x_{n+1}, \underbrace{\rho_1, \ldots, \rho_1}_{p^{k_1}}) \cdots (x_{n+1}, \underbrace{\rho_s, \ldots, \rho_s}_{p^{k_s}})$$

mod $D_{k+2}(F(n+1))$.

For a group commutator ρ in $x_1, \ldots, x_n$ with an arbitrary scheme of brackets let $\hat{\rho}$ denote the commutator in the free Lie algebra on $x_1, \ldots, x_n$ whose brackets are put in the same way as in ρ.

Consider the element

$$[x_{n+1}, \underbrace{\hat{\rho}_1, \ldots, \hat{\rho}_1}_{p^{k_1}}] + \cdots + [x_{n+1}, \underbrace{\hat{\rho}_s, \ldots, \hat{\rho}_s}_{p^{k_s}}]$$

of the free Lie algebra and its linearization f (see [2]). The Lie algebra $L_p(G)$ satisfies the identity $f = 0$. □

1.4. The reverse statement is not true. As we have seen above, the Nottingham group is infinitesimally-PI. A. Weiss, however, announced that an arbitrary finite p-group is embeddable into Nott(p) which rules out any pro-p identity for Nott(p).

1.5. It is an interesting open problem whether a p-adic analytic group necessarily satisfies a pro-p identity. Another form of the question is

Problem. ([25], [42]) Is it true that a nonabelian free pro-p group is not linear, that is, can not be embedded into $GL_n(\Lambda)$ for any n and a commutative ring Λ?

Zubkov [42] proved that the main congruence subgroup of $GL_2(\Lambda)$ satisfies a pro-p identity provided that $p \neq 2$. This implies that a nonabelian free pro-p group ($p \neq 2$) is not embeddable into $GL_2(\Lambda)$.

1.6. Residually-p groups are reasonably well behaved as compared to the Ol'shansky Monster (see [27]) and other recently emerged Monsters from Geometric Group Theory, but they still can display quite pathological properties. In 1964 E. Golod [5] showed that a finitely generated periodic residually-p group can be infinite. Other examples of such groups were constructed by Sushchansky [29], Grigorchuk [7], and Gupta-Sidki [9].

In a sense, a residually-p group can be as pathological as an associative (Lie) algebra can be. Golod constructed his examples from the examples of finitely generated nonnilpotent nil algebras over fields of positive characteristics. It is one of the deep theorems in Ring Theory that no such phenomenon is possible for an algebra satisfying a polynomial identity (see [13, 16]). We prove the same theorem for residually-p groups.

Theorem. *Let G be a finitely generated periodic residually-p group. If the Lie algebra $L_p(G)$ satisfies a polynomial identity then the group G is finite.*

1.7. The proof depends on (and, in fact, immediately follows from) the following theorem on Lie algebras. To formulate it we will need two more definitions.

An element a of a Lie algebra L is called ad-nilpotent (of degree n) if the adjoint operator $\mathrm{ad}(a) : L \to L$, $\mathrm{ad}(a) : x \to [x, a]$, is nilpotent (of degree n).

By a commutator in elements $a_1, \ldots, a_m$ we mean an arbitrary element that can be obtained from $a_1, \ldots, a_m$ by means of the (repeated) operation of commutation with an arbitrary system of brackets, the elements $a_1, \ldots, a_m$ included.

Theorem. *Let L be a Lie algebra that is generated by elements $a_1, \ldots, a_m$. Suppose that*

(i) L satisfies a polynomial identity,

(ii) every commutator in $a_1, \ldots, a_m$ is ad-nilpotent.

Then the algebra L is nilpotent.

1.8. To see the relation between Theorem (1.6) and Theorem (1.7) let us consider a finitely generated residually-p group G and an element $a \in D_k(G) \setminus D_{k+1}(G)$. If $a^{p^s} = 1$ then from Lazard's lemma it follows that

$$L_p(G)\mathrm{ad}(aD_{k+1}(G))^{p^s} = (0).$$

Suppose that the group G is generated by elements $a_1, \ldots, a_m$.

Again let us consider the free group $F(m)$ on the free generators $x_1, \ldots, x_m$. Let $\rho(x_1, \ldots, x_m)$ be an arbitrary group commutator and let $\hat{\rho}(x_1, \ldots, x_m)$ be the corresponding commutator in the free Lie algebra. If $\rho(a_1, \ldots, a_m)^{p^s} = 1$ then the element $\hat{\rho}(a_1D_2(G), \ldots, a_mD_2(G))$ is ad-nilpotent of degree $\leq p^s$.

Thus, for a periodic group G an arbitrary commutator in generators $a_1D_2(G),\dots,a_mD_2(G)$ of the algebra $L_p(G)$ is ad-nilpotent. If, in addition the algebra $L_p(G)$ is PI then by Theorem (1.7) $L_p(G)$ is nilpotent.

Let $L_p(G)^d=(0)$. This means that an arbitrary commutator in $a_1,\dots,a_m$ of length d can be represented as a product of powers $\rho_i^{p^{s_i}}$, where each ρ_i is a left-normed commutator in $a_1,\dots,a_m$ and for each i we have $\ell(\rho_i)p^{s_i}>d$.

Let $\rho_1,\dots,\rho_r$ be all left-normed commutators in $a_1,\dots,a_m$ of length $<d$. From what we said above it follows that for an arbitrary $n\geq 1$ an arbitrary element $g\in G$ can be represented as

$$g=\rho_1^{k_1}\cdots\rho_r^{k_r}g',$$

where $g'\in D_n(G)$.

Since the group G is periodic there exists the maximum of orders of elements $\rho_1,\dots,\rho_r$. Denote this maximum as p^s. Then without loss of generality we can assume that $k_i<p^s$, $i=1,\dots,r$. Hence, for any $n\geq 1$

$$|G/D_n(G)|<p^{sr}.$$

This implies that $|G|<p^{sr}$. So, the really difficult part of Theorem (1.6) is Theorem (1.7).

1.9. Here we will sketch some ideas from the proof of Theorem (1.7). Let L be a Lie algebra over a field F of characteristic $p>0$ which is generated by elements $a_1,\dots,a_m$ and such that

(i) every commutator in $a_1,\dots,a_m$ is ad-nilpotent,

(ii) L is PI.

Our aim is to prove that L is nilpotent. In view of the condition (i) we can use the well known Locally Nilpotent Radical argument: there exists the maximal locally nilpotent ideal $\mathrm{Loc}(L)$ of L such that the quotient algebra $L/\mathrm{Loc}(L)$ does not contain any nonzero locally nilpotent ideals.

If $\mathrm{Loc}(L)\neq L$ then factoring out the $\mathrm{Loc}(L)$ we shall assume that L does not contain any nonzero locally nilpotent ideas.

The crucial role in the proof of Theorem (1.7) is played by the notion of a sandwich.

Definition (A.I. Kostrikin [19, 20]) An element a of a Lie algebra L is called a sandwich if $[L,a,a]=[L,a,L,a]=(0)$.

Theorem on Sandwich Algebras. ([22]) *A Lie algebra generated by a finite collection of sandwiches is nilpotent.*

I know four proofs of this theorem. One of them is related to Jordan theory [38], two are entirely Lie theoretic [21, 22] and, finally, the fourth proof (Bachelin (see [30]), Chanyshev [3]) is entirely a combinatorics of words.

Without loss of generality we may assume the ground field F to be infinite. Suppose that we managed to find a polynomial $f(x_1,\ldots,x_r)$ such that f is not identically zero in L and for arbitrary elements $a_1,\ldots,a_r \in L$ the value $f(a_1,\ldots,a_r)$ is a sandwich of L.

Because of the infinity of the field F the linear span $Ff(L)$ of all values of f on L is an ideal of L. By the theorem on sandwich algebras the ideal $Ff(L)$ is locally nilpotent, giving a contradiction.

1.10. If a Lie algebra L satisfies the Engel identity

$$E_n : \ [x, \underbrace{y,\ldots,y}_{n}] = 0$$

then it is PI and at the same time an arbitrary element of L is ad-nilpotent.

If L satisfies (E_2) then an arbitrary element in L is a sandwich. If L satisfies (E_3) then an arbitrary value of the polynomial $[x, y, y]$ is a sandwich.

If L is a nonzero Lie algebra which satisfies E_n, $n < p$, then such a sandwich-valued polynomial f exists ($f(L) \neq (0)$ and an arbitrary value of f on L is a sandwich), but it may be rather complicated (see [21]).

1.11. If L is just a Lie algebra with a polynomial identity the situation is even worse: there may be no sandwich-valued polynomials for it. For example, if L is a finite dimensional simple Lie algebra then there are no polynomials f such that $f(L) \neq (0)$ and every value of f on L is a sandwich.

To use our scheme of the proof, that is, to construct a sandwich-valued polynomial we have to change the algebra L and even to change the notion of a polynomial. We will introduce generalized polynomials. Along with ordinary powers $\mathrm{ad}(x)^n$ these polynomials will involve divided powers of the operators ad.

1.12. Consider the associative commutative F-algebra E on generators e_i, $i = 1, 2, \ldots$ and relations $e_i^2 = 0$, $e_ie_j = e_je_i$ for all $i, j = 1, 2, \ldots$. The set of elements $e_\pi = e_{i_1} \cdots e_{i_r}$, where $\pi = \{i_1 < i_2 < \cdots < i_r\}$ runs over all finite sets of positive integers, is a basis of E.

Consider the tensor product $\tilde{L} = L \otimes_F E$. An arbitrary element a of the algebra $\tilde{L}$ has the unique representation as a finite sum

$$a = \sum_\pi \alpha_\pi \otimes e_\pi = \sum_\pi a_\pi.$$

1.13. Let V_i be the ideal of $\tilde{L}$ consisting of the elements a such that $a_\pi = 0$ whenever i does not occur in π. It is clear that $V_i^2 = (0)$ and $\tilde{L} = \sum_{i=1}^\infty V_i$.

Let A be a finite family of elements of $\tilde{L}$ such that

(a) every element a in A lies in one of the ideals V_i, and

(b) $[a, b] = 0$ for every pair a, b of elements of A.

Consider the following linear operator on L:

$$U_k(A) = \sum \mathrm{ad}(a_1)\cdots \mathrm{ad}(a_k),$$

where the summation extends over all k-elements subsets of A. It is clear that

$$U_1(A) = \mathrm{ad}\left(\sum_{a\in A} a\right).$$

We further set $U_0(A) =$ Id, the identity operator.

If the characteristic of the ground field exceeds k then we have

$$U_k(A) = \frac{1}{k!}\,\mathrm{ad}\left(\sum_{a\in A} a\right)^k,$$

so the operators U_k play the role of divided powers of the operators ad.

To demonstrate the advantages of these divided powers we shall mention the following two formulas:

$$\mathrm{ad}(aU_m(A)) = \sum_{i=0}^{m}(-1)^i U_i(A)\mathrm{ad}(a)U_{m-i}(A)$$

with all the coefficients being ± 1, and

$$U_i(A)U_j(A) = \binom{i+j}{i} U_{i+j}(A).$$

1.14. Now we shall try to be very formal in our definitions. For a given Lie algebra L we shall define a set of words in the alphabet x_i, U_k, (,), [,], ad. We shall call these words U-words relative to L. If L is fixed, we shall speak simply of U-words. The fact, that no letter x_i other than $x_1, \cdots, x_r$ occurs in the expression for a U-word W will be expressed as $W = W(x_1, \cdots, x_r)$.

If the variables x_i are given values a_i in $\tilde{L}$ then the U-word W is associated with the linear operator $W(a_1, \cdots a_r)$ acting on $\tilde{L}$.

By definition we make the following assumptions:

1) if ρ is a commutator in $x_1, \cdots, x_r$ then the word $W = ad(\rho)$ is a U-word. If the letters $x_1, \cdots, x_r$ are given values $a_1, \cdots, a_r \in \tilde{L}$ then the word W is associated with the linear operator $ad(\rho(a_1, \cdots, a_r)$,
2) if $W(x_1, \ldots, x_r)$ is a U-word then $ad(x_i W)$ is a U-word. If the letters $x_1, \cdots, x_r, x_i$ are given values $a_1, \cdots, a_r, a_i \in \tilde{L}$ then $ad(x_i W)$ is associated with the linear operator $ad(a_i W(a_1, \cdots, a_r))$,
3) if W, V are U-words then WV is a U-word. The linear operator associated with WV is just the product of linear operators associated with W and V respectively,

4) assume that $W = W(x_1, \cdots x_r)$ is a U-word such that for arbitrary elements $a, b, a_1, \cdots a_r \in \tilde{L}$

$$[aW(a_1, \cdots, a_r), bW(a_1, \cdots, a_r)] = 0,$$

and x_i does not occur in the expression for W. Then $W' = U_k(x_i W)$ is a U-word. If the letters $x_1, \cdots, x_r$ and x_i are given values $a_1, \cdots, a_r$ and a respectively and $a = \sum a_\pi$ is the canonical decomposition of a, then the set $A = \{a_\pi W(a_1, \cdots, a_r)\}$ satisfies conditions (a), (b) of (1.13). By definition the linear operator $U_k(A)$ is the value of W'.

By a U-word relative to L we shall mean a word which is a U-word because of 1) - 4).

1.15. By a U-polynomial relative to L we understand a system consisting of three objects: a homogeneous Lie polynomial $f(t_i; y_j; z_k)$ whose variables are divided into three disjoint subsets $T = \{t_i\}$, $Y = \{y_j\}$, $Z = \{z_k\}$; and two U-words $W = W(x_1, \cdots, x_r)$, $W' = W'(x_1, \cdots, x_r)$. If the variables t_i, y_j, z_k are given values a_i, b_j, c_k respectively in $\tilde{L}$, and $x_1, \cdots, x_r$ are given values $d_1, \cdots, d_r$ in $\tilde{L}$, then the following element is called a value of the U-polynomial $(f(t_i; y_j; z_k); W; W')$:

$$f(a_i W(d_1, \cdots, d_r),\ b_j W'(d_1, \cdots, d_r), c_k).$$

Now we are ready to formulate a proposition on sandwich valued polynomials.

Proposition. *Let L be a nonzero Lie algebra over a field F of positive characteristic and suppose that L satisfies a PI. Then there exists a U-polynomial $f(t_i W, y_j W', z_k)$ which is not identically zero on $\tilde{L}$ but every value of $f(t_i W, y_j W', z_k)$ on $\tilde{L}$ is a sandwich of $\tilde{L}$.*

The full linearization of a U-polynomial is an ordinary polynomial. Let $f = f(x_1, \cdots, x_r)$ be a U-polynomial having degrees N_i with respect to variables $x_i, i = 1, \cdots r$. Let $\tilde{f}$ be the full linearization of f. Then every value of $\tilde{f}$ on $\tilde{L}$ is a linear combination of $\leq N = N_1 \cdots N_r$ values of the U-polynomial f on L.

Thus, instead of a polynomial every value of which on L is a sandwich we constructed a multilinear polynomial $\tilde{f}$ every value of which on $\tilde{L}$ is a linear combination of $\leq N$ sandwiches. This is the crucial part of the proof of Theorem (1.7).

1.16. One of the consequences of Theorems (1.6), (1.7) is the solution of the so called Restricted Burnside Problem:

For any $m, n \geq 1$ there are only finitely many finite m-generated groups of exponent n.

Indeed, P. Hall and G. Higman [12] reduced the problem to the case when $n = p^k$ is a power of a prime. A finite group G of exponent p^k is nilpotent. Clearly, it is sufficient to find an upper bound for the class of nilpotency of G in terms of m and p^k. We have,

$$\text{class}(G) = \text{class}(L(G)),$$
$$\text{class}(G) \leq p^k \cdot \text{class}(L_p(G)),$$

The Lie algebra $L_p(G)$ is generated by m elements, satisfies the identify $(\tilde{E}_{p^k-1})$ and every commutator of generators is ad-nilpotent of degree $\leq p^k$. By Theorem (1.7) there exists a function N of two arguments such that

$$\text{class}(L_p(G)) \leq N(m, p^k).$$

Not much is known about the order of magnitude of the function $N(m, p^k)$ (see [33]).

Conjecture (a) If a Lie algebra L over a field F_p satisfies the Engel identity (E_{p-1}) then an arbitrary element $a \in L$ generates a nilpotent ideal in L.

The proof of this conjecture would imply that the function $N(m, p^k)$ grows linearly with respect to m. This is known for $p = 5$ (Higman, [15]). It is conceivable though that the Conjecture (a) is false but a weaker assertion is valid.

Conjecture (b) If a Lie algebra L over F_p satisfies (E_{p-1}) then there exists a number $d = d(p)$ such that every commutator $[a_1, \cdots, a_d], a_i \in L$, generates a nilpotent ideal.

This assertion, if true, would imply that the growth of $N(m, p^k)$ in m is polynomial. Recently M. R. Vaughan-Lee [32] proved it for $p = 7$.

Conjecture (c) If a Lie algebra L over F_p satisfies (E_{p-1}) then the adjoint-algebra of L (that is the subalgebra of $\text{End}_{F_p}(L)$ generated by all adjoint operators $ad(a), a \in L$) is PI.

The assertion (a) implies (c). In particular, the adjoint-algebra of a Lie algebra of characteristic 5 satisfying (E_4) is PI. On the other hand, by the theorem of S. Amitsur [1] (c) implies (b).

All known proofs of the Restricted Burnside Problem (both for $n = p$ and $n = p^k, k > 1$) are based on the analysis of the associated Lie algebras. Thus, unless some new proof is found, the question of getting bounds for classes of nilpotency boils down to the question, how much can we get from the linearized Engel identity $(\tilde{E}_n)$, $p < n$. I suspect that $(\tilde{E}_n)$ is not stronger than just an arbitrary identity.

Conjecture. If a Lie algebra over a field of positive characteristic satisfies a polynomial identity then it satisfies an identity $(\tilde{E}_n)$ for some sufficiently large n.

A. Kemer [17] proved the analogous assertion for associative PI-algebras.

It is still conceivable that classes of nilpotency of finite m-generated groups of exponent p^k are bounded from above by a polynomial in m.

2. Golod-Shafarevich inequalities

2.1. In 1964 E. Golod and I. Shafarevich [6] suggested a method to establish infinity of pro-p groups and associative algebras presented by generators and relators. Shafarevich immediately applied this method to solve the long standing problem of an infinite tower of class fields (see [6]) and Golod immediately used it to solve the General Burnside Problem on periodic groups [5].

Consider the free group $F(d)$ on d free generators $x_1, \cdots, x_d$ and its pro-p completion $F(d)_{\hat{p}}$, that is the free pro-p group. As we have remarked in §1 an arbitrary element c of $F(n)_{\hat{p}}$ can be represented as an infinite product $c = c_1 c_2 \cdots$, where $c_i \in D_i(F(d)) \backslash D_{i+1}(F(d))$ or $c_i = 1$. The minimal i such that $c_i \neq 1$ is called the degree of the element c.

Let $\{f_\alpha, \alpha \in \mathfrak{A}\}$ be a (possibly infinite) family of elements of $F(d)_{\hat{p}}$ and let N be the smallest closed normal subgroup containing $\{f_\alpha\}$. We say that a pro-p group $G = F(d)_{\hat{p}}/N$ is presented by generators $x_1, \cdots, x_d$ and relators $f_\alpha = 1$, $\alpha \in \mathfrak{A}$.

2.2. Now consider the free associative algebra A over the field F_p, $|F_p| = p$, on the free generators $a_1, \cdots, a_d$. The algebra A is graded: $A = \sum_{i=0}^{\infty} A_i$, where A_i is spanned by all monomials of length i while $A_0 = F_p \cdot 1$. Consider the ideal $I = \sum_{i=1}^{\infty} A_i$ of A, $A = F_p \cdot 1 + I$. The ideals $I^m = \sum_{i \geq m}^{\infty} A_i$, $m = 1, 2, \cdots$ define a topology on A. Let $\hat{A}$ be the completion of A with respect to this topology. Clearly, $\hat{A}$ is the algebra of infinite series over F_p on n noncommuting variables.

By a degree of an element $f \in \hat{A}$ we mean the smallest length of a monomial involved in f.

Let $\{g_\alpha, \alpha \in \mathfrak{A}\}$ be a family of elements in $\hat{A}$ of degrees ≥ 1 and let J be the smallest closed ideal of $\hat{A}$ that contains all the elements $g_\alpha, \alpha \in \mathfrak{A}$. We say that the quotient pro-p algebra

$$B = \hat{A}/J$$

is presented by generators $a_1, \cdots, a_d$ and relations $f_\alpha = 0, \alpha \in \mathfrak{A}$.

Clearly, $B = F_p \cdot 1 + B'$, where $B' = I/J$. The infinite series $H_B(t) = 1 + \sum_{i=1}^{\infty} \dim(B'^i / B'^{i+1}) t^i$ is called the Hilbert series of the algebra B.

2.3. Without loss of generality we will assume that all the relators $g_\alpha, \alpha \in \mathfrak{A}$, have degrees ≥ 2 since having a relator of degree 1 just means that some of the (topological) generators of the algebra B can be expressed by other generators.

Suppose that r_2 relators f_α have degree 2, r_3 relators g_α have degree 3, and so on. Let

$$H_R = r_2 t^2 + r_3 t^3 + \cdots .$$

For two infinite series $a(t) = \sum_{i=0}^{\infty} a_i t^i$ and $b(t) = \sum_{i=0}^{\infty} b_i t^i$ we say that $a(t) \geq b(t)$ if $a_i \geq b_i$ for all $i \geq 0$.

Golod and Shafarevich [6] proved that if all the relators $g_\alpha, \alpha \in \mathfrak{A}$, are homogeneous then

$$H_B(t)(1 - dt + H_R(t)) \geq 1.$$

If the relators g_α are not necessarily homogeneous then the inequality above is still valid if divided by $1-t$ (Roquette [28], Vinberg [34], Koch [18]),

$$\frac{H_B(t)}{1-t}(1 - dt + H_R(t)) \geq \frac{1}{1-t}$$

The second inequality is weaker than the first one since $\frac{1}{1-t} = 1 + t + t^2 + \cdots$ and $\frac{a(t)}{1-t} \geq \frac{b(t)}{1-t}$ means exactly that for an arbitrary n we have

$$\sum_{i=0}^{n} a_i \geq \sum_{i=0}^{n} b_i.$$

Now suppose that for some $t_0, 0 < t_0 < 1$, the series $1 - dt + H_R(t)$ converges and $1 - dt_0 + H_R(t_0) < 0$. Then the series $H_B(t)$ can not converge at $t = t_0$. Indeed, if the series $H_B(t)$ converges at $t = t_0$ then the formal inequality above would have implied

$$\frac{H_B(t_0)}{1-t_0}(1 - dt_0 + H_R(t_0)) \geq \frac{1}{1-t_0}$$

whereas $\frac{H_B(t_0)}{1-t_0}$ is positive and $1 - dt_0 + H_R(t_0)$ is negative, a contradiction.

In particular, the algebra B is infinite-dimensional because otherwise $H_B(t)$ would be a polynomial.

2.4. Definition. We say that a set of relators $g_\alpha = 0, \alpha \in \mathfrak{A}$, of an algebra B is *small in the sense of Golod-Shafarevich* if $\deg g_\alpha \geq 2$ for each α and there exists t_0, $0 < t_0 < 1$, such that $1 - dt_0 + H_R(t_0) < 0$.

As we have seen above, an algebra represented by a small set of relators is necessarily infinite-dimensional.

Any system of r relators of degree ≥ 2, where $r < \frac{d^2}{4}$, is small.

2.5. Now let us go back to pro-p groups. Consider the group of the invertible elements from $\hat{A}$ (we denote it by $\hat{A}^*$) with the induced topology. The closed subgroup of $\hat{A}^*$ generated by $1+a_1, \cdots, 1+a_d$ is the free pro-p group having $1+a_i, 1 \leq i \leq d$, as free generators.

Hence, an arbitrary element f of the free pro-p group $F(d)_{\hat{p}}$ on d generators $x_1, \cdots, x_d$ can be viewed as an element of the algebra $\hat{A}$ upon substituting $x_i = 1 + a_i, 1 \leq i \leq d$,

$$f(1+a_1, \cdots, 1+a_d) = 1 + f'(a_1, \cdots a_d),$$

where $f'(a_1, \cdots a_d)$ is an element of degree ≥ 1.

It is easy to see that the degree of the element $f(x_1, \cdots, x_d)$ in $F(d)_{\hat{p}}$ coincides with the degree of the element $f'(a_1, \cdots a_d)$ in the algebra $\hat{A}$.

Consider a d-generated pro-p group G presented by generators and relations $G = \langle x_1, \cdots x_d | f_\alpha = 1, \alpha \in \mathfrak{A} \rangle$. Substituting $1 + a_i$ for $x_i, 1 \leq i \leq d$, we get

$$f_\alpha(1+a_1, \cdots, 1+a_d) = 1 + f'_\alpha(a_1, \cdots a_d),\ f'_\alpha \in I.$$

The algebra

$$B = \langle a_1, \cdots, a_d | f'_\alpha(a_1, \cdots a_d) = 0, \alpha \in \mathfrak{A} \rangle$$

is closely related to the group G in the following way.

For an open normal subgroup H of G let $\omega(H)$ denote the ideal of the group algebra F_pG generated by elements $1-h, h \in H$. In other words, $\omega(H)$ is the kernel of the homomorphism $F_pG \to F_p(G/H)$ induced by the natural homomorphism $G \to G/H$.

The ideals $\omega(H)$ define a topology on the group algebra F_pG. The closure of F_pG with repect to this topology is isomorphic to the algebra B (see [10]). Hence,

$$H_B(t) = 1 = \sum_{i=1}^{\infty} \dim(\omega(G)^i / \omega(G)^{i+1}) t^i.$$

2.6. Suppose that all relators $f_\alpha, \alpha \in \mathfrak{A}$, have degrees ≥ 2. Suppose further that there are r_2 relators f_α of degree 2, there are r_3 relators of degree 3 and so on. As above, denote

$$H_R(t) = r_2 t^2 + r_3 t^3 + \cdots.$$

We have

$$\frac{H_B(t)}{1-t}(1 - dt + H_R(t)) \geq \frac{1}{1-t}.$$

Definition. We say that a set of relators $f_\alpha = 1, \alpha \in \mathfrak{A}$, of a a pro-$p$ group G is *small in the sense of Golod-Shafarevich* if all the f_α's have degrees ≥ 2 and there exists a number t_0, $0 < t_0 < 1$, such that $1 - dt_0 + H_R(t_0) < 0$.

From what we said above it follows that a pro-p group defined by a small set of relators is infinite. Any set of r relators of degree ≥ 2, where $r < \frac{d^2}{4}$, is small.

Let G be a finite p-group and let d be the minimal number of generators of G. It is known that $d = \dim_{F_p} H^1(G)$ (see [8]). Let r be the minimal number of relators that are needed to define G. Then $r = \dim_{F_p} H^2(G)$ (see [8]) and because of the minimality of d every such relator is automatically of degree ≥ 2. Thus, we see that

$$r \geq d^2/4.$$

This inequality is referred to as the Golod-Shafarevich inequality.

2.7. A. Lubotzky [24] discovered the connection between presentations of discrete and pro-p groups and formulated the appropriate version of the Golod-Shafarevich inequality for discrete groups.

Theorem. (Lubotzky, [24]) *Let G be a discrete group with a presentation $\langle X; R\rangle$, where X is a finite set of generators, and R is a set of relators. Let $G_{\hat{p}}$ be the pro-p completion of G. Then*

(1) $\langle X; R\rangle$ is a presentation of $G_{\hat{p}}$ in the category of pro-p groups.

(2) If the minimal number of generators of $G_{\hat{p}}$ as a pro-p group $d_p(G) = \dim_{F_p}(G/D_2(G))$ is less than $|X|$ then $G_{\hat{p}}$ has a presentation with $d_p(G)$ generators and $|R| - (|X| - d_p(G))$ relators.

This theorem suggests a proper version of the Golod-Shafarevich inequality for discrete groups.

Theorem. (Lubotzky, [24])

(1) Let G be a discrete group with a finite presentation $\langle X; R\rangle$. Suppose that the pro-p completion $G_{\hat{p}}$ of G can not be presented by a small set of relators. Then

$$|R| \geq |X| - d_p(G) + d_p(G)^2/4.$$

(2) Let $d_{ab}(G)$ be the Hirsch rank of the abelinization $G/(G,G)$. If for every prime p the pro-p completion $G_{\hat{p}}$ can not be defined by a small set of relators and $G_{\hat{p}}$ is not isomorphic to the group $Z_{\hat{p}}$ of p adic integers then

$$|R| \geq |X| - d_{ab}(G) + d_{ab}(G)^2/4.$$

2.8. After the original work of Golod and Shafarevich it was shown that many infinite pro-p groups also can not be defined by a small set of relators. H. Koch [18] and A. Lubotzky [24] noticed that the assertion is true for p-adic analytic pro-p groups. Indeed, as we have mentioned in (2.3), if there exists

$0 < t_0 < 1$ such that $1 - dt + H_R(t)$ converges and is negative at $t = t_0$ then the series $H_B(T) = 1 + \sum_{i=1}^{\infty} \dim(\omega(G)^i/\omega(G)^{i+1})t^i$ can not converge at $t = t_0$. But M. Lazard [23] proved that a finitely generated pro-p group G is p-adic analytic if and only if $H_B(t)$ is a polynomial.

A Lubotzky (see [24]) found a beautiful application of these results to the congruence subgroup problem in arithmetic groups. He proved that if Γ is an arithmetic group which has the congruence subgroup property then for any prime p the pro-p completion of Γ is p-adic analytic and the same assertion is true for every subgroup of Γ of finite index. Hence Theorem (2.7) is applicable. For arithmetic lattices in $SL_2(\mathbb{C})$ this leads to a contradiction. Thus, every arithmetic lattice in $SL_2(\mathbb{C})$ fails to have the congruence subgroup property.

2.9. As we have mentioned above E. Golod [5] used the Golod-Shafarevich property to construct infinite finitely generated residually -p groups. Roughly speaking he argued in the following way.

The free group $F(d)$ on $d \geq 2$ free generators $x_1, \cdots, x_d$ obviously is defined by a small set of generators (actually, by none). For any $\frac{1}{d} < t_0 < 1$ we have $1 - dt_0 = -\epsilon < 0$.

Lemma. (E. Golod) *For any $0 < t_0 < 1$ and any $\epsilon > 0$ there exists a subset $S \subset F(d)$ such that*

(i) the group $G = \langle X; S\rangle$ is periodic,

(ii) $H_S(t)$ converges at $t = t_0$ and $H_s(t_0) < \epsilon$.

Now, $1 - dt_0 + H_s(t_0) < 0$, so the pro-p group $G_{\hat{p}}$ is infinite.

2.10. J. Wilson [36] noticed that this argument works in both directions: one can use the positive solution of the Burnside's problem in a class of groups to deduce the Golod-Shafarevich inequality.

Let G be a pro-p group which is (topologically) generated by $d \geq 2$ elements $x_1, \cdots, x_d$ and defined by a small set of generators $R \subset F(d)_{\hat{p}}$. So, there exists a number t_0, $0 < t_0 < 1$, such that the series $H_R(t_0)$ converges at $t = t_0$ and $1 - dt_0 + H_R(t_0) = -\epsilon < 0$. Consider the discrete subgroup Γ of G generated by $x_1, \cdots, x_d$. By Golod's Lemma there exists a subset $S \subseteq F(d)$ such that the series $H_s(t)$ converges at $t = t_0$ and $H_s(t_0) < \epsilon$.

Consider the pro-p group

$$\overline{G} = \langle x_1, \cdots, x_d | R \cup S\rangle.$$

The set $R \cup S$ is still small since

$$1 - dt_0 + H_{R\cup S}(t_0) \leq 1 - dt_0 + H_R(t_0) + H_S(t_0) = -\epsilon + H_S(t_0) < 0.$$

Hence, the group $\overline{G}$ is infinite. On the other hand the discrete subgroup of $\overline{G}$ generated by the elements $x_1, \cdots, x_d$ is a periodic homomorphic image of the group Γ.

If the group Γ has some property that guarantees that all its periodic homomorphic images are finite (for example, of Γ is solvable) then $\overline{\Gamma}$ is finite and so is $\overline{G}$, a contradiction.

2.11. The most general theorem of this kind is:

Theorem. (Wilson-Zelmanov, [37]) *An infinitesimally PI pro-p group can not be defined by a small set of relators.*

Indeed, the property of a residually-p group to be infinitesimally PI is inherited by homomorphic images and by Theorem (1.6) every finitely generated periodic infinitesimally PI group is finite.

Remark, that all known (up to now) classes whose groups can not be defined by small sets of relators happen to be infinitesimally PI.

2.12. All these results show that a pro-p group presented by a small set of relators not only must be infinite, but it must be infinite in some very strong sense. In [36] it was conjectured that such a group necessarily contains a nonabelian free pro-p group. A partial confirmation of this conjecture – the existence of nonabelian abstract free subgroups – was established in [37]. Now we are ready to claim the conjecture as a theorem.

Theorem. *A pro-p group presented by a small set of relators contains a nonabelian free pro-p subgroup.*

The results of §1,2 suggest the following conjecture that I do not really believe is true but nevertheless is interesting.

Conjecture. A finitely generated pro-p group either contains a nonabelian free pro-p subgroup or is infinitesimally PI.

References

[1] S. Amitsur, Nil semigroups of rings with a polynomial identity, *Nagoya Math. J.* **27**(1966), 103–111.

[2] Yu.A. Bakhturin, *Identities in Lie algebras* (Nauka, Moscow, 1985).

[3] A.D. Chanyshev, personal communication.

[4] J.D. Dixon, M.P.F. du Satoy, A. Mann and D. Segal, *Analytic pro-p Groups* (London Math. Soc. Lecture Notes **157**, Cambridge Univ. Press, 1991).

[5] E.S. Golod, On nil-algebras and residually finite p-groups, *Izvestia Akad. Nauk SSSR, Ser. Mat.* **28**(1964), 273–276.

[6] E.S. Golod and I.R. Shafarevich, On the class field tower. *Izvestia Akad. Nauk SSSR, Ser. Mat.* **28**(1964), 261–272.

[7] R.I. Grigorchuk, On the Burnside problem for periodic groups. *Funct. Anal. Appl.* **14**(1980), 53–54.

[8] C. Gruenberg, *Some Cohomological Topics in Group Theory* (Queen Mary College Math. Notes, London, 1967).

[9] N. Gupta and S. Sidki, On the Burnside problem for periodic groups, *Math. Z.* **182**(1983), 385–386.

[10] K. Haberland, *Galois Cohomology of Algebraic Number Fields* (VEB Deutscher Verlag der Wissenschaften, Berlin, 1978).

[11] M. Hall, *The Theory of Groups* (Macmillan, New York, 1959).

[12] P. Hall and G. Higman, On the p-length of p-soluble groups and reduction theorems for Burnside's problem, *Proc. London Math. Soc.* **6**(1956), 1–42.

[13] I. Herstein, *Noncommutative rings* (Carus Mathematical Monographs No. 15, MAA, 1968).

[14] G. Higman, Lie ring methods in the theory of finite nilpotent groups, in *Proc. Intern. Congr. Mat. Edinburgh, 1958* (Cambridge Univ. Press, 1960), 307–312,

[15] G. Higman, On finite groups of exponent five, *Proc. Camb. Phil. Soc.*, **52**(1956), 381–390.

[16] N. Jacobson, *PI-algebras, an Introduction* (Lecture Notes in Math. **441**, Springer-Verlag, Berlin-Heidelberg-New York, 1975).

[17] A. Kemer, The standard identity in characteristic p. A conjecture of I.B. Volichenko, *Israel J. Math.* **81**(1993), 343–356.

[18] H. Koch, Zum Satz von Golod-Šafarevič, *Math. Nachr.* **42**(1969), 321–333.

[19] A.I. Kostrikin, The Burnside problem, *Izvestia Akad. Nauk SSSR, Ser. Mat.* **23**(1959), 3–34.

[20] A.I. Kostrikin, Sandwiches in Lie algebras, *Mat. Sb.* **110**(1979), 3–12.

[21] A.I. Kostrikin, *Around Burnside*, (Ergebnisse der Mathematik und ihrer Grenzgebiete, Springer–Verlag, Berlin, 1990).

[22] A.I. Kostrikin and E.I. Zelmanov, A theorem on sandwich algebras, *Trudy Mat. Inst. Steklov* **183**(1988), 142–149.

[23] M. Lazard, Groupes analytiques p-adiques, *Inst. Hautes Études Sci. Publ. Math.* **26**(1965), 389–603.

[24] A. Lubotzky, Group presentations, p-adic analytic groups and lattices in $SL_2(C)$, *Ann. Math.* **118**(1983), 115–130.

[25] A. Lubotzky and A. Shalev, On some Λ-analytic pro-p groups, *Israel J. Math.* **85**(1994), 307–337.

[26] W. Magnus, Über Gruppen und zugeordnete Liesche Ringe, *J. Reine Angew. Math.* **182**(1940), 142–159.

[27] Yu.A. Ol'shansky, *Geometry of defining relations in groups* (Nauka, Moscow, 1989).

[28] P. Roquette, On class field towers, in *Algebraic Number Theory* (Academic

Press, London, 1967), 231–249.

[29] V.I. Sushchansky, Periodic p-groups of permutations and the general Burnside problem, *Dokl. Akad. Nauk SSSR* **247**(1979), 447–361.

[30] V.A. Ufnarovsky, Combinatorial and asymptotic methods in algebra, in *Fundamental Problems in Mathematics* 57 (Viniti, Moscow, 1990), 5–177.

[31] M.R. Vaughan-Lee, *The Restricted Burnside problem*, 2nd edition (Oxford University Press, Oxford, 1993).

[32] M.R. Vaughan-Lee, The Nilpotency Class of Finite Groups of Exponent p, *Trans. Amer. Math. Soc.*, to appear.

[33] M.R. Vaughan-Lee and E.Zelmanov, Upper bounds in the restricted Burnside problem, *J. Algebra* **162**(1993), 107–145.

[34] E.B. Vinberg, On the theorem concerning the infinite dimension of an associative algebra, *Izv. Akad. Nauk SSSR* Ser. Mat. **29**(1965), 209–214.

[35] G.E. Wall, Lie methods in group theory, in *Topics in Algebra* (Lecture Notes in Math. **697**, Springer-Verlag, 1978), 137–173.

[36] J.S. Wilson, Finite presentations of pro-p groups and discrete groups, *Invent. Math.* **105**(1991), 177–183.

[37] J.S. Wilson and E. Zelmanov, Identities for Lie algebras of pro-p groups, *J. Pure Appl. Algebra* **81**(1992), 103–109.

[38] E. Zelmanov, Absolute zero divizors in Jordan pairs and Lie algebras, *Mat. Sb.* **112**(1980), 611–629.

[39] E. Zelmanov, The solution of the restricted Burnside problem for groups of odd exponent, *Math. USSR Izv.* **36**(1991), 41–60.

[40] E. Zelmanov, The solution of the restricted Burnside problem for 2-groups, *Mat. Sb.* **182**(1991), 568–592.

[41] E. Zelmanov, *Nil Rings and Periodic Groups* (The Korean Math. Soc. Lecture Notes in Math. Seoul, 1992).

[42] A. Zubkov, Non-abelian free pro-p groups can not be represented by 2-by-2 matrices, *Siberian Math. J.* **28**(1987), 742–747.

SOME NEW RESULTS ON ARITHMETICAL PROBLEMS IN THE THEORY OF FINITE GROUPS

JIPING ZHANG[1]

DMI, Ecole Normale Supérieure, Paris, France;
Mathematical Institute, Beijing University, Beijing, China

Abstract

In this paper we present a survey of the recent progress on arithmetical problems in the theory of finite groups concerning orders of elements, lengths of conjugacy classes, Sylow numbers, etc. We will also pose some open problems and conjectures.

1. Introduction

In recent years there has been considerable interest in studying the influence of arithmetical conditions on finite groups, mainly motivated by the famous conjecture of Huppert on character degrees, which is now referred to as the $\rho-\sigma$ conjecture. The reader is referred to [H2] for the conjecture and progress on it. Here we are concerned with some progress made recently on relevant interesting topics, such as orders of elements, lengths of conjugacy classes and Sylow numbers. As already shown by the work on Huppert's $\rho-\sigma$ conjecture, and as we will see again in this survey arithmetical conditions do, indeed, severely restrict the structure of finite groups.

We need first to fix some notations. For finite groups G we denote by G_p a Sylow p-subgroup of G, and by G_π any Hall π-subgroup of G if there exists such a subgroup and the trivial subgroup otherwise, where $\pi \subseteq \pi(G) = \{p : p \mid |G|\}$. For positive integers n with the prime-factor-decomposition $n = \prod_{i=1}^{k} p_i^{\alpha_i}$ (p_i's are different primes and $\alpha_i > 0$) we define

$$\omega(n) = \Sigma_{i=1}^{k}\alpha_i,\ \sigma(n) = k,\ \tau(n) = \max\{\alpha_i : i = 1, 2, ...k\}.$$

For finite groups G we put

$$O(G) = \{o(x) : x \in G\},\ C(G) = \{|G : C_G(x)| : x \in G \setminus Z(G)\}$$
$$S(G) = \{|G : N_G(G_p)| : p \mid |G|\},\ \Pi(G) = \{|G : N_G(G_\pi)| : \pi \subseteq \pi(G)\}$$

---

[1]Supported by the Commission of European Communities, via a CEC Research Fellowship.

and set

$$\sigma_o(G) = \max\{\sigma(n) : n \in O(G)\}, \ \rho_o(G) = \sigma(\prod_{n \in O(G)} n)$$

with analogous definitions for $\sigma_c(G), \sigma_s(G), \sigma_\pi(G), \rho_c(G), \rho_s(G)$ and $\rho_\pi(G)$.

For a finite group G we define the Sylow-graph $\Gamma_s(G)$ of G as follows: the vertices are prime divisors of the order of G and two vertices p and q are joined by an edge if there is a prime r such that $pq || G : N_G(G_r)|$, and define similarly the element-graph $\Gamma_o(G)$ and the class-graph $\Gamma_c(G)$.

2. Element orders

The study of the arithmetical structure of the set of element orders dates at least from Higman [Hi]. He proved that if G is a finite solvable group with $\sigma_o(G) = 1$ then G is a p-group or Frobenius group; he also determined largely the nonsolvable case, which was completed by Suzuki [Sz] and Brandl [B]. Shi [Sh] proved that if two finite simple groups G and H have the same order and the same set of element orders then they are in fact isomorphic. The element-graphs have already been investigated by J.Williams [Wi] and Iiyori and Yamaki [IY] and a combination of their results produces the following theorem.

Theorem 1. (Willams, Iiyori and Yamaki) *For any finite group G the number of components of the element-graph of G is at most 6.*

Is there a bound from above on $\rho_o(G)$ in terms of $\sigma_o(G)$? The following theorem answers this question.

Theorem 2. (Zhang [Z1]) *Let G be any finite group then we have*

$$\rho_o(G) \leq 16\sigma_o(G)^3(\sigma_o(G) + 3)e^{\sigma_o(G)}$$

The proof of this theorem depends on the classification of finite simple groups. If we assume in addition that G is solvable then we have a much better bound.

Theorem 3. (Zhang [Z1]) *If G is a finite solvable group then*

$$\rho_o(G) \leq \sigma_o(G)\ (\sigma_o(G) + 3)/2.$$

In [Z1] we constructed a finite solvable group G such that $\sigma_o(G) = 2$ and $\rho_o(G) = 5 = 2(2+3)/2$, so the bound is sharp in this case, but it is quite plausible that a linear bound may also work. So we make the following

Conjecture 1. (J. Zhang) There exists a positive number c such that $\rho_o(G) \leq c\sigma_o(G)$ for any finite solvable group G.

One can not expect to bound the nilpotent length of a finite solvable group G in terms of $\sigma_o(G)$, since there exists for any integer m a finite group G of order $p^a q^b$ such that the nilpotent length $F_\ell(G) \geq m$. But in fact $F_\ell(G)$ is bounded from above by a function of $\sigma_o(G)$ and $\tau_o(G)$.

Corollary 4. (Zhang [Z1]) *If G is a finite solvable group then*

$$F_\ell(G) \leq 4(\tau_o(G)+1)^{\sigma_o(G)(\sigma_o(G)+3)/2} - 1.$$

For further results on the set $O(G)$ the reader is referred to [Sh], and also the final remark in [CH].

3. Lengths of conjugacy classes

Conjugacy classes are fundamental in the theory of finite groups as well as in modular representation theory. The following two properties of conjugacy classes shared by the symmetric group Σ_3 are typical arithmetical ones.

(1) $C(G) = S \cup T$ with $S, T \neq C(G)$ and $(s,t) = 1$ for any $s \in S, t \in T$, or in other word, the class-graph $\Gamma_c(G)$ is not connected.

(2) Any two distinct conjugacy classes of G have different lengths.

Finite groups with property (1) have been determined by Bertram, Herzog and Mann [BHM]. Ito also proved some interesting results on such groups [I1].

Theorem 5. ([BHM]) *A finite group G has a non-connected class-graph $\Gamma_c(G)$ if and only if G is a quasi-Frobenius group with abelian kernel and complement.*

Property (2) is very involved and intensive studies have been made by Markel [Ma], Hayashi [Hy], Ward [War], Zhang [Z2], etc. The following result is proved by Zhang [Z2] and confirms a long standing conjecture.

Theorem 6. (Zhang [Z2]) *If G is a finite solvable group such that any two distinct non-central conjugacy classes of G have different lengths then G is either abelian or isomorphic to Σ_3.*

Problem 2. Determine the finite nonsolvable groups in which any two distinct conjugacy classes have different lengths.

In [CH], Chillag and Herzog characterize the finite group with squarefree class-lengths, using the classification of finite simple groups.

Theorem 7. (Chillag and Herzog [CH]) *If every conjugacy class of a finite group G has a square-free length then G is supersolvable and both $|G/F(G)|$ and $|F(G)'|$ are squarefree numbers. In particular, $G' \leq F(G), G/F(G)$ is cyclic, $F(G)$ has nilpotent class at most 2 and G has derived length at most 3.*

In fact, the classification of finite simple groups is not necessary here, since we see that for any Sylow subgroup P of G, the Frattini subgroup $\Phi(P)$ of P is contained in the center $Z(G)$ of G, and also it is easy to verify that $F(\overline{G}) =< x > N$, where $\overline{G} = G/\Phi(G)$, $N = Z(\overline{G}), < x >$ is of squarefree order.

The following conjecture and its study have some remarkable similarities with the character degree problems.

Conjecture 3. (Huppert) For any finite group G, $\rho_c(G) \leq 2\sigma_c(G)$.

Recently, Ferguson has proved the following

Theorem 8. (Ferguson [F]) *If G is a finite solvable group then $\rho_c(G) \leq 4\sigma_c(G) + 6$, and if in addition $(|G|, 6) = 1$ then $\rho_c(G) \leq 3.25\sigma_c(G)$.*

Casolo considered the conjecture for arbitrary finite groups and was able to prove the following remarkable result.

Theorem 9. (Casolo) *If G is a finite group which is p-nilpotent with abelian Sylow p-subgroup for at most one prime p dividing the order of G then $\rho_c(G) \leq 2\sigma_c(G)$.*

Thus the conjecture is almost true. It is easy to construct for any natural number n a finite group G of squarefree order such that $\rho_c(G) = 2\sigma_c(G) = 2n$, so the bound is best-possible.

Considering Casolo's result, we pose the following

Problem 4. Prove Conjecture 3 for finite supersolvable groups.

As for finite simple groups G, the set $C(G)$ probably characterizes them. Ito studied such problems for groups G with $|C(G)| \leq 5$.

Conjecture 5. (J. Thompson) Suppose G is a finite simple group and M a finite group with trivial center. If $C(G) = C(M)$ then $G \cong M$.

More generally, for any finite group G let $N(G)$ (resp. $\mathcal{N}(\mathcal{G})$) be the set of the numbers of elements (resp. subgroups) of the same order, so each number (> 1) in $N(G)$ is a sum of numbers in $C(G)$. Two finite groups G and H are said to be Gassmann (resp. semi-Gassmann) equivalent if $N(G) = N(H)$ (resp. $\mathcal{N}(\mathcal{G}) = \mathcal{N}(\mathcal{H})$). It is proved in [P] that G and H are

Gassmann equivalent if and only if for any regular embeddings $g : G \longmapsto \Sigma_n$ and $h : H \longmapsto \Sigma_n$ we have $Ind_{G^g}^{\Sigma_n}(F) = Ind_{H^h}^{\Sigma_n}(F)$, where $n = |G| = |H|$ and F is an algebraically closed field of characteristic zero and is considered at the same time as the trivial module of proper groups. It is trivial that (semi-) Gassmann equivalence preserves nilpotence. However, the equivalences do not preserve the isomorphism. For further information concerning Gassmann equivalence we refer to [Gu] and [P].

Problem 6. Investigate whether or not (super)solvability is invariant under (semi-)Gassmann equivalence.

X. Zhou has obtained some related results on the set of the lengths of conjugacy classes of cyclic subgroups.

4. Sylow numbers

Undoubtedly the most important basic result in the theory of finite groups is the theorem of Sylow, so it is very interesting to study the arithmetical problems related to Sylow subgroups. For convenience we call a natural number n a Sylow number for a finite group G if there is a prime p such that $n = |G : N_G(G_p)|$. We know that $n \equiv 1(\mod p)$. P. Hall studied in [Ha] the structure of Sylow numbers and proved that for any finite solvable group G the number of Sylow p-subgroups of G has the form $p_1^{\alpha_1} p_2^{\alpha_2} \cdots p_m^{\alpha_m}$, where α_i's are positive integers such that $p_i^{\alpha_i} \equiv 1 \pmod p$, $i = 1, 2, \cdots\cdots, m$. We will see in the following that arithmetical restrictions on Sylow numbers lead also to strong restrictions on the structure of finite groups, though probably not as strong as those on class-lengths do.

Theorem 10. (Zhang [Z3]) *Let G be a finite group. If all Sylow numbers of G are squarefree then either G is supersolvable or $G = HK < x >$, where H and K are normal subgroups of G with $H \cap K = Z(H)$ and $H \cong PSL(2,p)$ or $SL(2,p)$ for some prime $p = 8k + 5$, and $K < x >$ is supersolvable.*

Corollary 11. *Finite solvable groups with squarefree Sylow numbers are supersolvable.*

Theorem 12. (Zhang [Z3]) *If every Sylow number of a finite group G is a prime power then G is solvable. Moreover, there exists a normal subgroup H of G, such that H is a direct product of Hall subgroups of G of order $p^a q^b$ for some primes p and q with $p \neq q$ and G/H is of nilpotent length at most 2.*

From this theorem we know that $\rho_s(G)$ can not be bounded in terms of $\sigma_s(G)$.

The following theorem is an interesting new criterion for p-nilpotency.

Theorem 13. (Zhang [Z3]) *A finite group G is p-nilpotent if and only if each Sylow number of G is prime to p.*

This theorem and its proof prompt us to pose the following

Problem 7. For any finite non-abelian simple group G is it true that for any prime divisor p of the order of G there exists a prime q such that G has no Hall $\{p, q\}$-subgroups ?

Remark. Arad and Ward proved in [AW] the conjecture of Hall, which claims that a finite group is solvable if and only if there exists a Hall $\{p, q\}$-subgroup in the group for any primes p and q .

For any given graph F it is easy to construct a finite group G such that the Sylow-graph $\Gamma_s(G)$ of G is isomorphic to F. But we have the following result.

Theorem 14. (Zhang [Z3]) *If $\Gamma_s(G)$ has more than one component then G is not simple.*

Theorem 15. (Zhang [Z3]) *For any finite solvable group G the nilpotent length $F_\ell(G) \leq 4 + 2\log_3(\tau_s(G)/2)$.*

A dual result of Theorem 11 is the following.

Theorem 16. (Zhang [Z3]) *If G is a finite solvable group such that for any prime divisor p of the order of G the integer $|G : N_G(G_{p'})|$ is squarefree (so it is 1 or p) then G is supersolvable.*

We conclude by posing

Problem 8. Prove that $\rho_\pi(G) \leq 2\sigma_\pi(G)$ for any finite group G.

References

[AW] Z. Arad and M.B. Ward, New criterion for the solvability of finite groups, *J. Algebra* **77**(1982), 234–246.

[BHM] E. Bertram, M. Herzog and A. Mann, On a graph related to conjugacy classes of groups, *Bull. London Math. Soc.* **22**(1990), 569–575.

[BS] M. Bianchi, D. Chillag, A. Mauri, M. Herzog and S.M. Scoppola, Applications of a graph related to conjugacy classes in finite groups, *Arch. Math.* **58**(1992), 126–132.

[B] R. Brandl, Finite groups all of whose elements are of prime power order, *Boll. Un. Mat. Ital. A* **18**(1981), 491–493.

[BM] M. Broué and G. Malle, Théorèmes de Sylow génériques pour les groupes réductifs sur les corp finis, *Math. Ann.* **292**(1992), 241–262.

[CH] D. Chillag and M. Herzog, On the lengths of conjugacy classes of finite groups, *J. Algebra* **131**(1990), 110–125.

[CNP] J.H. Conway, R.T. Curtis, S.P. Norton, R.A. Parker and R.A. Wilson, *Atlas of Finite groups* (Clarendon Press, Oxford, 1985).

[F] P.A. Ferguson, Lengths of conjugacy classes of finite solvable groups II, *J. Algebra* **154**(1993), 223–227.

[G] D. Gluck, Primes dividing character degrees and character orbit sizes, *Proc. Sympos. Pure Math.* **47**(2)(1987), 45–46.

[Gu] R.M. Guralnick, Zeroes of permutation characters with applications to prime splitting and Brauer groups, *J. Algebra* **131**(1990), 294–302.

[HH] P. Hall and G. Higman, The *p*-length of a solvable group, and reduction theorems for Burnside's problem, *Proc. London Math. Soc.* **6**(3)(1956), 1–42.

[Ha] P. Hall, A note on solvable groups, *J. London Math. Soc.* **3**(1928), 98–105.

[H] T. Hawkes, On the Fitting length of a solvable linear group, *Pacific J. Math.* **44**(1973), 537–540.

[Hy] M. Hayashi, On a generalization of F.M. Markel's theorem, *Hokkaido Math. J.* **4**(1975), 278–280.

[Hi] G. Higman, Finite groups in which every element has prime power order, *J. London Math. Soc.* **32**(1957), 335–342.

[H1] B. Huppert, Inequalities for character degrees of solvable groups, *Arch. Math. (Basel)* **46**(1986), 387–392.

[H2] B. Huppert, *Research in Representation Theory at Mainz* (Progress in Math. **95**, Birkhäuser Verlag Basel, 1991).

[HM] B. Huppert and O. Manz, Orbit sizes of *p*-groups, *Arch. Math.* **54**(1990), 105–110.

[IY] N. Iiyori and H. Yamaki, Prime graph components of the simple groups of Lie type over the field of even characteristic, *J. Algebra* **155**(1993), 335–343.

[IP] I.M. Isaacs and D. Passman, A characterization of groups in terms of the degrees of their characters II, *Pacific J. Math.* **24**(1968), 467–510.

[I1] N. Ito, Some studies of group characters, *Nagoya Math. J.* **2**(1951), 17–28.

[M] O. Manz, Arithmetical conditions on character degrees and group structure, *Proc. Sympos. Pure Math.* **47**(2)(1987), 65–69.

[Ma] F.M. Markel, Groups with many conjugate elements, *J. Algebra* **26**(1973), 69–74.

[P] R. Perlis, On the equation $\zeta_K(s) = \zeta_{K'}(s)$, *J. Number Theory* **9**(1977), 342–360.

[Sh] W.J. Shi, Characterization of finite simple groups and related topics, *Adv. in Math. (Beijing)* **20**(1991), 135–141.

[Sz] M. Suzuki, Finite groups with nilpotent centralizers, *Trans. Amer. Math. Soc.* **99**(1961), 425–470.

[Wa] M. B. Ward, Finite groups in which no two distinct conjugacy classes have the same order, *Arch. Math.* **54**(1990), 111–116.

[W] J.S. Williams, Prime graph components of finite groups, *J. Algebra* **69**(1981), 487–513.

[Z1] J.P. Zhang, Arithmetical conditions on element orders and group structure, *Proc. Amer. Math. Soc.*, to appear.

[Z2] J.P. Zhang, Finite groups with many conjugate elements, *J. Algebra*, to appear.

[Z3] J.P. Zhang, Studies on Sylow numbers, to appear.

[Z4] J. P. Zhang, A note on the length of conjugacy classes, to appear.

GROUPS THAT ADMIT PARTIAL POWER AUTOMORPHISMS

JAY ZIMMERMAN

Department of Mathematics, Towson State University, Towson, MD 21204, U.S.A.

The influence of the existence of a group automorphism of a certain kind is the subject of much research. This paper will attempt to summarize one small corner of this research. We survey automorphisms that map most of the elements of a finite group to powers of themselves and the groups that support such automorphisms. The types of groups which have such automorphisms is limited in some cases, and in such cases, the proportion of elements taken to their powers can have only certain values.

The following notation will be used in this paper.

- $Z(G)$: The centre of the group G.
- G': The commutator subgroup of G.
- G^p: The group generated by all of the pth powers of elements in G.
- $(g)\phi$: The image of the group element g under the automorphism ϕ.
- $|G : A|$: The index of the subgroup A in the group G.
- $|T|$: The number of elements in the set T.
- $T_n(\phi) : \{g \in G : (g)\phi = g^n\}$ for an automorphism ϕ of G.
- $\mathcal{L}_p$: The set of all finite groups with order divisible by a prime p, but by no smaller prime.

It is well-known that if the map $\phi : x \to x^2$ for all $x \in G$ is an automorphism of the group G, then G is abelian and has odd order if it is finite. The existence of automorphisms of a group G of the form $\phi : x \to x^{-1}$ or $\phi : x \to x^3$ where G is finite also force G to be abelian. These facts were first shown by G.A. Miller [16] in 1927. Miller also showed that there are infinitely many finite non-abelian groups with automorphisms of the form $\phi : x \to x^n$ for any integer n, except $n = -1, 0, 1, 2, 3$. Miller wrote over 10 articles on these topics and a close examination of his collected works is highly recommended [17].

Automorphisms that keep every subgroup of G invariant are called *power automorphisms.* It is easy to see that every power automorphism has the form $\phi : x \to x^{n(x)}$ where $n(x)$ is an integer depending on $x \in G$. If $n(x)$ is a constant, then ϕ is called a ***universal power automorphism.*** We will call automorphisms which keep $< g >$ invariant for a certain fraction r of the elements g in G ***r-partial power automorphisms.***

Groups G which have a homomorphism $\sigma : G \to G$ given by $\sigma : x \to x^n$ for some positive integer n are called *n-abelian groups* (see [1], [22]). Hence any group which has a universal power automorphism for some positive integer n is n-abelian. J.L. Alperin [1] showed that if a group G has an automorphism $\phi : x \to x^n$ for some $n > 1$, then G is a quotient group of a subgroup of a direct product of an abelian group and a group of exponent dividing $(n-1)$ where n is an integer greater than one. This theorem and the result of Miller listed above show that there is almost no hope of classifying those groups which support a universal power automorphism, $\phi : x \to x^n$, unless n is either -1, 2, or 3, in which case the group is abelian.

However, it is possible to deduce properties of power automorphisms. It has been shown (Cooper [2]) that every power automorphism is central (i.e. acts as the identity on the central quotient, $G/Z(G)$). The following partial converse is also true (Pettet [20]). Let G be a group with periodic centre $Z(G)$, then every central automorphism of G which commutes with all central automorphisms induces a power automorphism on a subgroup of G.

In addition, more may be said if we restrict the group theoretic class of G. For example, if G is a locally finite group whose Sylow subgroups are regular, then every power automorphism is universal on every finite group (see [2]). If G is a p-group, it is well-known that the group of all power automorphisms of a finite non-abelian p-group is also a p-group. This result has been significantly generalized by both Cooper [2] and Meixner [15].

It is worth mentioning that an examination of power endomorphisms gives useful information on power automorphisms. (See [9] and [19]). However, we are more interested in considering groups which support an r - partial power automorphism. Suppose that a group G has an automorphism ϕ that sends a certain proportion r of its elements to their n-th powers. Is ϕ a universal power automorphism? The following theorem was proven by D. MacHale in 1985.

Theorem 1. (MacHale [14]) *If G belongs to $\mathcal{L}_p$ and $|T_n(\phi)| > |G|/p$ for some automorphism ϕ of G where $gcd(|G|, n-1) = 1$, then $T_n(\phi) = G$. Moreover, by Alperin's result [1], G is abelian.*

MacHale also gave several examples to show that this theorem is the best possible. A generalisation of this result is given by Laffey and Newell [9], where $T_n(\phi)$ is replaced by a subset S of it and the condition $gcd(|t|, n(n-1)) = 1$ for all $t \in S$ replaces $gcd(|G|, n-1) = 1$. In addition Laffey and Newell [9] also proved the following theorems. Let p be a prime.

Theorem 2. *Assume that G is not a p-group and there is an automorphism ϕ satisfying $|T_p(\phi)| > k|G|$ where*

(1) if $gcd(|G|, p(p-1)) = 1, k = 1/q$ where q is the least prime divisor of $|G|$;

(2) *if* $gcd(|G|, p(p-1)) \neq 1$, $k = (p-1)/p + 1/pq_0^{m_0}$ *where* $q_0^{m_0} = min\{q^m | q \neq p$ *is a prime divisor of* $|G|$ *and* m *is the order of* q *mod* $p\}$.

Then $G = A \times B$ *where* B *is a p-group and* A *is a group such that*

$$|\{u \in A/Z(A) | u^{p-1} = 1\}| > k|A/Z(A)|.$$

The groups A *and* B *are* ϕ*-invariant. Furthermore, if* $p = 3$*, then* A *is abelian.*

Theorem 3. *Assume that* ϕ *is an automorphism of* G *and that* G *is not a p-group. Suppose that* $|T_{p+1}(\phi)| > k|G|$ *where*

(1) *if* p *does not divide* $|G|$, $k = 1/q$ *where* q *is the least prime divisor of* $|G|$;

(2) *if* p *divides* $|G|$, $k = (p-1)/p + 1/pq_0^{m_0}$ *where* $q_0^{m_0} = min\{q^m | q \neq p$ *is a prime divisor of* $|G|$ *and* m *is the order of* q *mod* $p\}$.

Then $G = A \times B$ *where* A *is a p-group and* B *is an abelian* p'*-group. Also* ϕ *maps* x *to* x^{p+1} *for* $x \in B$.

Theorem 4. *Let* $n = 3$, $n = 4$, *or* $n = 5$ *and* $T_5(\phi)$ *has no elements of order* 4. *Suppose that* $|T_n(\phi)| > (7/9)|G|$. *Then* $T_n(\phi) = G$.

We are also interested in what types of groups support a partial power automorphism ϕ with $|T_n(\phi)| < |G|/p$. It seems to be a hopeless task to classify such groups unless n = -1, 2 or 3 for reasons stated previously. However, much is known about groups with r - partial power automorphisms which take the specified fraction r of the group elements to their n^{th} power where n = -1, 2, or 3.

Let us examine groups G with an automorphism ϕ that inverts a proportion r of the elements of G. Miller [17] wrote papers on this for $r = 3/4, 2/3$, and $1/2$. Suppose $G \in \mathcal{L}_p$ where p is an odd prime. By Theorem 1, if $r > 1/p$, then ϕ is a universal power automorphism. It follows that G is abelian. Liebeck and MacHale [12] classify those groups in $\mathcal{L}_p$ (p odd) where $r = 1/p$. Any such group G must have one of the following three forms:

(1) $G/Z(G)$ is elementary abelian of even rank, the commutator subgroup, G' is a direct factor of $Z(G)$ and $|G'| = p$,

(2) $G/Z(G)$ is non-abelian of order p^3,

(3) G is the semidirect product of an abelian group by the cyclic group of order p.

Groups of type (1) must also satisfy certain commutator relations in order to support an automorphism inverting $1/p$ of the elements.

If G is a group with even order, then ϕ is a universal power automorphism if $r > 3/4$. Liebeck and MacHale [11] also classified all groups where $r > 1/2$. Such groups, G must have one of the following three forms:

(1) G has an abelian subgroup A of index 2 and if $x \notin A$ then $r = (q+1)/2q$ where $q = |A : C_A(x)|$,

(2) $G/Z(G)$ is an elementary abelian 2-group of rank $2k$, G' is a subgroup of order 2 in $Z(G)$, and $r = (2^k + 1)/2^{k+1}$,

(3) G' and $G/Z(G)$ are both elementary abelian 2-groups with rank 2 and 4 respectively, and $r = 9/16$.

Groups of types (2) and (3) must also satisfy some commutator relations in order to support an automorphism inverting at least 1/2 of the elements of G. In addition, Fitzpatrick [4] showed that if G is not a 2-group and $r = 1/2$, then the central quotient group $G/Z(G)$ is isomorphic to either the alternating group on 4 letters, Alt(4) or the direct product of the symmetric group on 3 letters and two copies of the cyclic group with two elements, $\mathrm{Sym}(3) \times C_2 \times C_2$. Finally, Potter [21] showed that if $r > 4/15$, then G must be a solvable group and if G is a non-solvable group with $r > 1/4$, then G is isomorphic to the direct product of an abelian group and the alternating group on 5 letters, and in this case $r = 4/15$.

The situation for groups G having an automorphism ϕ squaring a proportion r of the elements of G is less satisfactory. By Theorem 1 if $G \in \mathcal{L}_p$ and $r > 1/p$, then ϕ is a universal power automorphism and hence G is abelian. Liebeck [10] classified all finite non-abelian groups $G \in \mathcal{L}_p$ with $r = 1/p$ as one of the following three types:

(1) G is nilpotent of class 2 with $|G'| = p$ and $G^p \cap G' = 1$ where p is an odd prime, or

(2) G is the semidirect product of an abelian group of odd order by a cyclic group of order p, or

(3) $p = 3$ and $G/Z(G)$ is the non-abelian group of order 27 and exponent 3.

In addition, Zimmerman [23] showed that if $r \geq 5/12$, then G is either abelian, of type (2) as above with $p = 2$, or $G/Z(G)$ is the alternating group on 4 letters and $r = 5/12$.

Next, we consider the case where G has an automorphism cubing a proportion r of the elements of G. By Theorem 1 if $G \in \mathcal{L}_p$ where p is an odd prime and $r > 1/p$, then ϕ is a universal power automorphism and hence G is abelian. MacHale [13] showed that if G is a group of even order and $r > 3/4$, then once again ϕ is a universal power automorphism and so G is abelian. Subsequently, Deaconescu and MacHale [3] gave the following classification.

Theorem. *Let G be an odd order non-abelian group with $G \in \mathcal{L}_p$ where $r = 1/p$ and let $\phi \in Aut(G)$.*

(1) If $|T_1(\phi)| = 1$, then G is nilpotent of class 2, $|G'| = p$, $G^p \cap G' = 1$ and $p \geq 5$.

(2) *If* $|T_1(\phi)| \neq 1$, *then* $T_3(\phi)$ *is an abelian subgroup of index p in G and there exists* $f \in T_1(\phi), f \notin T_3(\phi)$ *such that f has order p. Moreover* $gcd(|T_3(\phi)|, 3) = 1$.

MacHale [13] also showed that G is a non-abelian group with $r = 3/4$ if and only if the central quotient group $G/Z(G)$ has order 4 and the centre has no elements of order 3.

The following theorem of P. Hegarty [5] completes our survey of the case where G has an automorphism cubing a proportion of the elements of G.

Theorem. *If the finite group G admits an automorphism* ϕ *satisfying* $|T_3(\phi)| > (1/2)|G|$, *then G has one of the following structures:*

1. *G is abelian and 3 does not divide* $|G|$.
2. *G has a normal Sylow 3-subgroup S satisfying the following conditions:*
   (a) $S \subseteq K$ *where* $|G : K| = 2$ *and K is abelian*
   (b) $S \cap Z(G) = \{1\}$.

   In particular, if 3 does not divide $|G|$, *then it is sufficient that G have an abelian subgroup of index 2.*
3. *G is nilpotent of class 2 and 3 does not divide* $|G|$. *All Sylow p-subgroups* $(p > 3)$ *are abelian, and the Sylow 2-subgroup* S_2 *of G has one of the following structures:*
   (i) $S_2' \cong C_2$ *and* $S_2/Z(S_2)$ *is elementary abelian, generated by* $Zx_1, \ldots, Zx_k, Za_1, \ldots, Za_k$ *subject to the following commutator relations:* $[a_i, x_i] = z \neq 1, [a_i, x_j] = 1$ *for* $i \neq j$ *and* $[x_i, x_j] = [a_i, a_j] = 1$ *for* $i, j = 1, \ldots, k$.
   (ii) $S_2' \cong C_2 \times C_2$ *and* $S_2/Z(S_2)$ *is elementary abelian, generated by* Zx_1, Zx_2, Za_1, Za_2 *subject to the following commutator relations:* $[a_1, x_1] = z_1, [a_2, x_2] = z_2, [x_1, x_2] = [a_1, a_2] = [a_1, x_2] = [a_2, x_1] = 1$ *where* $S_2' = < z_1, z_2 >$. *Conversely, all groups described above admit an automorphism sending more than 1/2 of the group elements to their cubes.*

Finally, there has been some work on the special case where the partial power automorphism is the identity. Thomas J. Laffey ([7], [8], and [6]) examined how many elements in a group G that does not have exponent n can satisfy the equation $x^n = 1$ (for $n = 3$, 4, or a prime p).

We see that Theorem 1 is the starting point for much work on this subject. An obvious way to generalise Theorem 1 is to consider partial power automorphisms of G where the power depends on the element of G. This is embodied in the following conjecture.

Conjecture 1. Let ϕ be an automorphism of a group G with $G \in \mathcal{L}_p$ and define $T_\phi = \{g \in G : (g)\phi = g^{n(g)}$ for some integer $n(g)\}$. Suppose that $|T_\phi| > |G|/p$ and that for every $g \in T$ we have that $gcd(|G|, n(g) - 1) = 1$, then $G = T_\phi$.

We intend to prove this conjecture false. Further, we will show that the condition $|T|_\phi > |G|/p$ can not be replaced by a condition of the form $|T_\phi| > f(p) \cdot |G|$ where $f(p)$ is any function of p alone. In other words, we intend to exhibit infinitely many counterexamples where ϕ is not a power automorphism and where $(|T_\phi|/|G|)$ gets arbitrarily close to one.

Our first example is as follows.

Example A. Let G be the direct product of a cyclic group of order $p, < a >$, and a cyclic group of order $p^{k+1}, < x >$, for some natural number k. Define an automorphism ϕ of the group G by

$$\phi : \left\{ \begin{array}{l} x \to x^m \\ a \to a^m x^{p^k} \end{array} \right\} \quad \text{where } p \text{ doesn't divide } m.$$

It follows that $|T_\phi|/|G| = 1 - ((p-1)/p^{k+1})$. Thus ϕ is not a power automorphism, but $|T_\phi| > |G|/p$.

PROOF. Let $(x^s a^t)$ be an element of G where $0 \le s < p^{k+1}$ and $0 \le t < p$. The element $(x^s a^t)$ is in T_ϕ if and only if there exists a natural number n such that

$$(x^s a^t)^n = (x^s a^t)\phi = x^{ms+tp^k} \cdot a^{mt}.$$

This is equivalent to the system of congruences

$$ms + tp = sn (\bmod p^{k+1})$$
$$mt = nt (\bmod p).$$

The second congruence is equivalent to $n = m + vp$ for some natural number v. Substituting this in the first congruence we see that

$$ms + tp = ms + svp (\bmod p^{k+1}).$$

It follows that $(x^s a^t)$ is an element of T_ϕ if and only if there exists a natural number v such that $sv = tp^{k-1} (\bmod p^k)$.

Case 1. Suppose p divides s. Then $(x^s a^t)$ is in T_ϕ if and only if $t = 0$.

Case 2. Suppose p does not divide s. We see that $s = p^e q$ where e is a natural number satisfying $0 \le e < k$ and p doesn't divide q. It follows that $(x^s a^t)$ is in T_ϕ if and only if

$$qv = tp^{k-e-1} (\bmod p^{k-e}) \text{ for some natural number } v.$$

This congruence always has a solution since q is relatively prime to p.

Now we count the number of ordered pairs (s,t) where $(x^s a^t)$ is not an element of T_ϕ. This occurs when p^k divides s and $t \neq 0$. There are $(p-1)$ possible values of t and p possible values of s and so there are $p(p-l)$ ordered pairs (s,t) where $(x^s a^t)$ is not in T_ϕ. It follows that

$$1 - (|T_\phi|/|G|) = p(p-1)/(p^{k+2}).$$

□

Similar examples can be given for non-abelian groups G.

Example B. Define the group H_n by

$$H_n =< x, y : x^{p^{n+1}} = y^{p^{n+1}} = 1, x^p = y^p, [x,y] = x^{p^n} > .$$

Let $G_n =< c > \times H_n$ where $< c >$ is cyclic of order p and define the automorphism ϕ of G by

$$\phi : \begin{array}{l} x \to x^{p+1} \\ y \to y^{p+1} \\ c \to cx^{p^n}. \end{array}$$

An argument similar to the one given for Example A will show that

$$(|T_\phi|/|G|) = 1 - (p-1)/(p^k)$$

where $k = 2$ if $n = 1$ and $k = n + 2$ if $n \geq 2$. These examples show that Conjecture 1 is not only false, but cannot be fixed in any obvious way.

Finally, we observe that the automorphism ϕ can be restricted to the subgroup H_n of G_n. The automorphism induced on H_n is a power automorphism, but not a universal power automorphism, in general.

Acknowledgement. The author would like to thank the referee for many helpful suggestions.

References

[1] Alperin, J.L., A classification of n-abelian groups, *Canad. J. Math.* **21**(1969), 1238–1244.

[2] Cooper, C.D.H., Power automorphisms of a group, *Math. Z.* **107**(1968), 335–356.

[3] Deaconescu, M. and MacHale, D., Odd order groups with an automorphism cubing many elements, *J. Austral. Math. Soc. Ser. A* **46**(1989), 281–288.

[4] Fitzpatrick, P., Groups in which an automorphism inverts precisely half the elements, *Proc. Roy. Irish Acad.* **86A**(1)(1986), 81–89.

[5] Hegarty, P., *Proc. Roy. Irish Acad.*, submitted.

[6] Laffey, T.J., The number of solutions of $x^p = 1$ in a finite group, *Math. Proc. Cambridge Philos. Soc.* **80**(1976), 229–231.

[7] Laffey, T.J., The number of solutions of $x^3 = 1$ in a 3-group, *Math. Z.* **149**(1976), 43-45.

[8] Laffey, T.J., The number of solutions of $x^4 = 1$ in finite groups, *Proc. Roy. Irish Acad.* **79A**(4)(1979), 29–36.

[9] Laffey, T.J. and Newell, M.L., Group endomorphisms which are almost power mappings, *Proc. Roy. Irish Acad.* **83A**(2)(1983), 145 - 155.

[10] Liebeck, H., Groups with an automorphism squaring many elements, *J. Austral. Math. Soc.* **16**(1973), 33–42.

[11] Liebeck, H. and MacHale, D., Groups with automorphisms inverting most elements, *Math. Z.* **124**(1972), 51–63.

[12] Liebeck, H. and MacHale, D., Odd order groups with automorphisms inverting many elements, *J. London Math. Soc. (2)* **6**(1973), 215–223.

[13] MacHale, D., Groups with an automorphism cubing many elements, *J. Austral. Math. Soc. Ser A* **20**(1975), 253–256.

[14] MacHale, D., Universal power–automorphisms in finite groups, *J. London Math. Soc. (2)* **11**(1975), 366–368.

[15] Meixner, T., Power automorphisms of finite p-groups, *Israel J. Math.* **38**(1981), 345–360.

[16] Miller, G.A., Possible α-automorphisms of non-abelian groups, *Proc. Nat. Acad. Sci.* **15**(1929), 89–91.

[17] Miller, G.A., *The Collected Works of George Abram Miller*, Vol. I–V (University of Illinois, Urbana, Illinois, 1959).

[18] Newell, M.L., Normal and power endomorphisms of a group, *Math Z.* **151**(1976), 139–142.

[19] Newell, M.L. and Dark, R.S., On certain groups with a fourth power endomorphism, *Proc. Roy. Irish Acad.* **80A**(2)(1980), 167–172.

[20] Pettet, M.R., Central automorphisms of periodic groups, *Arch. Math.* **51**(1988), 20–33.

[21] Potter, W.M., Non-solvable groups with an automorphism inverting many elements, *Arch. Math.* **50**(1988), 292–299.

[22] Weichsel, P., On p-abelian groups, *Proc. Amer. Math. Soc.* **18**(1967), 736–737.

[23] Zimmerman, J., Groups with automorphisms squaring most elements, *Arch. Math.* **54**(1990), 241–246.

Problems

Problem 1. Let F be the free group of finite rank $n \geq 2$. Suppose an endomorphism ϕ of F takes every primitive element to a primitive one. Is it true that ϕ is actually an automorphism of F?

For $n = 2$, the answer was proved to be "yes" by S. Ivanov.

Vladimir Shpilrain

Problem 2. (Closure operators and finite solvable groups) Let G be a finite group, let $L(G)$ be its lattice of subgroups and let $\sigma : L \to L(G)$ be a closure operator on $L(G)$. This means that:

(i) if $H, K \leq G$, then $\sigma(H) \leq \sigma(K)$;
(ii) if $H \leq G$, then $H \leq \sigma(H)$; and
(iii) if $H \leq G$, then $\dot{\sigma}(H) = \sigma(\sigma(H))$.

Assume now that σ is a *non–trivial closure operator*, i.e. σ can be defined for every $L(G)$, where G is a finite group and there exists a finite group X such that $\sigma : L(X) \to (X)$ is not trivial.

The following problem arises: if $\sigma : L(G) \to L(G)$ is as above and if every subgroup of G is closed, i.e. $\sigma(H) = H, \ \forall H \leq G$, is it true that G is solvable?

Arguments for the "yes" answer:

1. If $\sigma(H) := H^G = \langle H^g | g \in G \rangle$ and if $\sigma(H) = H, \ \forall H \leq G$, then G is Dedekind, so nilpotent.
2. If $\sigma(H) := C_G(C_G(H))$ and if $\sigma(H) = H, \ \forall H \leq G$, then G is supersolvable (Gaschütz).
3. If $\sigma(H) := \bigcap\{M | M$ is a maximal subgroup of G and $H \leq M\}$ and if $\sigma(H) = H, \ \forall H \leq G$, then G is supersolvable (Menegazzo, di Martino *et. al.*)

Marian Deaconescu

Problem 3. Suppose that $G = HK$ is a finite group, with H and K subgroups, and let X be a subnormal subgroup of H and K. Then X is subnormal in G, by a theorem of Wielandt. However, if the subnormal defects of X in H and K are m and n, it does not appear to be known whether the subnormal defect of X in G is bounded by some function of m and n.

Stewart Stonehewer

Problem 4. (Is any solvable group a FE–group?) A group $\{G\cdot\}$ is called a (finite embedding) FE–group if for any finite subset X of G there exists a finite group $\{H*\}$, such that $X \subseteq H$ and $x \cdot y = x * y$ for any $x, y \in X$ with $x \cdot y \in X$. It is known that residually finite groups are FE. P. Hall has constructed an example of three generator solvable group which is not residually finite. Lately, it appears that this group is FE. Can this result be extended to all solvable groups?

Andrej Strojnowski

Problem 5. (Are *f.g.* power transitive groups Hopfian?) Given a group G, we say $x, y \in G$ are *power related* if there exists a $z \in G$ such that $x = z^k$ and $y = z^m$ for $k, m \in \mathbf{Z}$. The group G is *power transitive* (p.t.) if the relation of being power related is transitive on the non-identity elements of G.

Some examples of p.t. groups are:

(i) locally cyclic groups;

(ii) free groups;

(iii) free Abelian groups;

(iv) direct products of torsion–free p.t. groups;

(v) free products of p.t. groups;

(vi) dihedral groups;

(vii) elementary Abelian p–groups.

Although it is clear that one can get an example of an infinitely generated p.t. group which is NOT Hopfian, we have not been able to find a finitely generated non–Hopfian group which is p.t. Note that it is not hard to show that the Baumslag–Solitar groups are not p.t. Thus we pose the above question.

A.M. Gaglione and M.E. Hoffman

Problem 6. (Is every f.g. fully residually free group f.p.?) A group G is *fully residually free* (in the terminology of B. Baumslag) provided that to every finite set $S \subseteq G - \{1\}$ there is a free group F_S and a homomorphism $h_S : G \to F_S$ such that $h_S(g) \neq 1$ for all $g \in S$. The authors have heard that every finitely generated fully residually free is finitely presented, but they have no reference nor proof.

A.M. Gaglione and D. Spellman

Problem 7. The definition of fully residually free groups is given in Problem 6. Is every rank three 2–free fully residually free group free?

A.M. Gaglione and D. Spellman

Problem 8. Consider a double factorized group $G = AB$, where A and B are abelian and

(1) A is not normal in G,

(2) if $1 < H \leq B$, then H is not normal in G.

Under which conditions on B does it follow that $Z(G) \neq 1$?

For example: if B is (finite) cyclic, does it follow that $Z(G) \neq 1$? We are naturally interested in examples and counterexamples. This problem is important for the classification of the multiplication groups of quasigroups.

M. Niemenmaa

Problem 9. (Liftable Algebras) Let G be a finite group and $\mathbf{Z}G$ its integral group ring. For any prime p, $\mathbf{Z}G/p\mathbf{Z}G$ is a finite dimensional algebra over $\mathbf{F}_p$ which is of course isomorphic to the group algebra $\mathbf{F}_pG$. Now let Γ be an order in $\mathbf{Q}G$, or more generally, in any finite dimensional semi-simple $\mathbf{Q}$-algebra Σ. Again $A = \Gamma/p\Gamma$ is a finite dimensional $\mathbf{F}_p$-algebra, which we call "liftable".

Which are the liftable $\mathbf{F}_p$-algebras? In other words, characterize such liftable $\mathbf{F}_p$-algebras among all finite dimensional $\mathbf{F}_p$-algebras without reference to characteristic zero.

I know a necessary condition for liftability, which may or may not be useful. Of course, one might also replace $\mathbf{Q}$ by a number field (perhaps a splitting field for Σ) and ask the question for the appropriate finite fields $\mathbf{F}_q$. Or one might go p-adic.

Jan R. Strooker

Problem 10. Is there a description of the varieties of groups with residually finite groups?

D. Dikranjan

Problem 11. For this and the following question: a group G is *maximally almost periodic* if G has enough finite-dimentional unitary representations.

Is a variety consisting of only maximally almost periodic groups necessarily abelian?

D. Dikranjan

Problem 12. If G is a maximally almost periodic group, do there exist $\log|G|$ finite-dimensional unitary representations of G which separate the points of G?

D. Dikranjan

Problem 13. Let p be a prime and $\mathbf{Z}_p$ be the additive group of p-adic integers. For subsets G and C of $\mathbf{Z}_p$ let $[G : C] = \{\xi \in \mathbf{Z}_p : \xi C \subset G\}$, where ξC is the set C "multiplied" by ξ in the ring $\mathbf{Z}_p$. It is known, that for every cardinal number τ satisfying $1 < \tau \leq \log c^+$, where $\log c^+ = \min\{\lambda : 2^\lambda > c = 2^{\aleph_0}\}$, there exists a subgroup G_τ of $\mathbf{Z}_p$ such that $[G_\tau : C] \neq 0$ for each $C \subset \mathbf{Z}_p$ with $|C| < \tau$ and there exists $C_0 \subset \mathbf{Z}_p$ with $|C_0| = \tau$ and $[G_\tau : C_0] = 0$. Assume that $\log c^+ < c$. Is it possible to find such a G_τ for τ satisfying $\log c^+ < \tau \leq c$?

D. Dikranjan

Problem 14. The properties defined below arose several years ago in work with R. Alperin in computing the second homology group $H_2(G, \mathbf{Z})$.

A group satisifes the property C_n if for any collection of n elements $x_1, x_2, \ldots, x_n \in G$, there exist elements $z, y_1, \ldots, y_n \in G$ such that $x_i = [z, y_i]$ for $i = 1, \ldots, n$.

Thus a group satisfies C_1 precisely when every element of G is a commutator. It is well–known that for $n \geq 5$, A_n satisfies C_1. Several years ago, Ulf Rehmann (Bielefeld) and I checked that A_n satisfies C_2 for $5 \leq n \leq 10$. As I recall, A_5 does not satisfy C_3. If U is the multiplicative group of real quaternions of norm 1, then one can show that U satisfies C_3.

One might reasonably conjecture that A_n satisfies C_2 for $n \geq 5$. One might also ask if there is a function $c(n)$ such that A_m satisfies C_n precisely when $m \geq c(n)$. The same questions can be asked for other groups such as $PSL_n(F)$ and the finite non–abelian simple groups.

Keith Dennis

Problem 15. Let R be any ring. Let $GL(R)$ denote the infinite general linear group over R (the direct limit of the $GL_n(R)$), let $E(R)$ denote the subgroup generated by the elementary matrices and let $St(R)$ denote the Steinberg group over R. Dennis and Vaserstein have shown that every element of $E(R)$ or $St(R)$ is a product of 2 commutators. Can 2 be reduced to 1? For R a field, the answer is "yes". One case of particular interest is $E(\mathbf{Z}) = SL(\mathbf{Z})$. The theorem of Dennis and Vaserstein also applies to many other groups such as $Q = \varinjlim \operatorname{Aut}(F_n)$ where F_n is the free group of rank n. Is every element of Q a single commutator?

Keith Dennis

Problem 16. Let $\{1, p^a, \ldots, p^z\}$ be a set of powers of the prime p. Does there exist a finite p-group whose conjugacy class sizes are exactly these numbers? (For the corresponding problem regarding character degrees the answer is "yes".)

A. Mann

Problem 17. What are the subgroups of powerful p-groups? (A. Lubotzky has shown that all finite p-groups are sections of powerful groups. H. Heineken gave examples of p-groups that are not subgroups of powerful groups.)

A. Mann

Problem 18. Let $b_n(G)$ be the number of subgroups of G of index at most n. For a set X of subgroups, let $x_n(G)$ be the number of subgroups from X of index at most n. The lower limit of the ratio $x_n(G)/b_n(G)$, as $n \to \infty$, is the *lower density* of X.

Let G be finitely generated and residually finite, and let $X_d(G)$ be the set of d-generated subgroups of G. If $X_d(G)$ has a positive lower density, does G have a bounded rank (i.e. is there a bound on the number of generators of finitely generated subgroups)? This is known to be true if G is soluble.

A. Mann

Problem 19. Let G be a profinite group, and denote by $Q(G,k)$ the probability that k random elements of G generate a subgroup of finite index. Polynomial subgroup growth groups satisfy $Q(G,k) = 1$, for some k. Is the converse true? ($Q(G,k) > 0$, for some k, iff $P(G,d) > 0$, for some d. Here $P(G,d)$ is the probability that d random elements of G generate G. It is impossible that $P(G,k) = 1$. If F_d is a free pro-p group of rank d, where $2 \leq d < \infty$, then $0 < Q(G,k) < 1$, for $k \geq d$.)

A. Mann

Problem 20. Let G be a finite group. Consider the following property (WM). For every irreducible representation π of G there exists a subgroup H of G and a linear character G of H such that ind_H^G is a multiple of π. Does (WM) imply solvability?

Günter Schlichting

Problem 21. Let G be a (finite) transitive permutation group acting on a set Ω, $\alpha \in \Omega$. Denote by $\Omega : G_\alpha$ the partition of Ω into G_α-orbits.

Set $S(\alpha) := \langle G_{\alpha \cup \Gamma} | \{\alpha\} \neq \Gamma \in \Omega : G_\alpha \rangle$, where $G_{\alpha \cup \Gamma}$ denotes the pointwise stabiliser of the "suborbit" Γ in G_α. (Clearly $S(\alpha) \trianglelefteq G_\alpha$ and for all $g \in G$ $S(\alpha^g) = S(\alpha)^g$.)

What can be said about the set $\Omega_{S(\alpha)}$ of fixed-points in Ω of $S(\alpha)$? In particular, are there *primitive* groups G such that $\Omega_{S(\alpha)} = \{\alpha\}$?

There is no such primitive group with $|\Omega| \leq 50$ as a GAP search shows.

Wolfgang Knapp

Problem 22. Let G be a finite group and $G_{p_1}, G_{p_2}, \ldots, G_{p_n}$ its Sylow subgroups. Let $k(G)$ denote the number of conjugacy classes of G. We conjecture that $k(G) \leq k(G_{p_1})k(G_{p_2})\cdots k(G_{p_n})$.

Let G be a primitive subgroup of S_n. Is it true that $|N_{S_n}(G)/G| < n$? (This is true if G is simple by $CFSG$.) Does at least $|N_{S_n}(G)/G| < n^c$ hold for some constant c? (There is a bound of the form $n^{c \log \log n}$.)

L. Pyber

Problem 23. (Modules for infinite groups) Does there exist a finitely generated residually finite group G and a finitely presented $\mathbf{Z}G$–module M such that either (i) M is infinite simple or (ii) $M \neq 0$ and M has no non–trivial finite quotient?

B. Hartley

Problem 24. Is there an effective algorithm to decide whether two elements of $GL_n(R)$ generate a free group? Here R might be $\mathbf{Z}$, or, more preferably, $\mathbf{Z}[t_1, t_1^{-1}, \ldots, t_m, t_m^{-1}]$, or even $\mathbf{R}$.

H. Bass

Problem 25. (Automorphism groups of free groups) Let F be the free group on $x_1, \ldots, x_r$ $(r \geq 2)$ and $G = \mathrm{Aut}(F)$.

1. Does G satisfy the "Tits Alternative" i.e. does every non virtually solvable subgroup of G contain a non–abelian free group?
2. Let $L = \bigoplus_{d \geq 1} L_d$ be the free Lie algebra on $\overline{x}_1, \ldots, \overline{x}_r$ over a field k of characteristic 0: $L_1 = k\overline{x}_1 \oplus \cdots \oplus k\overline{x}_r$. Let $D = \mathrm{Der}_k(L)$, the Lie algebra of derivations of L, graded by $D = \bigoplus_{d \geq 0} D_d$, where $\delta \in D_d$ iff $\delta(L_n) \subset L_{n+d}$ $\forall n$. Put $D_+ = \bigoplus_{d>0} D_d$. Does D_+ satisfy a "Lie algebra Tits Alternative" i.e. if $E \subset D_+$ is a graded Lie subalgebra which contains no non–abelian free subalgebra, then must E be solvable (and hence nilpotent)? (An affirmative answer to 2 implies an affirmative answer to 1.)
3. Assume now that $r \geq 3$.
   (a) Is $G = \mathrm{Aut}(F)$ "rigid?" i.e. does G have only finitely many irreducible $\mathbf{C}$–representations in each dimension?
   Some subproblems: Let $\rho : G \to GL_n(\mathbf{C})$.
   (b) Is $\rho(F)$ virtually solvable?
   (Motivation: A result of Formanack–Procesi shows that $\rho(F_1)$ is virtually solvable if F_1 is generated by $x_1, \ldots, x_{r-1}$.)

(c) If ρ is irreducible, is $\rho(F)$ finite?

(d) Let $R \triangleleft_f F$ be a characteristic subgroup of finite index; hence $R \triangleleft G$. Let $H \leq_f G$ be a finite index subgroup containing R. Is $R/[H,R]$ finite?

Remarks. (1) $R^{ab} = R/[R,R]$ is a finitely generated free **Z**-module on which G acts, and $R/[H,R]$ trivializes the action of H on R^{ab}. For $R = F$, $R^{ab} = \mathbf{Z}^r$, on which G acts via its quotient $GL_r(\mathbf{Z})$. Trivializing the action of a finite index subgroup of $GL_r(\mathbf{Z})$ makes $\mathbf{Z}^r$ finite. (2) Affirmative answers to (b) and (d) imply an affirmative answer to (c).

H. Bass

Problem 26. (Enumeration of amalgams) Fix integers $m_0, m_1 \geq 2$. A *finite* (m_0, m_1)*–amalgam* is a configuration of finite groups

$$G_0 \geq H \leq G_1$$

with $|G_i : H| = m_i$ $(i = 0, 1)$. It is called *faithful* if H contains no non trivial subgroup that is normal in each G_i. Consider two amalgams the same if they are isomorphic, in the obvious sense.

If m_0 and m_1 are prime, then conjecturally there are only finitely many faithful finite (m_0, m_1)–amalgams. For example, D. Goldschmidt determined all 15 faithful finite $(3,3)$–amalgams.

If, on the other hand, m_0 or m_1 is composite, then there are infinitely many, even infinite ascending chains, of faithful finite (m_0, m_1)–amalgams. (Cf. Bass–Kulkarni, Uniform tree lattices, *J. Amer. Math. Soc.* (1991).)

Let $f(n) (= f_{m_0,m_1}(n))$ denote the number of faithful finite (m_0, m_1)–amalgams with $|G_0| = n$. What are the asymptotics of $f(n)$ as $n \to \infty$? For example, what type of growth and variation can it have? Is there an interesting generating function?

Motivation: Let X be a locally finite tree, and let Γ be a "uniform lattice" on X, i.e. a group of automorphisms with finite stabilizers Γ_X, and finite quotient graph $\Gamma \backslash X$. Let $f_\Gamma(n)$ denote the number of "over lattices of index n," i.e. the number of $\Gamma_1 \leq \mathrm{Aut}(X)$ such that $\Gamma \leq \Gamma_1$ and $|\Gamma_1 : \Gamma| = n$. This is known to be finite (*loc. cit.*). We would like to understand the asymptotics of $f_\Gamma(n)$.

Suppose that Γ has a fundamental domain in X that is an edge $0 \; e \; 1$. Then

$$\Gamma_0 \geq \Gamma_e \leq \Gamma_1 \qquad (*)$$

is a finite faithful (m_0, m_1)–amalgam, where $m_i = |\Gamma_i : \Gamma_e|$, and every finite faithful (m_0, m_1)–amalgam "containing" (*) defines an over lattice of Γ. Thus, the studies of $f_\Gamma(n)$ and of $f_{m_0,m_1}(n)$ are essentially equivalent.

H. Bass

Problem 27. (Does linear + rigid $\Longrightarrow$ super rigid?) *Explanation:* Let Γ be a finitely generated group. Consider finite dimensional complex representations

$$\rho : \Gamma \longrightarrow GL_N(\mathbf{C}).$$

Call Γ *linear* if there is a faithfull ρ. Call Γ *rigid* if, for each N, there are only finitely many irreducible ρ (up to conjugacy). Call Γ *super rigid* if the Zariski closures $\overline{Zar}(\rho(\Gamma))$ of $\rho(\Gamma)$ have dimensions which are bounded (independent of ρ).

Remarks:

1. Super rigid $\Longrightarrow$ rigid.
2. There are many residually finite rigid Γ which are not super rigid. However, all known examples are not linear.
3. Super rigid residually finite groups need not be linear, e.g. Golod–Shafarevich type groups, for which $\rho(\Gamma)$ is always finite.

H. Bass

Problem 28. (Integral Tanaka duality) For a group Γ and commutative ring K let $\text{Rep}_K(\Gamma)$ denote the category of $K\Gamma$—modules finitely presented as K-modules. Let

$$\Phi : \text{Rep}_K(\Gamma) \Longrightarrow \text{Rep}_K(1)$$

denote the functor "forget the Γ-action". Put

$$A_K(\Gamma) = \text{Aut}_{\bigotimes\text{-functor}}(\Phi)$$

Define

$$\varepsilon_{K,\Gamma} : \Gamma \longrightarrow A_K(\Gamma)$$

by

$$\varepsilon_{K,\Gamma}(g)_V = \rho_V(g).$$

Fact: $\varepsilon_{\mathbf{Z},\Gamma}$ is an isomorphism for $\Gamma = SL_n(\mathbf{Z})$ with $n \geq 3$. (This uses the congruence subgroup theorem, and rigidity.) What about the case $n = 2$?
Remark: $\varepsilon_{\mathbf{Z},SL_2(\mathbf{Z})}$ is an isomorphism iff $\varepsilon_{\mathbf{Z},F_r}$ is an isomorphism when F_r is a free group of rank $r \geq 2$.

H. Bass

Printed in the United States
By Bookmasters